The Workflow of Data Analysis Using Stata

The Workflow of Data Analysis Using Stata

J. SCOTT LONG
Departments of Sociology and Statistics
Indiana University–Bloomington

A Stata Press Publication
StataCorp LP
College Station, Texas

Published by Stata Press, 4905 Lakeway Drive, College Station, Texas 77845
Typeset in $\LaTeX\,2_\varepsilon$
Printed in the United States of America
10 9 8 7 6 5 4 3 2 1

ISBN-10: 1-59718-047-5
ISBN-13: 978-1-59718-047-4

To Valerie

Contents

Tables

Figures

Examples

Preface

This book is about methods that allow you to work efficiently and accurately when you analyze data. Although it does not deal with specific statistical techniques, it discusses the steps that you go through with any type of data analysis. These steps include planning your work, documenting your activities, creating and verifying variables, generating and presenting statistical analyses, replicating findings, and archiving what you have done. These combined issues are what I refer to as the *workflow of data analysis*. A good workflow is essential for replication of your work, and replication is essential for good science.

My decision to write this book grew out of my teaching, researching, consulting, and collaborating. I increasingly saw that people were *drowning* in their data. With cheap computing and storage, it is easier to create files and variables than it is to keep track of them. As datasets have become more complicated, the process of managing data has become more challenging. When consulting, much of my time was spent on issues of data management and figuring out what had been done to generate a particular set of results. In collaborative projects, I found that problems with workflow were multiplied. Another motivation came from my work with Jeremy Freese on the package of Stata programs known as SPost (Long and Freese 2006). These programs were downloaded more than 20,000 times last year, and we were contacted by hundreds of users. Responding to these questions showed me how researchers from many disciplines organize their data analysis and the ways in which this organization can break down. When helping someone with what appeared to be a problem with an SPost command, I often discovered that the problem was related to some aspect of the user's workflow. When people asked if there was something they could read about this, I had nothing to suggest.

A final impetus for writing the book came from Bruce Fraser's *Real World Camera Raw with Adobe Photoshop CS2* (2005). A much touted advantage of digital photography is that you can take a lot of pictures. The catch is keeping track of thousands of pictures. Imaging experts have been aware of this issue for a long time and refer to it as *workflow*—keeping track of your work as it flows through the many stages to the final product. As the amount of time I spent looking for a particular picture became greater than the time I spent taking pictures, it was clear that I needed to take Fraser's advice and develop a workflow for digital imaging. Fraser's book got me thinking about data analysis in terms of the concept of a workflow.

After years of gestation, the book took two years to write. When I started, I thought my workflow was very good and that it was simply a matter of recording what I did. As writing proceeded, I discovered gaps, inefficiencies, and inconsistencies in what I did.

Sometimes these involved procedures that I knew were awkward, but where I never took the time to find a better approach. Some problems were due to oversights where I had not realized the consequences of the things I did or failed to do. In other instances, I found that I used multiple approaches for the same task, never choosing one as the best practice. Writing this book forced me to be more consistent and efficient. The advantages of my improved workflow became clear when revising two papers that were accepted for publication. The analyses for one paper were completed before I started the workflow project, whereas the analyses for the other were completed after much of the book had been drafted. I was pleased by how much easier it was to revise the analyses in the paper that used the procedures from the book. Part of the improvement was due to having better ways of doing things. Equally important was that I had a consistent and documented way of doing things.

I have no illusions that the methods I recommend are the best or only way of doing things. Indeed, I look forward to hearing from readers who have suggestions for a better workflow. Your suggestions will be added to the book's web site. However, the methods I present work well and avoid many pitfalls. An important aspect of an efficient workflow is to find one way of doing things and sticking with it. Uniform procedures allow you to work faster when you initially do the work, and they help you to understand your earlier work if you need to return to it at a later time. Uniformity also makes working in research teams easier because collaborators can more easily follow what others have done. There is a lot to be said in favor of having established procedures that are documented and working with others who use the same procedures. I hope you find that this book provides such procedures.

Although this book should be useful for anyone who analyzes data, it is written within several constraints. First, Stata is the primary computing language because I find Stata to be the best, general-purpose software for data management and statistical analysis. Although nearly everything I do with Stata can be done in other software, I do not include examples from other packages. Second, most examples use data from the social sciences, because that is the field in which I work. The principles I discuss, however, apply broadly to other fields. Finally, I work primarily in Windows. This is not because I think Windows is a better operating system than Mac or Linux, but because Windows is the primary operating system where I work. Just about everything I suggest works equally well in other operating systems, and I have tried to note when there are differences.

I want to thank the many people who commented on drafts or answered questions about some aspect of workflow. I particularly thank Tait Runfeldt Medina, Curtis Child, Nadine Reibling, and Shawna L. Rohrman whose detailed comments greatly improved the book. I also thank Alan Acock, Myron Gutmann, Patricia McManus, Jack Thomas, Leah VanWey, Rich Watson, Terry White, and Rich Williams for talking with me about workflow. Many people at StataCorp helped in many ways. I particularly want to thank Lisa Gilmore for producing the book, Jennifer Neve for editing, and Annette Fett for designing the cover. David M. Drukker at StataCorp answered many of my questions. His feedback made it a better book and his friendship made it more fun to write.

Some of the material in this book grew out of research funded by NIH Grant Number R01TW006374 from the Fogarty International Center, the National Institute of Mental Health, and the Office of Behavioral and Social Science Research to Indiana University–Bloomington. Other work was supported by an anonymous foundation and The Bayer Group. I gratefully acknowledge support provided by the College of Arts and Sciences at Indiana University.

Without the unintended encouragement from my dear friend Fred, I would not have started the book. Without the support of my dear wife Valerie, I would not have completed it. Long overdue, this book is dedicated to her.

Bloomington, Indiana
October 2008

Scott Long

A word about fonts, files, commands, and examples

The book uses standard Stata conventions for typography. Items printed in a typewriter-style typeface are Stata commands and options. For example, `use mydata, clear`. Italics indicate information that you should add. For example, `use` *dataset-name*`, clear` indicates that you should substitute the name of your dataset. When I provide the syntax for a command, I generally show only some of the options. For full documentation, you can type `help` *command-name* or check the reference manual. Manuals are referred to with the usual Stata conventions. For example, [R] **logit** refers to the `logit` entry in the *Base Reference Manual* and [D] **sort** refers to the `sort` entry in the *Data Management Reference Manual*.

Within the text, the commands or output for some examples will trail off the right side of the page; see page 59 for an example. This is intentional to show you the consequence of not controlling the length of commands and output.

The book includes many examples that I encourage you to try as you read. If the name of a file begins with `wf`, you can download that file. I use (file: *filename*`.do`) to let you know the name of the do-file that corresponds to the example being presented. With few exceptions (e.g., some ado-files), if the name of a file does not begin with `wf` (e.g., `science2.dta`), the file is not available for download. To find where a downloaded file is used in the text, check the index under the entry for Workflow package files.

To download the examples, you must be in Stata and connected to the Internet. There are two Workflow packages for Stata 10 (`wf10-part1` and `wf10-part2`) and two for Stata 9 (`wf09-part1` and `wf09-part2`). To find and install the packages, type `findit workflow`, choose the packages you need, and follow the instructions. Although two packages are needed because of the large number of examples, I refer to them simply as the Workflow package. Before trying these examples, be sure to update your copy of Stata as described in [GS] **20 Updating and extending Stata—Internet functionality**. Additional information related to the book is located at http://www.indiana.edu/~jslsoc/workflow.htm.

1 Introduction

This book is about methods for analyzing your data effectively, efficiently, and accurately. I refer to these methods as the *workflow of data analysis*. Workflow involves the entire process of data analysis including planning and documenting your work, cleaning data and creating variables, producing and replicating statistical analyses, presenting findings, and archiving your work. You already have a workflow, even if you do not think of it as such. This workflow might be carefully planned or it might be ad hoc. Because workflow for data analysis is rarely described in print or formally taught, researchers often develop their workflow in reaction to problems they encounter and from informal suggestions from colleagues. For example, after you discover two files with the same name but different content, you might develop procedures (i.e., a workflow) for naming files. Too often, good practice in data analysis is learned inefficiently through trial and error. Hopefully, my book will shorten the learning process and allow you to spend more time on what you really want to do.

Reactions to early drafts of this book convinced me that both beginners and experienced data analysts can benefit from a more formal consideration of how they do data analysis. Indeed, when I began this project, I thought that my workflow was pretty good and that it was simply a matter of writing down what I routinely do. I was surprised and pleased by how much my workflow improved as a result of thinking about these issues systematically and from exchanging ideas with other researchers. Everyone can improve their workflow with relatively little effort. Even though changing your workflow involves an investment of time, you will recoup this investment by saving time in later work and by avoiding errors in your data analysis.

Although I make many specific suggestions about workflow, most of the things that I recommend can be done in other ways. My recommendations about the best practice for a particular problem are based on my work with hundreds of researchers and students from all sectors of employment and from fields ranging from chemistry to history. My suggestions have worked for me and most have been refined with extensive use. This is not to say that there is only one way to accomplish a given task or that I have the best way. In Stata, as in any complex software environment, there are a myriad of ways to complete a task. Some of these work only in the limited sense that they get a job done but are error prone or inefficient. Among the many approaches that work well, you will need to choose your preferred approach. To help you do this, I often discuss several approaches to a given task. I also provide examples of ineffective procedures because seeing the consequences of a misguided approach can be more effective than hearing about the virtues of a better approach. These examples are all real, based on mistakes

I made (and I have made lots) or mistakes I encountered when helping others with data analysis. You will have to choose a workflow that matches the project at hand, the tools you have, and your temperament. There are as many workflows as there are people doing data analysis, and there is no single workflow that is ideal for every person or every project. What is critical is that you consider the general issues, choose your own procedures, and stick with them unless you have a good reason to change them.

In the rest of this chapter, I provide a framework for understanding and evaluating your workflow. I begin with the fundamental principle of replicability that should guide every aspect of your workflow. No matter how you proceed in data analysis, you must be able to justify and reproduce your results. Next I consider the four steps involved in all types of data analysis: preparing data, running analysis, presenting results, and preserving your work. Within each step there are four major tasks: planning the work, organizing your files and materials, documenting what you have done, and executing the analysis. Because there are alternative approaches to accomplish any given aspect of your work, what makes one workflow better than another? To answer this question, I provide several criteria for evaluating the way you work. These criteria should help you decide which procedures to use, and they motivate many of my recommendations for best practice that are given in this book.

1.1 Replication: The guiding principle for workflow

Being able to reproduce the work you have presented or published should be the cornerstone of any workflow. Science demands replicability and a good workflow facilitates your ability to replicate your results. How you plan your project, document your work, write your programs, and save your results should anticipate the need to replicate. Too often researchers do not worry about replication until their work is challenged. This is not to say that they are taking shortcuts, doing shoddy work, or making decisions that are unjustified. Rather, I am talking about taking the steps necessary so that all the good work that has been done can be easily reproduced at a later time. For example, suppose that a colleague wants to expand upon your work and asks you for the data and commands used to produce results in a published paper. When this happens, you do not want to scramble furiously to replicate your results. Although it might take a few hours to dig out your results (many of mine are in notebooks stacked behind my file cabinets), this should be a matter of retrieving the records, not trying to remember what it was you did or discovering that what you documented does not correspond to what you presented.

Think about replication throughout your workflow. At the completion of each stage of your work, take an hour or a day if necessary to review what you have done, to check that the procedures are documented, and to confirm that the materials are archived. When you have a draft of a paper to circulate, review the documentation, check that you still have the files you used, confirm that the do-files still run, and double check that the numbers in your paper correspond to those in your output. Finally, make sure that all this is documented in your research log (discussed on page 37).

If you have tried to replicate your own work months after it was completed or tried to reproduce another author's results using only the original dataset and the published paper, you know how difficult it can be to replicate something. A good way to understand what is required to replicate your work is to consider some of the things that can make replication impossible. Many of these issues are discussed in detail later in the book. First, you have to find the original files, which gets more difficult as time passes. Once you have the files, are they in formats that can be analyzed by your current software? If you can read the file, do you know exactly how variables were constructed or cases were selected? Do you know which variables were in each regression model? Even if you have all this information, it is possible that the software you are currently using does not compute things exactly the same way as the software you used for the original analyses. An effective workflow can make replication easier.

A recent example illustrates how difficult it can be to replicate even simple analyses. I collected some data that were analyzed by a colleague in a published paper. I wanted to replicate his results to extend the analyses. Due to a drive failure, some of his files were lost. Neither of us could reproduce the exact results from the published paper. We came close, but not close enough. Why? Suppose that 10 decisions were made in the process of constructing the variables and selecting the sample for analysis. Many of these decisions involve choices between options where neither choice is incorrect. For example, do you take the square root of publications or the log after adding .5? With 10 such decisions, there are $2^{10} = 1{,}024$ different outcomes. All of them will lead to similar findings, but not exactly the same findings. If you lose track of decisions made in constructing your data, you will find it very difficult to reproduce what you have done. By the way, remarkably another researcher who was using these data discovered the secret to reproducing the published results.

Even if you have the original data and analysis files, it can be difficult to reproduce results. For published papers, it is often impossible to obtain the original data or the details on how the results were computed. Freese (2007) makes a compelling argument for why disciplines should have policies that govern the availability of information needed to replicate results. I fully support his recommendations.

1.2 Steps in the workflow

Data analysis involves four major steps: cleaning data, performing analysis, presenting findings, and saving your work. Although there is a logical sequence to these steps, the dynamics of an effective workflow are flexible and highly dependent upon the specific project. Ideally, you advance one step at a time, always moving forward until you are done. But, it never works that way for me. In practice, I move up and down the steps depending on how the work goes. Perhaps I find a problem with a variable while analyzing the data, which takes me back to cleaning. Or my results provide unexpected insights, so I revise my plans for analysis. Still, I find it useful to think of these as distinct steps.

1.2.1 Cleaning data

Before substantive analysis begins, you need to verify that your data are accurate and that the variables are well named and properly labeled. That is, you clean the data. First, you must bring your data into Stata. If you received the data in Stata format, this is as simple as a single `use` command. If the data arrived in another format, you need to verify that they were imported correctly into Stata. You should also evaluate the variable names and labels. Awkward names make it more difficult to analyze the data and can lead to mistakes. Likewise, incomplete or poorly designed labels make the output difficult to read and lead to mistakes. Next verify that the sample and variables are what they should be. Do the variables have the correct values? Are missing data coded appropriately? Are the data internally consistent? Is the sample size correct? Do the variables have the distribution that you would expect? Once these questions are resolved, you can select the sample and construct new variables needed for analysis.

1.2.2 Running analysis

Once the data are cleaned, fitting your models and computing the graphs and tables for your paper or book are often the simplest part of the workflow. Indeed, this part of the book is relatively short. Although I do not discuss specific types of analysis, later I talk about ways to ensure the accuracy of your results, to facilitate later replications, and to keep track of your do-files, data files, and log files regardless of the statistical methods you are using.

1.2.3 Presenting results

Once the analyses are complete, you want to present them. I consider several issues in the workflow of presentation. First, you need to move the results from your Stata output into your paper or presentation. An efficient workflow can automate much of this work. Second, you need to document the provenance of all findings that you present. If your presentation does not preserve the source of your results, it can be very difficult to track them down later (e.g., someone is trying to replicate your results or you must respond to a reviewer). Finally, there are a number of simple things that you can do to make your presentations more effective.

1.2.4 Protecting files

When you are cleaning your data, running analyses, and writing, you need to protect your files to prevent loss due to hardware failure, file corruption, or unintentional deletions. Nobody enjoys redoing analyses or rewriting a paper because a file was lost. There are a number of simple things you can do to make it easier to routinely save your work. With backup software readily available and the cost of disk storage so cheap, the hardest parts of making backups is keeping track of what you have. Archiving is distinct from backing up and more difficult because it involves the long-term preservation of files so that they will be accessible years into the future. You need to consider if the file formats and storage media will be accessible in the future. You must also consider the operating system you use (it is now difficult to read data stored using the CP/M operating system), the storage media (can you read 5 1/4" floppy disks from the 1980s or even a ZIP disk from a few years ago?), natural disasters, and hackers.

1.3 Tasks within each step

Within each of the four major steps, there are four primary tasks: planning your work, organizing your materials, documenting what you do, and executing the work. While some tasks are more important within particular steps (e.g., organization while planning), each task is important for all steps of the workflow.

1.3.1 Planning

Most of us spend too little time planning and too much time working. Before you load data into Stata, you should draft a plan of what you want to do and assess your priorities. What types of analyses are needed? How will you handle missing data? What new variables need to be constructed? As your work progresses, periodically reassess your plan by refining your goals and analytic strategy based on the work you have completed. A little planning goes a long way, and I almost always find that planning saves time.

1.3.2 Organization

Careful organization helps you work faster. Organization is driven by the need to find things and to avoid duplication of effort. Good organization can prevent you from searching for lost files or, worse yet, having to reconstruct them. If you have good documentation about what you did, but you cannot find the files used to do the work, little is gained. Organization requires you to think systematically about how you name files and variables, how you organize directories on your hard drive, how you keep track of which computer has what information (if you use more than one computer), and where you store research materials. Problems with organization show up when you have not been working on a project for a while or when you need something quickly. Throughout the book, I make suggestions on how to organize materials and discuss tools that make it easier to find and work with what you have.

1.3.3 Documentation

Without adequate documentation, replication is virtually impossible, mistakes are more likely, and work usually takes longer. Documentation includes a research log that records what you do and codebooks that document the datasets you create and the variables they contain. Complete documentation also requires comments in your do-files and labels and notes within data files. Although I find writing documentation to be an onerous task, certainly the least enjoyable part of data analysis, I have learned that time spent on documentation can literally save weeks of work and frustration later. Although there is no way to avoid time spent writing documentation, I can suggest things that make documenting your work faster and more effective.

1.3.4 Execution

Execution involves carrying out specific tasks within each step. Effective execution requires the right tools for the job. A simple example is the editor used to write your programs. Mastering a good text editor can save you hours when writing your programs and will lead to programs that are better written. Another example is learning the most effective commands in Stata. A few minutes spent learning how to use the `recode` command can save you hours of writing `replace` commands. Much of this book involves selecting the right tool for the job. Throughout my discussion of tools, I emphasize standardizing tasks and automating them. The reason to standardize is that it is generally faster to do something the way you did it before than it is to think up a new way to do it. If you set up templates for common tasks, your work becomes more uniform, which makes it easier to find and avoid errors. Efficient execution requires assessing the trade-off between investing the time in learning a new tool, the accuracy gained by the new tools, and the time you save by being more efficient.

1.4 Criteria for choosing a workflow

As you work on the various tasks in each step of your workflow, you will have choices of different ways to do things. How do you decide which procedure to use? In this section, I consider several criteria for evaluating your current workflow and choosing from among alternative procedures for your work.

1.4.1 Accuracy

Getting the correct answer is the *sine qua non* of a good workflow. Oliveira and Stewart (2006, 30) make the point very well, "If your program is not correct, then nothing else matters." At each step in your work, you must verify that your results are correct. Are you answering the question you set out to answer? Are your results what you wanted and what you think they are? A good workflow is also about making mistakes. Invariably, mistakes will happen, probably a lot of them. Although an effective workflow can prevent some errors, it should also help you find and correct them quickly.

1.4.2 Efficiency

You want to get your analyses done as quickly as possible, given the need for accuracy and replicability. There is an unavoidable tension between getting your work done and the need to work carefully. If you spend so much time verifying and documenting your work that you never finish the project, you do not have a viable workflow. On the other hand, if you finish by publishing incorrect results, both you and your field suffer. You want a workflow that gets things done as quickly as possible without sacrificing the accuracy of your results. A good workflow, in effect, increases the time you have to do your work, without sacrificing the accuracy of what you do.

1.4.3 Simplicity

A simpler workflow is better than a more complex workflow. The more complicated your procedures, the more likely you will make mistakes or abandon your plan. But what is simple for one person might not be simple for another. Many of the procedures that I recommend involve programming methods that may be new to you. If you have never used a loop, you might find my suggestion of using a loop much more complex than repeating the same commands for multiple variables. With experience, however, you might decide that loops are the simplest way to work.

1.4.4 Standardization

Standardization makes things easier because you do not have to repeatedly decide how to do things and you will be familiar with how things look. When you use standardized formats and procedures, it is easier to see when something is wrong and ensure that you do things consistently the next time. For example, my do-files all use the same structure for organizing the commands. Accordingly, when I look at the log file, it is easier for me to find what I want. Whenever you do something repeatedly, consider creating a template and establishing conventions that become part of your routine workflow.

1.4.5 Automation

Procedures that are automated are better because you are less likely to make mistakes. Entering numbers into your do-file by hand is more error prone than using programming tools to transfer the information automatically. Typing the same list of variables multiple times in a do-file makes it easy to create lists that are supposed to be the same but are not. Again automation can eliminate this problem. Automation is the backbone for many of the methods recommended in this book.

1.4.6 Usability

Your workflow should reflect the way you like to work. If you set up a workflow and then ignore it, you do not have a good workflow. Anything that increases the chances of maintaining your workflow is helpful. Sometimes it is better to use a less efficient approach that is also more enjoyable. For example, I like experimenting with software and prefer taking longer to complete a task while learning a new program than getting things done quicker the old way. On the other hand, I have a colleague who prefers using a familiar tool even if it takes a bit longer to complete the task. Both approaches make for a good workflow because they complement our individual styles of work.

1.4.7 Scalability

Some ways of work are fine for small jobs but do not work well for larger jobs. Consider the simple problem of alphabetizing 10 articles by author. The easiest approach is to lay the papers on a table and pick them up in order. This works well for 10 articles but is dreadfully slow with 100 or 1,000 articles. This issue is referred to as *scalability*—how well do procedures work when applied to a larger problem? As you develop your workflow, think about how well the tools and practices you develop can be applied to a larger project. An effective workflow for a small project where you are the only researcher might not be sustainable for a large project involving many people. Although you can visually inspect every case for every variable in a dataset with 25 measures of development in 80 countries, this approach does not work with the National Longitudinal Survey that has thousands of cases and thousands of variables. You should strive for a workflow that adapts easily to different types of projects. Few procedures scale perfectly. As a consequence you are likely to need different workflows for projects of different complexities.

1.5 Changing your workflow

This book has hundreds of suggestions. Decide which suggestions will help you the most and adapt them to the way you work. Suggestions for minor changes to your workflow can be adopted at any time. For example, it takes only a few minutes to learn how to use `notes`, and you can benefit from this command almost immediately. Other suggestions might require major changes to how you work and should be made only when you have the time to fully integrate them into your work. It is a bad idea to make major changes when a deadline is looming. On the other hand, make sure you find time to improve your workflow. Time spent improving your workflow should save time in the long run and improve the quality of your work. An effective workflow is something that evolves over time, reflecting your experience, changing technology, your personality, and the nature of your current research.

1.6 How the book is organized

This book is organized so that it can be read front to back by someone wanting to learn about the entire workflow of data analysis. I also wanted it to be useful as a reference for people who encounter a problem and who want a specific solution. For this purpose, I have tried to make the index and table of contents extensive. It is also useful to understand the overall structure of this book before you proceed with your reading.

Chapter 2 – Planning, organizing, and documenting your work discusses how to plan your work, organize your files, and document what you have done. Avoid the temptation of skipping this chapter so that you can get to the "important" details in later chapters.

Chapter 3 – Writing and debugging do-files discusses how do-files should be used for almost all your work in Stata. I provide information on how to write more effective do-files and how to debug programs that do not work. Both beginners and advanced users should find useful information here.

Chapter 4 – Automating Stata is an introduction to programming that discusses how to create macros, run loops, and write short programs. This chapter is not intended to teach you how to write sophisticated programs in Stata (although it might be a good introduction), rather it discusses tools that all data analysts should find useful. I encourage every reader to study the material in this chapter before reading chapters 5–7.

Chapter 5 – Names and labels discusses both principles and tools for creating names and labels that are clear and consistent. Even if you have received data that are labeled, you should consider improving the names and labels in the dataset. This chapter is long and includes a lot of technical details that you can skip until you need them.

Chapter 6 – Cleaning data and constructing variables discusses how to check whether your data are correct and how to construct new variables and verify that they were created correctly. At least 80% of the work in data analysis involves getting the data ready, so this chapter is essential.

Chapter 7 – Analyzing, presenting, and replicating results discusses how to keep track of the analyses used in presentations and papers, issues to consider when presenting your results, and ways to make replication simpler.

Chapter 8 – Saving your work discusses how to back up and archive your work. This seemingly simple task is often frustratingly difficult and involves subtle problems that can be easy to overlook.

Chapter 9 – Conclusions draws general conclusions about workflow.

Appendix A reviews how the Stata program operates; considers working with a networked version of Stata, such as that found in many computer labs; explains how to install user-written programs, such as the Workflow package; and shows you how to customize the way in which Stata works.

Additional information about workflow, including examples and discussion of other software, is available at http://www.indiana.edu/~jslsoc/workflow.htm.

2 Planning, organizing, and documenting

This chapter describes the three critical activities that occur at each step of data analysis: planning your work, organizing materials, and documenting what has been done. These tasks, which are closely related and equally irksome to many, are an essential part of your workflow. Planning is strategic, focusing on broader objectives and priorities. Organization is tactical, developing the structures and procedures needed to complete your plan. This includes deciding what goes where, what to name it, and how to find it. Documentation involves bookkeeping, recording what you have done, why you did it, when it was done, and where you put it. Without documentation, replication is effectively impossible.

All data analysts plan, organize, and document (PO&D) but to greatly differing degrees. When you begin your analysis, you have at least a basic idea of what you want to do (the plan), you know where things will be put (the organization), and you keep at least a few notes (the documentation). Most researchers will benefit from a more formal approach to these activities. Although this is true for all research, the importance of PO&D increases with the complexity of the project, the number of projects you are working on, and the frequency of interruptions while you work.

There is a huge temptation to jump into analysis and let planning, organization, and documentation come later. Crunching numbers is immensely more engaging than writing a plan, putting files in order, and documenting what you have done. However, even preliminary, exploratory analysis needs a plan, benefits from organization, and must be documented. Investing time in these activities makes you a better data analyst, speeds up your work, and helps you avoid mistakes. Critically, these activities make it easier to replicate your work.

One of the few advantages of working on a mainframe computer during the 1960s, 1970s, and 1980s was that card punches with 10-minute limits for use, queues to submit programs, delays in mounting tapes, and waits of hours or days for output encouraged and rewarded efficiency and planning. Although you waited for results, you had time to plan your next steps, to document what you were doing, and to organize earlier printout. Importantly, you also had the opportunity to watch how more experienced researchers did things. With delays built into the process, you did not want to forget a critical step in your program, incorrectly type a command, lose analyses that were completed, use the wrong variables, add unnecessary steps to the analyses, or forget what you had already done. Because computing was more expensive during the day (and you paid real dollars to compute), you used the day to plan the most efficient way to proceed and submitted your programs to run overnight. An unanticipated cost of cheap computing is that computation no longer imposes delays that encourage you to plan, organize, and document. Such planning is still rewarded, but the inducements are less obvious. With personal computers, there is less opportunity to watch and to learn from how others work.

The most impressive example of planning that I know of involves Blau and Duncan's (1967) masterpiece *The American Occupational Structure*. In the preface, the authors write (1967, 18–19)

> It should be mentioned here that at no time have we had access to the original survey documents or to the computer tapes on which individual records are stored. ... Consequently it was necessary for us to provide detailed outlines of the statistical tables we desired for analysis without inspecting the "raw" data, and to provide these, moreover, some 9 to 12 months ahead of the time when we might expect their delivery. ... We had to state in advance just which tables were wanted, out of the virtually unlimited number that conceivably might have been produced, and to be prepared to make the best of what we got. Cost factors, of course, put strict limits on how many tables we could request. We had to imagine in advance most of the analysis we would want to make, before having any advance indications of what any of the tables would look like. The general plan of the analysis had, therefore, to be laid out a year or more before the analysis actually began, ... We were conscious of the very real hazard that our initial plans would overlook relationships of great interest. However, some months of work were devoted to making rough estimates from various sources to anticipate as closely as possible how the tables might look.

I doubt if this exemplar of quantitative social science research would have been completed more quickly or better if the authors had been given full access to the data and complete control of a mainframe.

2.1 The cycle of data analysis

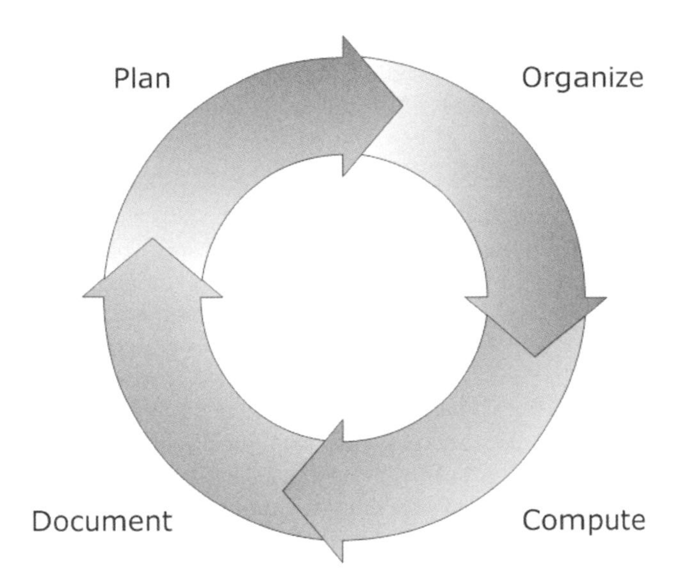

Figure 2.1. The cycle of data analysis

In an ideal world, planning, organizing, computing, and documenting occur in the sequence illustrated in figure 2.1. You begin by sketching a plan for analysis, setting up a folder for data and do-files, spending a week fitting models, and taking a few notes as you proceed. In practice, you are likely to go through this cycle many times, often moving among tasks in any order. On a large project, you begin with the master plan (e.g., the grant proposal, the dissertation proposal), set up an initial structure to organize your work (e.g., notebooks, files, a directory structure on disk drives), and examine the general characteristics of your datasets (e.g., how many cases, where data are missing). Once you have a sense of the complexities and problems with your data (e.g., inconsistent coding of missing data, problems converting the data into Stata), you develop a more detailed plan for cleaning the data, selecting your sample, and constructing variables. As analyses progress, you might reorganize your files to make them easier to find. At this point, you are ready to fit additional models. Preliminary results might uncover problems with variables that send you back to cleaning the data, perhaps requiring you to construct new variables; thus, starting the cycle again.

An effective workflow involves PO&D at different levels and in different ways. Broad plans consider your research within the context of the existing literature and determine where your research can make a contribution. More specific plans consider which variables to extract, how to select the sample, and what scales to construct. When data have been extracted and variables created, you need a plan for which models to fit, tests to make, and graphs and tables to summarize your results. Similarly, you need to organize materials including datasets, reprints, output, and budget sheets. You must decide where to locate files and where to archive them. During the analyses, you organize your do-files so that you can find what you need quickly and within the files you organize the commands in a logical sequence. Documentation also occurs on many levels. A research log keeps track of what you did and when. Codebooks, along with variable and value labels, document variables. Comments within do-files provide indispensable documentation of your analyses. When you write a paper, book, or presentation, you need to record where each number comes from should you need to revisit it later.

Planning, organizing, and documenting are ongoing tasks that affect everything that you do throughout the life of the project. At each new stage of data management and statistical analysis, you should reassess and extend your plan, decide how to incorporate new work into the existing organization, and update your documentation. Each of these tasks pays huge dividends in the quality and efficiency of your work. As you read this chapter, keep in mind that PO&D do not need to take a great deal of time and often save time. For example, I find that it takes much longer to search for one lost file than to create a directory structure that prevents losing a file. Plus many of the tasks are quite simple. For example, when I suggest that you "decide how to incorporate new work into the existing organization", this might simply involve looking at the directories you have and deciding everything is fine or it might require quickly adding one or two directories to hold new analyses.

2.2 Planning

Planning at the beginning of a project saves time and prevents errors. A plan begins with broad considerations and goals for the entire project, anticipating the work that needs to be completed and thinking about how to complete these tasks most efficiently. Data analysis often involves side trips to deal with unavoidable problems and to explore unanticipated findings. A good plan keeps your work on track. Michael Faraday, one of the greatest scientists of all time, seemed well aware of the need to stay focused until a project is complete. His laboratory had a sign that said simply (Cragg 1967): "Work. Finish. Publish." A plan is a reminder to stay on track, finish the project, and get it into print.

Although planning is important in all types of research, I find it particularly valuable in certain types of projects. First, in collaborative work, inadequate planning can lead to misunderstandings about who is doing what. This leads to a duplication of effort, to working at cross-purposes with one person undoing what someone else is doing, and to misunderstandings about access to data and authorship. Second, the larger the project and the more complex the analysis, the more important it is to plan. In projects such as a dissertation or book, it is impossible to remember all the details of what you have done. However, even if your analysis is exploratory and the project is small, your work will benefit from a plan. Third, the longer the duration of a project, the more important it is to plan and document your work. Finally, the more projects you work on, the greater the need to have a written plan.

In the rest of this section, I suggest issues to consider as you plan. This list is suggestive, not definitive. It includes topics that might be irrelevant to your work and excludes other topics that might be important. The list suggests the range of issues that should be considered as you plan. Ultimately, you have the best idea of what issues need to be addressed.

General goals and publishing plans

Begin with the broad objectives of your research. What papers do you plan to write and where will you submit them? Thinking about potential papers is a useful way to prioritize tasks so that initial writing is not held up by data collection or data management that has not been completed.

Scheduling

A plan should include a timeline with target dates for completing key stages of the project (e.g., data collection, cleaning and documenting data, and initial analysis). You might not meet the goals, but comparing what you have done with the timeline is useful for assessing your plan. If you are falling behind, consider revising the plan. You also want to note deadlines. If there are conferences where you want to present the results, when are the submission deadlines? If there is external funding, are there deadlines for progress reports or expending funds?

Size and duration

The size and duration of the project have implications for how much detail and structure is needed. If you are writing a research note, a simple structure suffices. A paper takes more planning and organization, whereas a book or series of articles makes it more important to think about how the structure you develop adapts as the research evolves.

Division of labor

Working in a group requires special considerations. Who is responsible for which tasks? Who coordinates data management? If multiple people have access to the data, how do you ensure that only one person is changing the data at a time? If the analysis begins while data collection continues, how do you make sure that people are working with the latest version of the data? Who handles backups and keeps the documentation up to date? What agreements do team members have about collaboration and joint authorship? Both the success of the project and interpersonal relationships depend on these considerations.

The enforcer

In collaborations, you need to agree on policies for documentation and organization, including many of the issues discussed in chapters 5–8. Even if everyone agrees, however, it is easy to assume (or hope) that somebody else is taking care of PO&D while you fit the models. By the time a problem is noticed, it can take longer to fix things than if the issue had been anticipated and resolved earlier. In collaborative research, you should decide who is responsible for enforcing policies on documenting, organizing, and archiving. This does not need to be the person who is doing the work, but someone has to make it their responsibility and a high priority.

Datasets

What data will be used? Do you need to apply for access to restricted datasets such as the National Longitudinal Study of Adolescent Health? What variables will be used? How many panels? Which countries? Anticipating the complexity of the dataset can prevent initial decisions that later cause problems. If you are extracting variables from a large dataset, reviewing the thousands of variables and deciding which you need to extract can prevent you from repeatedly returning to the dataset to get a few forgotten variables. If your research includes many variables, consider dividing the variables among multiple datasets. For example, in a study of work, health, and labor-force participation using the National Longitudinal Survey, we decided that keeping all variables in one file would not work because only one person could construct new variables at a time. We divided variables into groups and created separate datasets for each type of variable (e.g., demographic characteristics, health measures, and work history). We created analysis datasets by merging variables from these files (see page 279 for details on merging files).

Variable names and labels

Start with a master plan for naming and labeling variables, rather than choosing names and labels in an ad hoc manner. A simple example of the problems caused by careless names and labels occurred in a survey where the same question was asked early in the survey and again near the end. Unfortunately, the variables were named `ownsex` with the label `How good own sexuality?` and `ownsexu` with the label `Own sexuality is ....` Neither the names or the labels made it clear which variable corresponded to the question that was asked first. It took hours to verify which was which. When planning names, anticipate new variables that could be added later. For example, if you expect to add future panels, you need names that distinguish between variables in different panels (e.g., health status in panel 1, health status in panel 2). If you are using software that restricts names to eight characters, you should plan for this. Chapter 5 has an extended discussion on variable names (section 5.6) and labels (section 5.7).

Data collection and survey design

When collecting your own data, many things can go wrong. Before you start collecting data, I recommend that you create a codebook and write the do-files that create variable and value labels. This gives you one more chance to find problems when you can do something about them. Another survey gave respondents options for percentage of time that included the ranges 0–10%, 20–30%, 40–50%, and so on. After the data collection was complete, the person adding value labels noticed that 11–19%, 31–39%, and so on had been excluded.

Missing data

What types of missing data will be encountered, and how will these types be coded? Will a single code for missing values be sufficient, or will you need multiple codes that indicate why the data are missing (e.g., attrition, refusal, or a skip pattern in the survey)? Try to use the same missing-value codes for all variables. For example, letting `.n` stand for "not answered" for one variable and stand for "not applicable" in another is bound to cause confusions. See section 6.2.3 for details on missing data in Stata.

Analysis

What types of statistical analyses are anticipated? What software is needed, and is it locally available? Thinking about software helps you plan data formats, naming conventions, and data structures. For example, if you plan to use software that limits names to eight characters, you might want a simpler naming structure than if you plan to work exclusively in Stata, which allows longer names.

Documentation

What documentation is needed? Who will keep it? In what format? A plan for how to document the project makes it more likely that things will be documented.

Backing up and archiving

Who is going to make regular backups of the files? Long-term preservation should also be considered. If the research is funded, what requirements does the funding agency have for archiving the data? What sort of documentation do they expect and what data formats? If the research is not funded, would it not be a good idea to make the data available when you finish the research? Creating the documentation as you go makes this much simpler. See chapter 8 for further information on backing up and archiving files.

2.3 Organization

Organization involves deciding what goes where, what to name it, and how you will find it. A good plan makes it easier to create a rational structure to organize your work. Plans for the broader objectives help you define how complex your organization needs to be. Plans for more specific issues, such as how to name files, help you complete the work accurately and quickly. Thoughtful organization also makes it simpler to document your work because a clear logic to the organization makes it easier to explain what files are and where they are located.

2.3.1 Principles for organization

There are several principles that should guide how you organize your work. These principles apply to all aspects of your research, including budget sheets, reprints, computer files, and more. Because this book is about data management, I focus on issues related to data analysis.

Start early

The more organized you are when a project begins, the more organized you will be at the end. Organization is contagious. If things are disorganized, there is a temptation to leave them that way because it takes so much time to put them into order. If things start out organized, keeping them organized takes very little time.

Simple, but not too simple

More elaborate schemes for organization are not necessarily better. The goal is to be organized but to do this as simply as possible. A complex directory or folder structure is essential for large projects but makes things harder for simple projects. For example, if you have only one dataset and a few dozen do-files, a single directory should be fine. If you have hundreds of do-files and dozens of datasets, it can be difficult to find things in a single directory. Because I find that most projects end up more complicated than anticipated, I prefer more elaborate organization at the start. You can also start with a simple structure, and let it grow more complex as needed. Examples of how to organize directories are given in section 2.3.2.

Consistency

Consistency and uniformity pay dividends in organization as well as in documentation. If you use the same scheme for organizing all your projects, you will spend less time thinking about organization because you can take advantage of what you already know. For example, if all projects keep codebooks in a directory named `\Documentation`, you always know where to find this information. If you organize different projects differently, you are bound to confuse yourself and spend time looking for things.

Can you find it?

Always keep in mind how you will find things. This seems obvious but is easily overlooked. For example, how will you find a file that is not in the directory where it should be? Software that searches for files helps, but these programs work better if you plan your file naming and content so that search programs work more effectively. For example, suppose you have a paper about cohort effects on work and health that you refer to as the CWH paper. To take advantage of searching by name, filenames must include the right information (e.g., the abbreviation `cwh`). With search programs, you can look for a file with a specific name (e.g., `cwh-scale1.do`) or for a file with a name that matches some pattern (e.g., `cwh*.do` looks for all files that begin with `cwh` and end with `.do`). To search by content, you must include keywords within your files. For example, suppose that all do-files related to the project include the letters "CWH" within them. If you lose a file, you can let a search program run overnight to find all files that have the extension `.do` and contain the phrase "CWH". If you forget to include "CWH" inside a file, you will not find the file. Or, if you place different files with the same name in different directories (e.g., two projects each use a file called `extract-data.do`), searching by filename will turn up multiple files.

Document your organization

You are more likely to stay organized if you document your procedures. Written documentation helps you find things, prevents you from changing conventions midproject if you forget the original plan, and reminds you to stick to your plan for organization. In collaborations, written procedures are essential.

2.3.2 Organizing files and directories

> *It is easier to create a file than to find a file. It is easier to find a file than to know what is in the file. With disk space so cheap, it is tempting to create a lot of files.*

Do any of the following sound familiar?

- You have multiple versions of a file and do not know which is which.
- You cannot find a file and think you might have deleted it.
- You and a colleague are not sure which draft of your paper is the latest or find that there are two different "latest" drafts.
- You want the final version of the questionnaire and are not sure which file it is because two versions of the questionnaire include "final" in the name.

I find that these and similar problems are very common. One approach is to document the name, content, and location of each file in your research log. In practice, this takes too long. Instead, care in naming files and organizing their location is the key to keeping track of your files.

The easiest approach to organizing project files is to start with a carefully designed directory structure. When files are created, place them in the appropriate directory. For example, if you decide that all PDFs for readings associated with a project belong in the `\Readings` directory, you are less likely to have PDFs scattered across your hard drive, including duplicate copies downloaded after you misplaced the first copy. Another advantage of a carefully planned directory structure is that a file's location becomes an integral part of your documentation. If a file is located in the directory `\CWH` in the subdirectory `\Proposal`, you know the file is related to the research proposal for the CWH project. Section 2.3.3 discusses creating a directory structure. Approaches to naming files are discussed in chapter 5. Before proceeding, keep in mind, that if you create an elaborate directory structure but do not use it consistently, you will only make things worse.

What characters to use in names?

Not all names work equally well in all operating systems. Names are most likely to work across operating systems if you limit the characters used to `a`–`z`, `A`–`Z`, `0`–`9`, the underscore `_`, and the dash `-`. Macintosh names can include any character except a colon `:`. Windows names have more exceptions and should not use `/`, `[`, `]`, `;`, `=`, `"`, `\`, `:`, `|`, `*`, and `,` . In Linux, names can include numbers, letters, and the symbols `.`, `_`, and `-`. Although blank spaces can be used in file and directory names, some people feel strongly that spaces should never be used. For example, instead of having a directory called `\My Documents`, they prefer `\My-documents`, `\My_documents`, or simply `\Documents`. Blanks can make it more difficult to refer to a file. For example, suppose that I save `auto.dta` in `c:\Workflow\My data\auto.dta`. To use this dataset, I must include double quotes: `use "d:\Workflow\My data\auto.dta"`. If I forget the quotes, an error message is received:

```
. use d:\Workflow\My data\auto.dta
invalid `data'
r(198);
```

Similarly, if you name a do-file `my pgm.do` and need to search for the file, you need to search for `"my pgm.do"`, not simply `my pgm.do`. As a general rule, I avoid filenames that include spaces, but I use spaces in directory names when the spaces make it easier for me to understand what is in the directory or because I think it looks better. Thus, in the names of the directories that I suggest below, some directory names include spaces, although the most frequently used directories do not. If you want to avoid spaces, you can replace them with either a dash (`-`) or an underscore (`_`), or simply remove the space from the name.

Pick a mnemonic for each project

The first step in naming files and directories is to pick a short mnemonic for your project. For example, `cwh` for a paper on cohort, work, and health; `sdsc` for the project on sex differences in the scientific career; `epsl` for my collaboration with Eliza Pavalko.

This lets me easily add the project identifier to file and directory names. When choosing a mnemonic, pick a string that is short because you do not want your names to get too long. Avoid mnemonics that are commonly found in other contexts or as part of words. For example, do not choose the mnemonic `the` because "the" occurs in many other contexts, and do not use `ead` because these letters are part of many common words.

2.3.3 Creating your directory structure

Directories allow you to organize many files, just as file cabinets and file folders allow you to organize many papers. Indeed, some operating systems use the term folder instead of directory. When referring to a directory or folder, I start the name with \, such as `\Examples`. Directories themselves can contain directories, which are called *subdirectories* because they are "below" the parent directory. All the work related to a project should be contained within a single directory that I refer to as the *project directory* or the level-0 directory. For example, `\Workflow` is the project directory for this book. The project directory can be a subdirectory of some other directory or can be on a network, on your local hard drive, on an external drive, or on a flash drive. Under the project directory you can create subdirectories to organize your files. The term *level* indicates how far a directory is below the project directory. A level-1 directory is the first level under the project directory. For example, `\Workflow\Examples` indicates the level-1 directory `\Examples` contained within the level-0 directory `\Workflow`. A level-1 directory can have level-2 directories within it, and so on. For example, `\Workflow\Examples\SourceData` adds the level-2 directory `\SourceData`. When referring to a directory, I might indicate all levels (e.g., `\Workflow\Examples\SourceData`) or simply refer to the subdirectory of interest (e.g., `\SourceData`). With this terminology in hand, I consider several directory structures for use with increasingly complex projects.

A directory structure for a small project

Consider a small project that uses a single data source, only a few variables, and a limited number of statistical analyses. The project might be a research note about labor-force participation. I start by creating a project directory `\LFP` that will hold everything related to the project. Under the project directory, there are five level-1 subdirectories:

Directory	Content
`\LFP`	Project name
`\Administration`	Correspondence, budgets, etc.
`\Documentation`	Research log, codebooks, and other documentation
`\Posted`	Completed text, datasets, do-files, and log files
`\Readings`	PDF files with articles related to the project
`\Work`	Text and analyses that are being worked on

To make it easier to find things, all files are placed in one of the subdirectories, rather than in the project directory itself.

The \Work and \Posted directories

The folders `\Work` and `\Posted` are critical for the workflow that I recommend. The directory `\Work` holds work in progress. For example, the draft of a paper I am actively working on would be located here, as would the do-files that I am debugging. At some point I decide that a draft is ready to circulate to colleagues. Before sharing the paper, I move the text file to the `\Posted` directory. Or, when I think that a group of do-files is running correctly and I want to share the results with others, I move the files to `\Posted`. There are two essential rules for posting files:

> The share rule: Results are only shared after the associated files are posted.
>
> The no-change rule: Once a file is posted, it is never changed.

These simple rules prevent many problems and help assure that publicly available results can be replicated. By following these rules, you cannot have multiple copies of the "same" paper or results that differ because they were changed after they were shared. If you decide something is wrong in your analyses or you want to revise a paper that was circulated, you create new files with new names, but do not change the posted files. The distinction between the `\Work` and `\Posted` directories also helps me keep track of work that is not finished (e.g., I am still revising a draft of a paper; I am debugging programs to construct scales) and work that is finished. When I return to a project after an interruption, I check the `\Work` directory to see if there is work that I need to finish. For a detailed discussion of the idea of posting and why it is critical for your workflow, see page 125.

Expanding the directory structure

As my work develops, I might accumulate dozens or hundreds of do-files. When this happens, I could divide `\LFP\Posted` to include level-2 subdirectories for different aspects of data management and statistical analysis. For example,

Directory	Content
`\LFP`	Project name
`  \Posted`	Datasets, do-files, logs, and text files
`    \Analysis`	Do-files and logs for statistical analyses
`    \DataClean`	Do-files and logs for data management
`    \Datasets`	Datasets
`    \Text`	Drafts of paper

The idea is to add subdirectories when you have trouble keeping track of what is in a directory. The principle is the same as used when putting reprints in a file cabinet.

Initially, I might have sections A–F, G–K, L–P, and Q–Z. If you have a lot of papers in the L–P folder, I might divide that folder into L–M and N–P. Or, if I have lots of papers by R. A. Fisher, I might create a separate folder just for his papers.

A directory structure for a large, one-person project

Larger projects require a more elaborate structure. Suppose that you are the only person working on a paper, book, or grant. Collaborative projects are discussed below. Your project directory might begin with a structure like this:

Directory	Content
\Administration	Files for administrative issues
\Budget	Budget spreadsheets and billing information
\Correspondence	Letters and emails
\Proposal	Grant proposal and related materials
\Posted	Datasets, do-files, logs, and text files
\DataClean	Clean data and construct variables
\Datasets	Datasets
\Derived	Datasets constructed from the source data
\Source	Original, unchanged data sources
\DescStats	Descriptive statistics
\Figures	Programs to create graphs
\PanelModels	Panel models of discrimination
\Text	Drafts of paper
\Documentation	Project documentation (e.g., research log, codebooks)
\Readings	Reprints and bibliography
\Work	Text and analyses that are being worked on

Later in this section, I suggest other directories that you might want to add, but first I discuss changes needed for collaborative projects.

Directories for collaborative projects

A clear directory structure is particularly important for collaborative projects where things can get disorganized quickly. In addition to the directories from the prior section, I suggest a few more.

The mailbox directory

You need a way to exchange files among researchers. Sending files as attachments can fill up your email storage quota and is not efficient. I suggest a mailbox directory. Suppose that Eliza, Fong, and Scott are working on the project. The mailbox looks like this:

Directory	Content
`\Mailbox`	Files being exchanged
`\Eliza to Fong`	Eliza's files for Fong
`\Eliza to Scott`	Eliza's files for Scott
`\Fong to Eliza`	Fong's files for Eliza
`\Fong to Scott`	Fong's files for Scott
`\Scott to Eliza`	Scott's files for Eliza
`\Scott to Fong`	Scott's files for Fong

We exchange files by placing them within the appropriate directory.

Private directories

I also suggest private directories where you can put work that you are not ready to share with others. One approach is to create a level-1 directory `\Private` with subdirectories for each person:

Directory	Content
`\Private`	
`\Eliza`	Eliza's private files
`\Fong`	Fong's private files
`\Scott`	Scott's private files

With only a few team members, you might not need the `\Private` directory and could create the private directories in the first level of the project directory, such as `\epsl\Eliza` and `\epsl\Scott`. Each person can decide how they want to organize files within their private directory.

The data manager and transfer directories

Even if everyone agrees in principle on where the files should be put, you need a data manager to enforce the agreement. Otherwise, entropy creeps in and you will lose files, have multiple copies of some files, and have different files with the same name. The data manager makes sure that files are put in the right place. The principle is the same as used by libraries where librarians rather than users shelve the books. Each member of the team needs a way to transfer files to the data manager. To make this work, I suggest a data transfer directory called `\- To file` along with subdirectories for each member of the team. The directory name begins with `-` so that it appears at the top of a sorted list of files and directories. For our project, we set up this structure:

Directory	Content
`\- To file`	Files for the data manager to relocate
`\- To clean`	Files that need to be evaluated before filing
`\From Eliza`	Files Eliza wants to have relocated
`\From Fong`	Files Fong wants to have relocated
`\From Scott`	Files Scott wants to have relocated

The data manager verifies each file before moving it to the appropriate location. The `\- To clean` directory is for those files that invariably appear that nobody is sure who created or what they are.

Restricting access

For collaborations, you are probably using a local area network (LAN) where everyone can potentially access the files. If people store project files on their local hard drives, you risk having data scattered across multiple machines and it is difficult to find and to back up what you need. Although a LAN solves this problem, you might have files that you do not want everyone to use. For example, you might want to restrict access to the budget materials in `\Administration\Budget`. Or you might want some people to have only read access to datasets to avoid the possibility of accidental changes. You can work with your network administrator to set up *file permissions* that determine who gets what type of access to which files and directories.

Is the LAN backed up?

If you are using a LAN, you should not assume that it is backed up until you talk with your LAN manager. Find out how often the LAN is backed up, how long the backups are kept, where the backups are located, and how easy it is to retrieve a lost file from the backup. These issues are discussed in chapter 8.

Special-purpose directories

I also use several special-purpose directories for things such as holding work that needs to be done or holding backup copies of files. Although I begin the names of these directories with a dash (e.g., `\- To do`), you can remove the dash if you prefer (e.g., `\To do`).

The \- To do directory

Work that has not been started goes here as a subdirectory under `\Work`. These files are essentially a to-do list. If I think of something that needs to be done, a reprint I need to read, a do-file that needs to be revised, etc., it belongs here until I get a chance to do it. I begin the name with a dash so that it appears at the top of a sorted list of directories.

The \- To clean directory

Inevitably, I accumulate files that I am not sure about or that need to be moved to the appropriate directory. By having a special folder for these files, I am less likely to carelessly put them in the wrong directory. At some point, I review these files and move them to their proper location. This directory can be located immediately under the project directory or as a subdirectory elsewhere.

The \- Hold then delete directory

This directory holds files that I want to eventually delete and short-term copies of files as a fail-safe in case I accidentally delete or incorrectly change the original. For example, if I decide to abandon a set of do-files and logs for analyses that did not work, I move them here. This makes it easy to "undelete" the files if I change my mind. Or suppose that I am writing a series of do-files to create scales, select cases, merge datasets, and so on. These programs work, but before finalizing them I want to add labels and comments and perhaps streamline the commands. Making these improvements should not change the results, but there is a chance that I will make a mistake and break a program that was working correctly. When this happens, it is sometimes easiest to return to the version of the program that worked and start again rather than debugging the program that does not work. With this in mind, before I start revising the programs I copy them from `\Work` to `\- Hold then delete`. I might have subdirectories with the date on which the backup was made. For example,

Directory	Content
`\- Hold then delete`	Temporary copies of files
`\2006-01-12`	Files backed up on January 12, 2006
`\2006-02-09`	Files backed up on February 9, 2006

Or I might use subdirectories that indicate what the backups are for. For example,

Directory	Content
`\- Hold then delete`	Temporary copies of files
`\VarConstruct`	Files used in variable construction
`\REmodels`	Files used to fit random-effects models

When I have completed a major step in the project (e.g., submitted a paper for review), I might *copy* all the critical files to `\- Hold then delete`. For example,

Directory	Content
\- Hold then delete	Temporary copies of files
\2007-06-13 submitted	Do-files, logs, data, and text when paper was submitted
\2008-04-17 revised	Do-files, logs, data, and text when revisions were submitted
\2008-01-02 accepted	Do-files, logs, data, and text when paper was accepted

The critical files should already be in the `\Posted` directory, but before posting files, I often delete things that I do not expect to need. By keeping temporary copies of these files, I can easily recover a file if I made a mistake by deleting it. In many ways, this directory is like the Windows Recycle Bin or Mac OS Trash Can. I put files here that I do not expect to need again, but I want to easily recover them if I change my mind. When organizing files, it is important to keep track of the files you need and also the files which you do not need. If you do not keep track of files that can be deleted, you are likely to end up with lots of files that you do not know what to do with (sound familiar?). When I need disk space or the project is finished, I delete the files in the `\- Hold then delete` directory.

The \Scratch directory

When learning a new command or method of analysis, I often experiment to make sure that I understand how things work. For example, if I am importing data, I might verify that missing data codes are transferred the way I expect. If I am trying a new regression command, I might experiment with the command using data from a published source where I know what the estimates should be. These analyses are important, but I do not need to save the results. For this type of work, I use a `\Scratch` directory. When I need disk space or the project is finished, these files can be deleted. Generally, `\Scratch` is located within the `\Work` directory. But, wherever it appears, I know that the files are not critical.

Remembering what directories contain

You need a way to keep track of what a directory is for and which files go where. You could give each directory a long name that describes its contents, such as `\Text for workflow book`. However, if each directory name is long, you can end up with path names that are so long that some programs will not process the file. Long names are also tedious to type. To keep track of what a directory is for, I suggest a combination of the following approaches.

First, decide on a directory structure with short names and use the same structure for everything you do. Eventually, it will become second nature. For example, if every project directory contains a subdirectory `\Work`, you know where things you are currently working on are located when you return to the project. You can choose a different name than `\Work` but use the same name for all your projects.

Second, use a text file within the directory to explain what goes in the directory. For example, the `\Workflow\Posted\Text\Submitted` directory for the workflow project could have a file `Submitted.is` that contains

```
Project:     Workflow of Data Analysis
Directory:   \Workflow\Posted\Text\Submitted
Content:     Files submitted to StataCorp for production.
Author:      Scott Long
Created      2008-06-09
Note:        These files were submitted to StataCorp for copy
             editing and latexing.  Prior drafts are located
             in \Workflow\Posted\Text\Drafts.
```

The naming file can be as large as you like. Because you must open the file to read the information, this approach is not effective as a quick reminder.

Third, you can create *naming directories* whose sole purpose is to remind you of what is in the directory above it. For example,

Directory	Content
`\Private`	Private files
`    \- Private files for team members`	Description of the `\Private` directory

I use this approach to keep track of directories containing backup files. The naming directory tells me which external drive holds the backup copies. For example,

Directory	Content
`\- Hold then delete`	Backup files
`    \2006-01-12`	Date files were placed in this directory
`        \- Copied to EX02`	Reminder that files are on external drive `EX02`
`    \2007-06-03`	Date files were placed in this directory.
`        \- Copied to EX03`	Reminder that files are on external drive `EX03`

Finally, I use a directory named `\-  History` that contains naming directories with critical information about the files in the project. For example,

Directory
`\-  History`
`    \2006-01-12 project directory created`
`    \2006-06-17 all files backed up to EX02`
`    \2007-03-10 initial draft completed`
`    \2007-03-10 all files backed up to EX04`

I find these reminders to be very useful when returning to a project that has been put on hold. It also documents where backup copies of files have been put (e.g., `EX02` is the volume name of an external drive).

Planning your directory structure

You might prefer to use different directory names than I have suggested. Having names that make sense to you is an advantage, but there is also an advantage to using names that have been documented. This, I believe, is a good reason to stick with the names I suggest or versions of these names that replace spaces with dashes or underscores. If you add people to your project, they can read this chapter to find out what the directories are for. Still, even if you use my names, you will need to customize some things. A spreadsheet is a convenient way to plan your directory structure. For example (file: `wf2-directory-design.xls`),[1] see figure 2.2.

Project Directory	Level 1	Level 2	Level 3	Purpose
\AgeDisc				Project directory.
	\- To file			Files to examine and move to appropriate location.
	\Administration			Administration.
		\Budget		Budget sheets.
		\Correspondence		Letters and emails.
		\Proposal		Grant proposal and related materials.
	\Documentation			Documentation for project.
		\Codebooks		Codebooks for source and constructed variables.
	\Hold then delete			Delete when project is complete.
		\2007-06-13 submited		Do, data and text when paper was submitted.
		\2008-04-17 revised		Do, data and text when revisions are sumbitted.
		\2008-01-02 accepted		Do, data and text when paper is accepted.
	\Posted			Completed files that cannot be changed.
		\- Datasets		Datasets.
			\Derived	Dataset constructed from original data files.
			\Source	Original data without modifications.
		\- Text		Completed drafts of paper.
		\DataClean		Data cleaning and variable construction.
		\DescStats		Descriptive statistics and sample selection.
		\Figures		Graphs of data.
		\PanelModels		Panel models for discrimination.
	\Readings			Articles related to project; bibliography.
	\Work			Work directory.
		\- To do		Work that hasn't been started.
		\Text		Active drafts of paper.

Figure 2.2. Spreadsheet plan of a directory structure

This spreadsheet would be kept in the `\Documentation` directory.

1. This is the first time I have referred to a file that is available from the Workflow web site. Throughout the book, files that have names that begin with `wf` can be downloaded. See the *Preface* for further details.

Naming files

After you set up a directory structure, you should think about how to name the files in these directories. Just as you need a logical structure for your directories, you need a logical structure for how you will name files. For example, if you put reprints in the `\Readings` directory, but the files are not consistently named, it will be hard to find them. My PDF files with reprints are a good example of what not to do. Although I routinely filed paper reprints by author in a file cabinet, I often downloaded files and kept whatever names they had. As a result, here is a sample of files from my `\Readings` directory:

```
03-19Greene.pdf
0OWENS94.pdf
12087810.pdf
12087811.pdf
Chapter03.pdf
CICoxBM95.pdf
cordermanton.pdf
faiq-example.pdf
gllamm2004-12-10.pdf
long2.pdf
Muthen1999biometrics.pdf
```

It is not worth the effort to rename these files, but I name new PDFs with the first author's last name followed by year, journal abbreviation, and keyword (e.g., `Smith 2005 JASA missingdata.pdf`). Issues of naming, which are even more important when it comes to do-files and datasets, are discussed in chapter 5.

Batch files

I prefer to create the directory structure using a batch file in Windows or a script file in Mac OS or Linux rather than right-clicking, choosing **Create a new folder**, and typing the name. A batch file is a text file that contains instructions to your operating system about doing things such as creating directories. The first advantage of a batch file is that if you change your mind, you can easily edit the batch file to re-create the directories. Second, you can use the batch file from one project as a template for creating the directory structure for another project. For example, I use this file to create the directories for a project with Eliza (file: `wf2-dircollab.bat`):

```
md "- Hold then delete"
md "- To file\Eliza to data manager"
md "- To file\Scott to data manager"
md "- To file\- To clean"
md "Administration\Budget"
md "Administration\Correspondence"
md "Administration\Proposal"
md "Posted\Datasets"
md "Documentation\Codebooks"
md "Mailbox\Eliza to Scott"
md "Mailbox\Scott to Eliza"
md "Private\Eliza"
md "Private\Scott"
md "Readings"
```

To set up directories for a different project, I only need to make a few changes to the batch file. Details on batch files are beyond the scope of this book; ask your local computer support person for help.

2.3.4 Moving into a new directory structure (advanced topic)

Ideally, you create a directory structure at the start of a project and routinely place new files in the proper directory. However, even with the best intentions, you are likely to end up with orphan files created over several years and scattered across directories on several computers. At some point, these files need to be combined into one project directory. Or, perhaps this chapter has convinced you to reorganize your files. In this section, I discuss how to merge files from multiple locations into a unified directory structure. Reorganizing files is difficult, especially if you have lots of files. If you start the job but do not finish it, you are likely to make things worse. If you begin to reorganize files without a careful plan, you can make things worse and even lose valuable data.

Aside on software

When doing a lot of work with files, utility programs can save time and prevent errors. First, third-party file managers are often more efficient for moving and copying files than those built into the operating system. Second, when you copy a file, most programs do not verify that the copy is exactly like the original. For example, in Windows when Explorer copies a file, it only verifies that the copied file can be opened but it does not (contrary to what you sometimes read) verify that the new file is exactly like the source file. I highly recommend using a program that verifies the copy is exactly the same as the original by comparing every bit in the original file to every bit in the destination file. This is referred to as *bit verification*. Programs for backing up files and many file managers do this. Third, when combining files from many locations, you are likely to have duplicate files. It is slow and tedious to verify that files with the same names are in fact identical and that files with different names are not the same. I recommend using a utility to find duplicate files. Software for file management is discussed on the Workflow web site.

Example of moving into a new directory structure

To make the discussion of moving into a new directory structure concrete, I explain how I would do this for a collaborative project known as `epsl` (named with the initials of the two researchers).

Step 1. Inform collaborators

Before I start to reorganize files, I let everyone using the files know what I am doing. Others can still use files from their current locations, but they should not add, change, or delete files within the current directory structure. Instead, I create new directories (e.g., `\epsl-new-files\eliza` and `\epsl-new-files\scott`) where new or changed files can be saved until the new directory structure is completed.

Step 2. Take an inventory

Next I take an inventory of all files related to the project. The inventory is critical because I do not want to complete the reorganization and then discover that I forgot some files. I found files on the LAN directory `\epsl`; on Eliza's home, office, and laptop computers; and on my home and two work computers. I create a text file that lists each file and where it was found. This list is used to document where files were before they were reorganized and to help plan the new organization. I do not want to try to relocate 10,000 files without having a good idea of where I want to put things. Most operating systems have a way to list files; see the Workflow web site for further details.

Step 3. Copy files from all source locations

On an external drive, I create a holding directory with subdirectories for each source location. For example,

Directory	Content
`\epsl-to-be-merged`	Holding directory with copies of files to be merged
`\Eliza-home`	Files from Eliza's home computer
`\Eliza-laptop`	Files from Eliza's laptop
`\Eliza-office`	Files from Eliza's office computer
`\LAN`	Files from LAN
`\Scott-home`	Files from Scott's home computer
`\Scott-officeWin`	Files from Scott's Windows computer
`\Scott-officeMac`	Files from Scott's Mac computer

Using bit verification, I copy files from each source location to the appropriate directory in `\epsl-to-be-merged`. Do not delete the files from their original location until the entire reorganization is complete.

Step 4. Make a second copy of the combined files

After all the files have been copied to the external drive, I make a second backup copy of these files. If you do not have many files, you could copy the files to CDs or DVDs, although I prefer using a second external drive because hard drives are much faster and hold more. The copies are bit verified against the originals. The first portable drive will be used to move files into their new location, while the second backup copy is put in a safe place as part of the backups for the project.

Step 5. Create a directory structure for the merged files

Next I create the destination directory structure that will hold the merged source files. For example,

Directory	Content
\epsl-cleaned-and-merged	Destination directory with cleaned files
\- Hold then delete	Files that can be deleted
\- To file	Files to move to their proper folder
\- To clean	Files to clean before relocating
\From Eliza	
\From Scott	
\Administration	Administrative materials
\Budget	
\Correspondence	
\Documentation	Project documentation
\Codebooks	
\Mailbox	Location for exchanging files
\Eliza to Scott	
\Scott to Eliza	
\Posted	Posted datasets, do-files, etc.
\Datasets	Completed datasets
\Derived	
\Source	
\Text	Completed drafts of papers
\Private	Private files
\Eliza	
\Scott	
\Readings	PDFs related to project

I make the directory structure as complete as possible. For example, if there are a lot of analysis files, I would create subdirectories for each type of analysis. Creating the new directory structure takes careful planning but is critical for getting the job done efficiently.

Step 6. Delete duplicate files

There are likely to be multiple copies of some files. For example, Eliza and I might both have copies of the grant proposal or key datasets. Or my laptop and office machine might have copies of many of the same files. We could also have files with different names but identical content. Or worse, we could have files with the same name but different content. I need to delete these duplicate files, but the problem is finding them efficiently. For this, I use a utility that searches for duplicate files.

Step 7. Move files to the new directory structure

Next I move the files from the directory `\epsl-to-be-merged` to their new location in `\epsl-cleaned-and-merged`. Because I am moving the files, I cannot accidentally copy the same file to two locations and end up with more files than I started with. Moving the files to their new location can take a lot of time and I might encounter files that I am unsure about. I put these files in the `\- To file\- To clean` directory to relocate later.

Step 8. Back up and publish the new files and structure

When I am done moving files to their new location, I back up the newly merged files in `\epsl-cleaned-and-merged`. If I have room for these files on the portable drive that I used for the backup copy of `\epsl-to-be-merged`, I would put them there. Next I move `\epsl-cleaned-and-merged` to its new location on the LAN and start implementing new procedures for saving files.

Step 9. Clean up and let people know

I now either delete the original files or move them into a directory called `\- Hold and delete epsl`. It is essential that people stop using their old files or we will end up repeating the entire process, but next time we will need to deal with the files that were just cleaned. I inform collaborators that the new directory structure is available and ask them to move any new files they created to the `\- To file` directory.

Step 10. Documentation

I return to the list of files I created in step 2 and add details on where the files were moved. I also list problems that I encountered and assumptions that I made (e.g., I assumed that `mydataxyz.dta` was the most recent version of data even though it had an older date). I also add information to my research log that briefly discusses how the files were reorganized and where the archived copies of the original files are stored.

2.4 Documentation

> *Long's law of documentation*: It is always faster to document it today than tomorrow.

Documentation boils down to keeping track of what you have done and thought. It reminds you of decisions made, work completed, and plans for future work. Without documentation, replication is essentially impossible. Unfortunately, writing good documentation is hard and few enjoy the task. It is more compelling to discover new things by analyzing your data than it is to document how you scaled your variables, where you stored a file, or how you handled missing data. But, the time spent documenting your work saves time in the long run. When writing a paper or responding to reviews,

I often use analyses that were completed months or even years before. This is much easier when decisions and analyses are clearly documented. For example, a collaborator and I were asked by a reviewer to refit our models using the number of children under 18 years old in the family rather than the number of children under 6 years old, which we had used. Using our documentation and archived copies of the do-files, the new analyses took only an hour. Without careful documentation and archiving, it would have taken us much longer, perhaps days.

If you do not document your work, many of the advantages of planning and organization are lost. A wonderful directory structure is not much help if you forget what goes where. The most efficient plan for archiving is of no value if you forget what the plan is or you fail to document the location of the archived files. To ensure that you keep up with documentation, you need to include it as a regular part of your workflow. You can add the task to your calendar just like a meeting, although this does not work for me. Instead, I keep up with documentation by linking it to the completion of key steps in the project. For example, when a paper is sent for review, I check the documentation for the analyses used in the paper, add things that are missing, organize files, and verify that files are archived. When I finish data cleaning and am ready to start the analysis, I make sure that my documentation of the dataset and variables is up to date.

Ironically, the insights you gain through days, weeks, or years on a project make it harder to write documentation. When you are immersed in data analysis, it is difficult to realize that details that are second nature to you now are obscure to others and may be forgotten by you in the future. Was cohort 1 the youngest cohort or the oldest? Which is the latest version of a variable? What assumptions were made about missing data? Is `ownsex` or `ownsexu` the name of the variable for the question asked later in the survey? Does JM refer to Jack Martin or Janice McCabe? As you work on a project, you accumulate tacit knowledge that needs to be made explicit. Rather than thinking of documentation as notes for your own use, think of it as a public record that someone else could follow. Terry White, a researcher at Indiana University, refers to the "hit-by-a-bus" test. If you were hit by a bus, would a colleague be able to reconstruct what you were doing and keep the project moving forward?

Although documentation is central to training in some fields, it is largely ignored in others. In chemistry, a great deal of attention is given to recording what was done in the laboratory and publishers even sell special notebooks for this purpose. The American Chemical Society has published *Writing the Laboratory Notebook* (Kanare 1985), which is devoted entirely to this topic. A search of the web provides wonderful examples of how chemists document their work. For example, Oregon State University's Special Collection Library maintains a web site with scans of 7,680 pages from 46 volumes of research notes written by Nobel Laureate Linus Pauling (http://osulibrary.oregonstate.edu/specialcollections/rnb/index.html). A Google search turns up jobs descriptions that include statements like (http://ilearn.syr.edu/pgm_urp_project.htm): "Involvement in on-going chemical research toward published results. Act as junior scientist, not skilled technician. Maintain research log, attend weekly (evening) group meetings, present own results informally."

In my experience, documentation is rarely discussed in courses in applied statistics (if you know of exceptions, please let me know). This is not to say that skilled data analysts do not keep research records but rather that the training is haphazard and too many data analysts learn the hard way about the importance of documentation.

2.4.1 What should you document?

What needs to be documented varies by the nature of the research. The ultimate criterion for whether something should be documented is whether it is necessary for replicating your findings. Unfortunately, it is not always obvious what will be necessary. For example, you might not think of recording which version of Stata was used to fit your model, but this can be critical information (see section 7.6.2). Hopefully, the following list gives you an idea of the range of materials to consider for inclusion in your documentation.

Data sources

If you are using secondary sources, keep track of where you got the data and which release of the data you are using. Some datasets are updated periodically to correct errors, to add new information, or to revise the imputations for missing data.

Data decisions

How were variables created and cases selected? Who did the work? When was it done? What coding decisions were made and why? How did you scale the data and what alternatives were considered? If you dichotomized a scale, what was your justification? For critical decisions, also document why you decided not to do something.

Statistical analysis

What steps were taken in the statistical analysis, in what order, and what guided those analyses? If you explored an approach to modeling but decided not to use it, keep a record of that as well.

Software

Your choice of software can affect your results. This is particularly true with recent statistical techniques where competing packages might use different algorithms leading to different results. Moreover, newer versions of the same software package might compute things differently.

Storage

Where are the results archived? When you complete a project or put it aside to work on other projects, keep a record of where you are storing the files and other materials.

Ideas and plans

Ideas for future research and lists of tasks to be completed should be included in the documentation. What seems like an obvious idea for future analysis today might be forgotten later.

2.4.2 Levels of documentation

Documentation occurs on several levels that complement one another.

The research log

The research log is the cornerstone of your documentation. The log chronicles the ideas underlying the project, the work you have done, the decisions made, and the reasoning behind each step in data construction and statistical analysis. The log includes dates when work was completed, who did the work, what files were used, and where the materials are located. As the core of your documentation, the log should indicate what other documentation is available and where it is located. In section 2.4.4, I present an excerpt from one of my research logs and provide a template that makes it easier to keep a log.

Codebooks

A codebook summarizes information on the variables in your dataset. The codebook reflects the final decisions made in collecting and constructing variables, whereas the research log chronicles the steps taken and computer programs used to implement these decisions. The amount of detail in a codebook depends on a number of things. How many people will use the data? How much detail is in your research log? How much documentation was stored internally to the dataset, such as variable labels, value labels, and notes. Additional information on codebooks is provided in section 2.4.5. See also section 8.5 on preparing data for archival preservation.

Dataset documentation

If you have many datasets, you might want a registry of datasets. This will help you find a particular dataset and can help ensure that you are working with the latest data. An example is given below. You can also use Stata's `label` and `notes` commands to add metadata to your datasets as discussed in section 2.4.6 and chapter 5.

Documenting do-files

Although the research log should include information about your do-files, your do-files should also include detailed comments. These comments are echoed in the Stata log file and clarify what the output means, where it came from, and how it should be interpreted. You need to find a practical balance between how much information goes in the research log and how much goes in the do-file. My research log usually has limited information about each do-file, with fuller documentation located within the do-files. Indeed, for smaller projects, you might find that your do-files along with the variable labels, value labels, and notes in the dataset provide all the documentation you need for a project. This approach, however, requires that you include very detailed comments in your do-files and that you are able to fully replicate your results by rerunning the do-files in sequence.

Internally labeling documents

Every document should include the author's name, the name of the document file (so you can search for the file if you have a paper copy but want to edit the file), and the date it was created. One of the most frequent and easily remedied problems I see is documents that do not include this information. Worse yet, someone revises a document, but does not change the document's internal date and perhaps does not change the name of the file. (Have you ever been in a meeting where participants debate which version of a document is the latest?) On collaborative projects, it is easy to lose track of which version of a document is the latest. This can be avoided if you add a section at the end of each document that records a document's pedigree. With each revision, add a new entry indicating who wrote it, when, and what it was called. You might wonder why you cannot use the operating system's file date to determine when a file was created. Unfortunately, that date can be changed by the operating system even if the file has not changed. It is much safer to rely on a date that is internal to the file.

2.4.3 Suggestions for writing documentation

Although there are many ways to write documentation and I encourage you to find the method that works best for you, there are several principles of documentation that are worth remembering.

Do it today

When things are fresh in your mind, you can write documentation faster and more accurately.

Check it later

If you write documentation while doing the work, it is easy to forget information that is obvious now but that should be recorded for future reference. Ideally, write your documentation soon after the work is completed. Then either have someone else check the documentation or check it yourself at a later time.

Know where the documentation is

Decide where to keep your documentation. If you cannot find it, it does not do you any good! I keep electronic copies of my documentation in the `\Documentation` subdirectory of each project. I usually keep a printed copy in a project notebook that I update after each step of the project is completed.

Include full dates and names

When it comes to dates, the year is important. On February 26, it might seem inconceivable that the project will continue through the next calendar year, but even simple research notes can take years to finish. Include full names. "Scott" or the initials "sl" may be clear now, but at a later time, there might be more than one Scott or two people with the same initials.

Evaluating your documentation

Here is a thought experiment for assessing the adequacy of your documentation. Think of a variable or data decision that was completed early in the project. In a study of aging, this could be how the age of a respondent was determined. Imagine that you have finished the first draft of a paper and then discovered that age was computed incorrectly. This might seem far fetched, but the National Longitudinal Survey has revised the birth years of respondents several times. How long would it take to create a corrected dataset and redo the analyses? Could other researchers understand your documentation well enough to revise your programs to correct the variable and recompute all later analyses? If not, your documentation is inadequate. When teaching statistics, I require students to keep a research log. This log mimics what they should record if they were working on a research paper. The standard for assessing the adequacy of the log and the file organization is the following. During the last week of class, imagine returning to the second assignment, removing the first three cases in the dataset (i.e., `drop if _n < 4`), and rerunning the analyses. If the documentation and file organization are adequate, this should take less than five minutes.

2.4.4 The research log

The research log is the core of your documentation, serving as a diary of the work you have done on a project. Your research log should accomplish three things:

- The research log keeps your work on track. By including your research plan, the log helps you set priorities and complete work in an efficient way.
- The research log helps you deal with interruptions. Ideally, you start a project and work on it without interruption until it is finished. In practice, you are likely to move among projects. When you return to a project, the research log helps you pick up the work where it ended without spending a lot of time remembering what you did and what needs to be done.
- The research log facilitates replication. By recording the work that was done and the files that were used, the research log is critical for replicating your work.

As long as these objectives are met, your research log is a good one.

Researchers keep logs in many formats (e.g., bound books, loose-leaf notebooks, computer files) and refer to them by different names (e.g., project notes, think books, project diaries, workbooks). While writing this book, I asked several people to show me how they keep track of their research. I discovered that there are many styles and approaches, all of which do an admirable job of meeting the fundamental objective of recording what was done so that results could be reproduced at a later time. Several people conveyed stories of how their logs became more detailed as the result of a painful lesson caused by inadequate documentation in the past. Without question, keeping a research log involves considerable work. So, it is important to find an approach to keeping a log that appeals to you. If you prefer writing by hand to typing, use a bound volume. If you would rather type or your handwriting is illegible, use a word processor. The critical thing is to find a way that allows you to keep your documentation current.

My research log records what I have done, why I did it, and how I did it. It also records things that I decided not to pursue and why. Finally, it includes information on what I am thinking about doing next. To accomplish this, the log includes information on the source of data, problems encountered and solutions used, extracts from emails from coauthors, summaries of meetings, ideas related to the current analyses, and a lists of things I need to do. When I begin a project, I start with my research plan. The plan lays out the work for the following weeks or months. As work progresses, the plan is folded into the record of work that was completed. As such, the plan becomes a to-do list, whereas the research log is a diary of how the work was completed.

A sample page from a research log

To give you an idea of what my research logs look like, figure 2.3 shows an extract from the research log for a paper completed several years ago.

jslFLIMlog: 4/1/02 to 6/12/22 - Page 11

First complete set of analysis for FLIM measures paper

f2alt01a.do - 24May2002

Descriptive information on all rhs, lhs, and flim measures

f2alt01b.do - 25May2002

Compute bic' for each of four outcomes and all flim measures.

```
** Outcome: Can Work                    global lhs "qcanwrk95"
** Outcome: Work in three categories    global lhs "dhlthwk95"
** Outcome: bath trouble                global lhs "bathdif95"
** Outcome: adlsum95 - sum of adls      global lhs "adlsum95"
```

f2alt01c.do - 25May2002

Compute bic' for each of four outcomes and with only these restricted flim measures.

```
*    1.  ln(x+.5) and ln(x+1)
*    2.   9 counts: >=5=5  >=7=7  (50% and 75%)
*    3.   8 counts: >=4=4  >=6=6  (50% and 75%)
*    4.   18 counts: >=9=9 >=14=14  (50% and 75%)
*    5.   probability splits at .5; these don't work well in prior tests
```

f2alt01d.do - 25May2002

bic' for all four outcomes in models that include all raw flim measures (fla*p5; fll*p5); pairs of u/l measures; groups of LCA measures

f2alt01e.do - all LCA probabilities - 25May2002

:::

f2alt01j.do - use three probability measures from LCA - 29May2002

:::

f2alt02c.do - 29May2002

use three binary variables, not just LC class numbers.
: dummies work better than the class number;
: effects of lower and severe are not significantly different.

Redo f2 analyses - error in adlsum - 3Jun2002

ARGH! adlsum is incorrect -- it included going to bed twice.
All of the f2alt analyses need to be redone using the corrected dataset.

f3alt_qflim07.do: create qflim07.dta 3Jun2002

1) Correct aldsum: adlsum95b
2) Add binary indicators of Lmaxp5: LmaxNonep5, etc.

f3alt01a (redo f2alt01a.do) - 3Jun2002

f3alt01b.do (redo f2 job) - 3Jun2002

Figure 2.3. Sample page from a research log

This section of the log documents analyses completed after the data were cleaned and variables were constructed. The do-files from `f2alt01a.do` to `f2alt02c.do` complete the primary analyses for the paper. When reviewing these results, I discovered that a summated scale was incorrect, as it included the same variable twice. The program `f3alt_qflim07.do` fixed the problem and created the dataset `qflim07.dta`. The do-files `f3alt*.do` are identical to `f2alt*.do` except that the corrected scale is used. As I reread this research log, which was written four years ago, I found many things that were unclear. But, because the log pointed to the do-files that were used, it was simple to figure things out by checking those files. Thus the comments in the do-files were critical for making the log effective. The point is that your research log does not need to record every detail. Rather, it needs to provide enough detail so that you can find the information you need.

A template for research logs

Keeping a research log is easier if you start with a template. For example, I use the Word document `research-log-blank.docx` (see figure 2.4) to start a new research log (available at the Workflow web site):

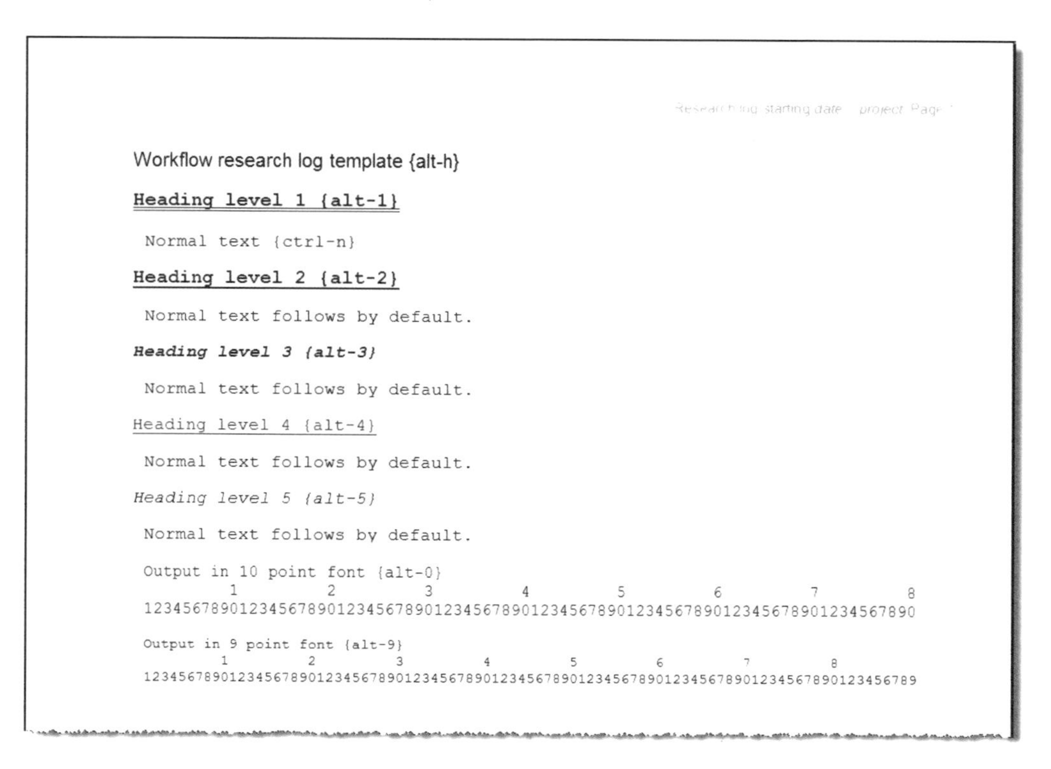

Figure 2.4. Workflow research log template

I load the file, change the header and title to correspond to the project, and delete the remaining lines in the file. These lines are included in the template to remind me of the keyboard shortcuts built into the document. For example, to add a level-1 heading, press Alt+1; to add output in a 9-point font, press Alt+9; and so on. The body of the document is in a fixed font, which I find easiest because I can paste output and it will line up properly. I change the name of the file and save it to the `\Documentation` directory for the project.

2.4.5 Codebooks

Codebooks describe your dataset. If you are collecting your own data, you need to create a codebook for all the variables. If you are using an existing dataset that has a codebook, you only need to create a codebook for variables that you add.

There is an excellent guide for preparing a codebook, *Guide to Social Science Data Preparation and Archiving: Best Practice Throughout the Data Life Cycle* (ICPSR 2005), which can be downloaded as a PDF. Here I highlight only some key points. The *Guide* suggests that you start by writing a plan for what the codebook will look like and think about how you can use the output from your software to help you write the codebook. For example, Stata's `codebook` command might have most of the information you want to include in the codebook. For each variable, consider including the following information:

- The variable name and the question number if the variable came from a survey.
- The text of the original question from the survey that collected the data or the details on the method by which the data for that variable were obtained. Include the variable label used in your data file.
- If the data are collected with a survey that includes branching (e.g., if the respondent answers A, go to question 2; if B, got to question 7), include information on how the branching was determined.
- Descriptive statistics including value labels for categorical variables.
- Descriptions of how missing data can occur along with codes for each type of missing data.
- If there was recoding or imputation, include details. If a variable was constructed from other variables in the survey (e.g., a scale), provide details, including how missing data were handled.
- An appendix with abbreviations and other conventions used.

A codebook based on the survey instrument

If you are collecting data, editing the original survey instrument is a quick and effective way to create a codebook. For example, figure 2.5 is an edited version of the survey instrument used for the SGC-MHS Study (Pescosolido et al. 2003). The vaiiable names and labels were added in bold. Other information on coding decisions, skip patterns, and so on was documented elsewhere.

	Not at all Important									Very Important
Q43. Turn to family for help	1	2	3	4	5	6	8	9	9	10
tcfam Q43 How Important: Turn to family for help										
Q44. Turn to friends for help	1	2	3	4	5	6	8	9	9	10
tcfriend Q44 How Important: Turn to friends for help										
Q45. Turn to a minister, priest, Rabbi or other religious leader	1	2	3	4	5	6	8	9	9	10
tcrelig Q45 How Important: Turn to a Minister, Priest, Rabbi or other religious leader										
Q46. Go to a general medical doctor for help	1	2	3	4	5	6	8	9	9	10
tcdoc Q46 How Important: Go to a general medical doctor for help										
Q47. Go to a psychiatrist for help	1	2	3	4	5	6	8	9	9	10
tcpsy Q47 How Important: Go to a psychiatrist for Help										
Q48. Go to a mental health professional for help	1	2	3	4	5	6	8	9	9	10
tcmhprof Q48 How Important: Go to a mental health professional										

ALLOWED DEFINITION – PSYCHOLOGIST, THERAPIST, SOCIAL WORKER, OR COUNSELOR

INTERVIEWER NOTE: CODE "DON'T KNOW" AS 98 ABOVE SEQUENCE.

The next few questions deal with the government's responsibility to help people like NAME. For each statement please tell me if you think the government definitely should, probably should, probably should not, or definitely should not be responsible for helping people with situations like NAME.

Figure 2.5. Codebook created from the survey instrument for the SGC-MHS Study

2.4.6 Dataset documentation

Your research log should include details on how each dataset was created. For example, the log might indicate that `cwh-data01a-scales.do` started with the dataset `cwh-01.dta`, created new scales, and saved the dataset `cwh-02.dta`. I also recommend including information inside the dataset. Stata's `label data` command lets you add a label that is displayed every time you load your data. For example,

```
. use jsl-ageism04
(Ageism data from NLS \ 2006-06-27)
```

The data label, listed in parentheses, reminds me that the file is for a project that is analyzing reports of age discrimination from the NLS and that the dataset was created on June 27, 2006. Stata's `notes` command lets you embed additional information in your dataset. When I create a dataset, I include a `notes` with the name of the do-file that created the dataset. When a file is updated or merged with another file, the notes are carried along. This means that internal to the dataset I have a history of how the dataset was created. For example, `jsl-ageism04.dta` is from a project with Eliza Pavalko that has been ongoing for five years. The project required dozens of datasets, thousands of variables, and hundreds of do-files. If I found a problem in `jsl-ageism04.dta`, I can use `notes` to track down what caused the problem. For example,

```
. notes _dta
_dta:
  1.  base01.dta: base vars with birthyr and cohort \ base01a.do jsl 2001-05-31.
  2.  base02.dta: add attrition info \ base01b.do jsl 2001-06-29.
  (output omitted)
 38.  jsl-ageism04.dta: add analysis variables \ age07b.do jsl 2006-06-27.
```

There were 38 steps that went into creating this dataset. If a problem was found with the attrition variable, the second note indicates that this variable was created by `base01b.do` on June 29, 2001. I can check the research log for that date or go to the do-file to find the problem. The advantage of internal documentation is that it travels with the dataset and saves me from searching research logs to track down the problem. Essentially, I use the `notes` command to index the research log. Details on Stata's `label data` and `notes` commands are given on page 138.

For large projects, you might want a registry of datasets. For example, I am working on a project in which we will be receiving datasets from 17 countries where each country has several files. We created a registry to keep track of the datasets. The data registry can be kept in a spreadsheet that looks like figure 2.6 (file: `wf2-data-registry.xls`):

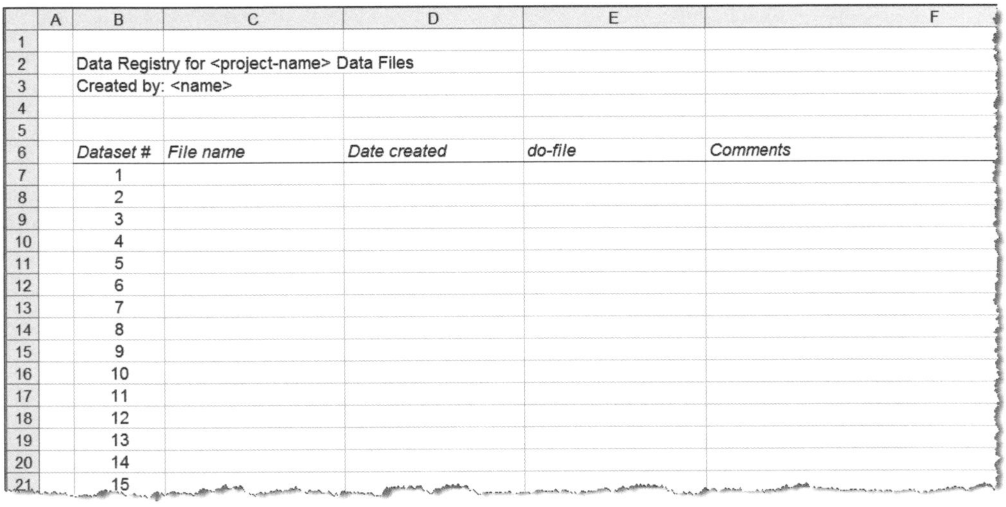

	A	B	C	D	E	F
1						
2		Data Registry for <project-name> Data Files				
3		Created by: <name>				
4						
5						
6		*Dataset #*	*File name*	*Date created*	*do-file*	*Comments*
7		1				
8		2				
9		3				
10		4				
11		5				
12		6				
13		7				
14		8				
15		9				
16		10				
17		11				
18		12				
19		13				
20		14				
21		15				

Figure 2.6. Data registry spreadsheet

2.5 Conclusions

The critical point of this chapter is that planning, organizing, and documenting are essential tasks in data analysis. Planning saves time. Organization makes it easier to find things. Documentation is essential for replication, and replication is fundamental to the research enterprise. Although I hope that my discussion will help you accomplish these tasks more effectively and convince you of their importance, any way you look at it PO&D are hard work. When you are tempted to postpone these tasks, keep in mind that it is almost always easier to do these tasks earlier than later. Make these tasks a routine part of your work. Get in the habit of checking your documentation at natural divisions in your work. If you find something confusing (e.g., you cannot remember how a variable was created) or if you have trouble finding something, take the time right then to improve your documentation and organization. When thinking about PO&D consider the worst case scenario when things go wrong and time is short, not the ideal situation when you have plenty of uninterrupted time to work on a project from start to finish. By the time you lose track of what you are doing, it often takes longer to create the plan, organize the files, and document the work than if you had started these tasks at the very beginning.

The next two chapters look at features of Stata that are critical for developing an effective workflow. Chapter 3 reviews basic tools and provides handy tricks for working with Stata. Chapter 4 introduces Stata features for automating your work. Time spent learning these tools really pays off when using Stata.

3 Writing and debugging do-files

Before discussing how to use Stata for specific tasks in your workflow, I want to talk about using Stata itself. Part of an effective workflow is taking advantage of the powerful features of your software. Although you can learn the basics of Stata in an hour, to work efficiently you need to understand some of its more advanced features. I am not talking about specific commands for transforming data or fitting a model, but rather about the interface of the program, the principles for writing do-files, and how to automate your work. The time you spend learning these tools will quickly be recovered as you apply these tools to your substantive work. Moreover, each of these tools contributes to the accuracy, efficiency, and replicability of your work. This chapter discusses writing and debugging do-files. Chapter 4 introduces powerful tools for automating your work. The tools and techniques from chapters 3 and 4 are used and expanded upon in chapters 5–7 where different parts of the workflow of data analysis are discussed.

I begin the chapter reviewing three ways to execute commands: submit them from the Command window, construct them with dialog boxes, or include them in do-files. Each approach has its advantages, but I argue that the most effective way to work is with do-files. Because the examples in the rest of the book depend on do-files, I discuss in section 3.2 how to write more effective do-files that are easier to understand and that will continue to work on different computers, in later versions of Stata, and after you change the directories on your computer. Although these guidelines can prevent many errors, sometimes your do-files will not work. Section 3.3 describes how to debug do-files, and section 3.4 describes how to get help when the do-files still do not work.

I assume that you have used Stata before, although I do not assume that you are an expert. If you have not used Stata, I encourage you to read [GS] *Getting Started with Stata* and those sections of the [U] *User's Guide* that seem most useful. Appendix A discusses how the Stata program works, which directories it uses, how to use Stata on a network, and ways to customize Stata. Even experienced users may find some useful information there.

3.1 Three ways to execute commands

There are three ways to execute commands in Stata. You can submit commands interactively from the command line. This is ideal for trying new things and exploring your data. You can use dialog boxes to construct and submit commands, which is particularly useful for finding the options you need when exploring new commands. You can also run do-files, which are text files that contain Stata commands. Each method has advantages, but I will argue that serious work requires do-files. Indeed, I only use the other methods to help me write do-files.

3.1.1 The Command window

You can type one command at a time in the Command window. Type the command and press Enter. When experimenting with how a command works or checking some aspect of my data, I often use this method. I try a command, press Page Up to redisplay the command in the Command window, revise it, press Enter to run it again, and so on. The disadvantage of working interactively is that you cannot easily rerun your commands at a later time.

Stata has a number of features that are very useful when working from the Command window.

Review window

The commands you submit from the Command window are echoed to the Review window. When you click on a command in the Review window, it is pasted into the Command window where you can revise it and then submit it by pressing Enter. If you double-click on a command in the Review window, it is sent to the Command window and automatically executed.

Page up and page down

The Page Up and Page Down keys let you scroll through the commands in the Review window. Pressing Page Up multiple times moves through multiple prior commands. Page Down moves you forward to more recent commands. When a command appears in the Command window, you can edit it and then rerun it by pressing Enter.

Copy and paste

You can highlight and copy text from the Command window or the Results window. This information can be pasted into other applications, such as your text editor. This allows you to debug a command interactively, then copy the corrected commands to your do-file.

Variables window

The Variables window lists the variables in the current dataset. If you click on a variable name in this window, the name is pasted into the Command window. This is often the fastest way to construct a list of variable names. You can then copy the list of names and paste it into your do-file.

Logging with log and cmdlog

If you want to reproduce the results you obtain interactively, you should save your session to a log file with the `log using` command. You can then edit the log file to create a do-file to rerun the commands. Suppose that you start an interactive session with the command

```
log using datacheck, replace text
```

After you are done with your session, you close the log file with `log close` to create the file `datacheck.log`. To create a do-file that will produce the same results, you can copy the log file to `datacheck.do`, remove the `.`'s in front of each command, and delete the output. This is tedious but sometimes quite useful. An alternative is to use `cmdlog` to save your interactive commands. For example, `cmdlog using datacheck.do, replace` saves all commands from the Command window (but no output) to a file named `datacheck.do`, which you can use to create your do-file. You close a `cmdlog` with the `cmdlog close` command.

3.1.2 Dialog boxes

You can use dialog boxes to construct commands using point-and-click. You open a dialog box from the menus in Stata by selecting the task you want to complete. For example, to construct a scatterplot matrix, you select **Graphics** (Alt+G) > **Scatterplot matrix** (s, Enter). Next you select options using your mouse. After you have selected your options, click on the **Submit** button to run the command. The command you submit is echoed to the Results window so that you can see how to type the command from the Command window or with a do-file. If you press Page Up, the command generated by the dialog box is brought into the Command window where you can edit it, copy it, or rerun it.

Although dialog boxes are easy to learn, they are slow to use. However, dialog boxes are very efficient when you are looking for an option used by a complex command. I use them frequently when creating graphs. I select the options I need, run the command by clicking on the **Submit** button, and then copy the command from the Results window to my do-file.

3.1.3 Do-files

Over 99% of the work I do in Stata uses do-files. Do-files are simply text files that contain your commands. Here is a simple do-file named `wf3-intro.do`.

```
log using wf3-intro, replace text
use wf-lfp, clear
summarize lfp age
log close
```

This program loads data on labor-force participation and computes summary statistics for two variables. If you have installed the Workflow package in your working directory, you can run this do-file by typing the command `do wf3-intro.do`.[1] The extension `.do` is optional, so you could simply type `do wf3-intro`. After submitting the file, I obtain these results:

```
      log:  e:\workflow\work\wf3-intro.log
 log type:  text
opened on:   3 Apr 2008, 05:27:01
. use wf-lfp, clear
(Workflow data on labor force participation \ 2008-04-02)
. summarize lfp age
    Variable |       Obs        Mean    Std. Dev.       Min        Max
-------------+--------------------------------------------------------
         lfp |       753    .5683931    .4956295          0          1
         age |       753    42.53785    8.072574         30         60
. log close
      log:  e:\workflow\work\wf3-intro.log
 log type:  text
closed on:   3 Apr 2008, 05:27:01
```

That is how simple it is to run a do-file. If you have avoided them in the past, this is a good time to take an hour and learn how they work. That hour will save you many hours later.

I use do-files for two major reasons. First, with do-files you have a record of the commands you ran, so you can rerun them in the future to replicate your results or to modify the program. Recall the research log on page 41 that documented a problem with how a variable was created. If I had not been using do-files, I would have needed to reconstruct weeks of work rather than changing a few lines of code and rerunning the do-files in sequence. Second, with do-files, you can use the powerful features of your text editor, including copying, pasting, global changes, and much more (see the Workflow web site for information on text editors). The editor built into Stata can be opened several ways: run the command `doedit`, select the Do-file Editor from the **Window** menu of Stata, or click on the Do-file Editor icon. For details on the Stata Do-file Editor, type `help doedit`, or see [R] **doedit**.

3.2 Writing effective do-files

The rest of the book assumes that you are using do-files to run commands, with the exceptions of occasionally testing commands from the Command window or using dialog boxes to track down options. In this section, I consider how to write do-files that are robust and legible. Here is what I mean by these terms:

1. Appendix A explains the idea of a working directory. The *Preface* has information on installing the Workflow package.

> *Robust do-files* produce exactly the same result when run at a later time or on another computer.
>
> *Legible do-files* are documented and formatted so that it is easy to understand what is being done.

Both criteria are important because they make it possible to replicate and correctly interpret your results. As a bonus, robust and legible do-files are easier to write and debug. To illustrate these characteristics of do-files, I use examples that contain basic Stata commands. Although you might encounter a command that you have not seen before, you should still be able to understand the general points I am making even if you do not follow the specific details.

3.2.1 Making do-files robust

A do-file is robust if it produces exactly the same result when it is rerun on your computer or run on a different computer. The key to writing robust do-files is to make sure that results do not depend on something left in memory (e.g., from another do-file or a command submitted from the Command window) or how your computer is set up (e.g., the directory structure you use). To operationalize this standard, imagine that after running a do-file you copy this file and all datasets used to a USB drive, insert the USB drive in another computer, and run the do-file again without any changes. If you cannot do this and get the same results, replication will be difficult or impossible. Here are my suggestions for making your do-files robust.

Make do-files self-contained

Your do-file should not rely on something left in memory by a prior do-file or commands run from the Command window. A do-file should not use a dataset unless it loads the dataset itself. It should not compute a test of coefficients unless it estimates those coefficients. And so on. To understand why this is important, consider a simple example. Suppose that `wf3-step1.do` creates new variables and `wf3-step2.do` fits a model. The first program loads a dataset and creates two variables indicating whether a family has young children and whether a family has older children:

```
log using wf3-step1, replace text
use wf-lfp, clear
generate hask5 = (k5>0) & (k5<.)
label var hask5 "Has children less than 5 yrs old?"
generate hask618 = (k618>0) & (k618<.)
label var hask618 "Has children between 6 and 18 yrs old?"
log close
```

The program `wf3-step2.do` estimates the logit of `lfp` on seven variables, including the two created by `wf3-step1.do`:

```
log using wf3-step2, replace
logit lfp hask5 hask618 age wc hc lwg inc, nolog
log close
```

If these programs are run one after the other, with no commands run in between, everything works fine. What if the programs are not run in sequence? For example, suppose that I run `wf3-step1.do` and then run other do-files or commands from the Command window. Or I might later decide that the model should not include `age`, so I modify `wf3-step2.do` and run it again without running `wf3-step1.do` first. Regardless of the reason, if I run the second do-file without running `wf3-step1.do` first, I get the following error:

```
. logit lfp hask5 hask618 age wc hc lwg inc, nolog
no variables defined
r(111);
```

The error occurs because the dataset is no longer in memory. I might change the program so that the original dataset is loaded

```
log using wf3-step2, replace
use wf-lfp, clear
logit lfp hask5 hask618 age wc hc lwg inc, nolog
log close
```

Now the error is

```
. logit lfp hask5 hask618 age wc hc lwg inc, nolog
variable hask5 not found
r(111);
```

This error occurs because `hask5` is not in the original dataset but was created by `wf3-step1.do`.

To avoid this type of problem, I can modify the two programs to make them self-contained. I change the first program so that it saves a dataset with the new variables (file: `wf3-step1-v2.do`):

```
log using wf3-step1-v2, replace
use wf-lfp, clear
generate hask5 = (k5>0) & (k5<.)
label var hask5 "Has children less than 5 yrs old?"
generate hask618 = (k618>0) & (k618<.)
label var hask618 "Has children between 6 and 18 yrs old?"
save wf-lfp-v2, replace
log close
```

I change the second program so that it loads the dataset created by the first program (file: `wf3-step2-v2.do`):

```
log using wf3-step2-v2, replace
use wf-lfp-v2, clear
logit lfp hask5 hask618 age wc hc lwg inc, nolog
log close
```

The do-file `wf3-step2-v2.do` still requires running `wf3-step1-v2.do` to create the new dataset, but it does not require running `wf3-step2-v2.do` immediately after `wf3-step1-v2.do` or even that it be run in the same Stata session.

There are a few exceptions of do-files that need to be run in sequence. For example, if I am doing postestimation analysis of coefficients from a model that takes a long time to fit (e.g., `asmprobit`), I do not want to refit the model repeatedly while I debug the postestimation commands. I would use one do-file to fit the model and a second do-file for postestimation analysis. The second do-file only works if the prior do-file was run. To ensure that I remember that the programs need to be run in tandem, I add a comment to the second do-file:

```
// Note: This do-file assumes that program1.do was run first.
```

After debugging the second program, I would combine the two do-files to create one do-file that is self-contained.[2]

Use version control

If you run a do-file at a later time, perhaps to verify a result or to modify some part of the program, you could be using a newer version of Stata. If you share a do-file with a colleague, she might be using a different version of Stata. Sometimes new versions of Stata change the way in which a statistic is computed, perhaps reflecting advances in computational methods. When this occurs, the same commands can produce different results in different versions of Stata. Newer versions of Stata might change the name of a command (e.g., `clear` in Stata 9 was changed to `clear all` in Stata 10). The solution is to include a `version` command in your do-file. For example, if your do-file includes the command `version 6` and you run the do-file in Stata 10, you will get exactly the same answer that you would obtain in Stata 6. This is true even if Stata 10 computes the particular statistic differently (e.g., the computations in some `xt` commands changed between Stata 6 and Stata 10). On the other hand, if your do-file includes the command `version 10` and you try to run the program in Stata 8.2, you get an error:

```
. version 10
this is version 8.2 of Stata; it cannot run version 10.0 programs
     You can purchase the latest version of Stata by visiting
     http://www.stata.com.
r(9);
```

You could rerun the program after changing the `version 10` command to `version 8.2`. There is no guarantee that programs written for newer versions of Stata will work in older versions.

Exclude directory information

I almost never specify a directory location in commands that read or write files. This lets my do-files run even if the directory structure of the computer I am using changes. For example, suppose that my do-file loads data with the command

2. With Stata 10, I might use the new **estimates save** command to save the estimates in the first do-file and then load them at the start of the second do-file that does postestimation analysis. This would allow each program to be self-contained, even when debugging the second program. For details, see [R] **estimates save**.

```
use c:\data\wf-lfp, clear
```

Later, when I rerun the do-file on a computer where the dataset is stored in `d:\data\`, I get an error:

```
. use c:\data\wf-lfp, clear
file c:\data\wf-lfp.dta not found
r(601);
```

To avoid such problems, I do not include a directory location. For example, to load `wf-lfp.dta`, I use the command

```
use wf-lfp, clear
```

When no directory is specified, Stata looks in the working directory.

The working directory is the directory you are in when you launch Stata.[3] In Windows, you can determine your working directory by typing `cd`. For example,

```
. cd
e:\data
```

In Mac OS or Unix, you use the `pwd` command. For example, on a Mac:

```
. pwd
~:data
```

You can change your working directory with the `cd` command. For example, when testing commands for this book, I used the `e:\workflow\work` directory. To make this my working directory, I would type

```
cd e:\workflow\work
```

To change to the working directory used for the CWH project, I would type

```
cd e:\cwh\work
```

If the directory name includes blanks or special characters, you need to put the name in quotes. For example,

```
cd "c:\Documents and Settings\jslong\Projects\workflow\work"
```

The advantage of not including directory locations in your do-file is that you can run your do-files on other computers without any changes. Although it is tempting to say that you will always keep your data in the same place (e.g., `d:\data`), this is unlikely for several reasons.

1. If you change computers or add a new drive to your computer, the drive letters might change.

3. Appendix A has a detailed discussion of the directories used by Stata.

2. If you keep data on external drives, including USB flash drives, the operating system will not always assign the drive the same drive letter.
3. If you reorganize your files, the directory structure could change.
4. When you restore files from your archive, you might not remember what the directory structure used to be.

If you share do-files with a collaborator or someone helping you debug your program, they will probably have a different directory structure than yours. If you hardcode the directory, the person you send the do-file to must either create the same directory structure or change your program to load data from a different directory. When the collaborator sends you the corrected do-file, you will have to undo the directory changes that were made, and so on. All things considered, I think that it is best practice to write do-files that do not require a particular directory structure or location for the data. There are two exceptions that are useful. First, if you are loading a dataset from the web, you need to specify the specific location of the file. For example, `use http://www.stata-press.com/data/r10/auto, clear`. Second, you can specify relative directories. Suppose there is a subdirectory `\data` located in your working directory. To keep things organized, you place all your datasets in this directory, while your do-files and log files remain in your working directory. You can assess the datasets by specifying the subdirectory. For example, `use data\wf-lfp, clear`.

Include seeds for random numbers

Random numbers are used in a variety of ways in data analysis. For example, if you are bootstrapping standard errors, Stata draws repeated random samples. If you try to replicate results that use random numbers, you need to use the same random numbers or you will obtain different results. Stata uses pseudorandom numbers that are generated by a formula in which one pseudorandom number is transformed to create the next number. This transformation is done in such a way that the sequence of numbers behaves as if it were truly random. With pseudorandom numbers, if you start with the same number, referred to as the *seed*, you will re-create exactly the same sequence of numbers. Accordingly, to reproduce exactly the same results when you rerun a program that uses pseudorandom numbers, you need to start with the same seed. To set the seed, use the command

```
set seed #
```

where # is a number you choose. For example, `set seed 11020`. For further details and an example, see section 7.6.3.

3.2.2 Making do-files legible

I use the term legible to describe do-files that are internally documented and carefully formatted. When writing a do-file, particularly one that does complex statistical analyses or data manipulations, it is easy to get caught up in the logic of what you are doing and forget about documenting your work and formatting the file to make the content clear. Applying uniform procedures for documenting and formatting your do-files makes them easier to debug and helps you and your collaborators understand what you did. There are many ways to make your do-files easier to understand. If you do not like my stylistic suggestions, feel free to create your own style. The important thing is to establish a style that you and others find legible. If you are collaborating, try to agree upon a common style for writing do-files that makes it simpler to share programs and results. Clear and well-formatted do-files are so important for working efficiently that one of the first things I do when helping someone debug a program is to reformat their do-file to make the code easier to read.

Use lots of comments

I have never returned to a do-file and regretted how many comments it had, but I have often wished that I had written more. Commands that seem obvious when I write them can be obscure later. I try to add at least a few comments when I initially write a do-file. After the program works the way I want, I add additional comments. These comments are used both to label the output and to explain commands and options that might later be confusing.

Stata provides three ways to add comments. The first two create comments on a single line, whereas the third allows you to easily write multiline comments. The method you use is largely a matter of personal preference.

* comments

If you start a line with a `*`, everything that follows on that line is treated as a comment. For example,

```
* Select sample based on age and gender
```

or

```
* The following analysis includes only those people
* who responded to all four waves of the survey.
```

You can temporarily stop a command from being executed:

```
* logit lfp wc hc age inc
```

// comments

You can add comments after a `//`. For example,

```
// Select sample based on age and gender
```

This method can also be used at the end of a command. For example,

```
logit lfp wc hc // includes only education, add wages later
```

/* and */ comments

Everything between an opening /* and a closing */ is treated as a comment. This is particularly useful for comments that extend over multiple lines. For example,

```
/*
   These analyses are preliminary and are based on those countries
   for which complete data were available by January 17, 2005.
*/
```

Comments as dividers

Comments can be used as dividers to distinguish among different parts of your program. For example,

```
*************************************
**  Descriptive statistics by gender
```

or

```
// ====================================================
// = Logit models of depression on genetic factors
```

Obscure comments

Comments are useful only when they are accurate and clear. When writing a complex do-file, I use comments to remind me of things I need to do. For example,

```
* check this. wrong variable?
```

or

```
* see ekp´s comment and model specification
```

After the program is written, these comments should be deleted because later they will be confusing.

Use alignment and indentation

It is easier to verify your commands if things line up. For example, here are two ways to format the same commands for renaming variables. Which is easier for spotting a mistake? This?

(Continued on next page)

```
rename dev origin
rename major jobchoice
rename HE parented
rename interest goals
rename score testscore
rename sgbt sgstd
rename restrict restrictions
```

Or this?

```
rename  dev       origin
rename  major     jobchoice
rename  HE        parented
rename  interest  goals
rename  score     testscore
rename  sgbt      sgstd
rename  restrict  restrictions
```

Most text editors, including Stata's Do-file Editor, allow tabbing that makes lining things up easier.

When commands take more than one line, I indent the second and later lines. I find it easier to read

```
logit y var01 var02 var03 var04 var05 var06  ///
        var07 var08 var09 var10 var11 var12  ///
        var13 var14 var15
```

than

```
logit y var01 var02 var03 var04 var05 var06 ///
var07 var08 var09 var10 var11 var12 ///
var13 var14 var15
```

Some text editors, including Stata's, can automatically indent. This means that if you indent a line, the next line is automatically indented. If you find that the Stata Do-file Editor does not do this, you need to turn on the **Auto-indent** option. While in the Editor, press Alt+e and then f to open the dialog box where you can set this option. You can also highlight lines in the Do-file Editor and indent them all by pressing Ctrl+i or outdent them by pressing Ctrl+Shift+i.

Use short lines

Mistakes are easy to make if you cannot see the complete command or all the output. To avoid problems with truncation or wrapping, I keep my command lines to 80 columns or less and set the line size to 80 (`set linesize 80`) because this works with most printers and on most screens. To illustrate why this is important, here is a problem I encountered when helping someone debug a program using the `listcoef` command, which is part of SPost. The do-file I received looked like this, where the line with `mlogit` that trails off the right-hand side of the page is 182 characters long (file: `wf3-longcommand.do`):

```
use wf-longcommand, clear
mlogit jobchoice income origin prestigepar aptitude siblings friends scale1_std demands interestlvl
listcoef
```

Because the outcome had three categories, `listcoef` should have listed coefficients comparing outcomes 1 to 2, 2 to 3, and 1 to 3 for each variable. For some variables, that was the case:[4]

```
Variable: income (sd=1.1324678)
Odds comparing
Alternative 1
to Alternative 2 |        b          z     P>|z|      e^b    e^bStdX
-----------------+--------------------------------------------------
2        -3      |   0.49569     0.825    0.409    1.6416    1.7530
2        -1      |   0.68435     2.483    0.013    1.9825    2.1706
3        -2      |  -0.49569    -0.825    0.409    0.6092    0.5704
3        -1      |   0.18866     0.377    0.706    1.2076    1.2382
1        -2      |  -0.68435    -2.483    0.013    0.5044    0.4607
1        -3      |  -0.18866    -0.377    0.706    0.8281    0.8076
--------------------------------------------------------------------
```

For other variables, some comparisons were missing:

```
Variable: female (sd=.50129175)
Odds comparing
Alternative 1
to Alternative 2 |        b          z     P>|z|      e^b    e^bStdX
-----------------+--------------------------------------------------
2        -1      |   1.25085     1.758    0.079    3.4933    1.8721
1        -2      |  -1.25085    -1.758    0.079    0.2863    0.5342
--------------------------------------------------------------------
```

Initially, I did not see a problem with the model and began looking for a problem in the code for the `listcoef` command. Eventually, I did what I should have done from the start—I reformatted the do-file so that it looked like this:

```
mlogit jobchoice income origin prestigepar aptitude siblings friends      ///
    scale1_std demands interestlvl jobgoal scale3 scale2_std motivation ///
    parented city female, noconstant baseoutcome(1)
```

Once reformatted, I immediately saw that the problem was caused by the `noconstant` option. Although `noconstant` is a valid option for `mlogit`, it was inappropriate for the model as specified. While this problem did not show up in the `mlogit` output, it did lead to misleading results from `listcoef`.

Having output lines that are too long also causes problems. Because you can control line length of output in your do-file, this is a good place to talk about it. Suppose that your line size is set at 132 and you create a table (file: `wf3-longoutputlines.do`):

```
set linesize 132
tabulate occ ed, row
```

4. The real example had comparisons among six categories, so the output took dozens of pages.

When you print the results they are truncated on the right:

```
+-------------------+
| Key               |
|-------------------|
|     frequency     |
|  row percentage   |
+-------------------+

           |                             Years of education
Occupation |         3          6          7          8          9         10         11
-----------+-----------------------------------------------------------------------------
   Menial  |         0          2          0          0          3          1          3
           |      0.00       6.45       0.00       0.00       9.68       3.23       9.68
-----------+-----------------------------------------------------------------------------
  BlueCol  |         1          3          1          7          4          6          5
           |      1.45       4.35       1.45      10.14       5.80       8.70       7.25
-----------+-----------------------------------------------------------------------------
    Craft  |         0          3          2          3          2          2          7
           |      0.00       3.57       2.38       3.57       2.38       2.38       8.33
-----------+-----------------------------------------------------------------------------
 WhiteCol  |         0          0          0          1          0          1          2
           |      0.00       0.00       0.00       2.44       0.00       2.44       4.88
-----------+-----------------------------------------------------------------------------
     Prof  |         0          0          1          1          0          0          2
           |      0.00       0.00       0.89       0.89       0.00       0.00       1.79
-----------+-----------------------------------------------------------------------------
    Total  |         1          8          4         12          9         10         19
           |      0.30       2.37       1.19       3.56       2.67       2.97       5.64
```

Depending on how you print the log file, the results might wrap and look like this:

```
           |                             Years of education
Occupation |         3          6          7          8          9         10
        11         12         13 |     Total
-----------+------------------------------------------------------------------
-----------------------------------+----------
   Menial  |         0          2          0          0          3          1
         3         12          2 |        31
           |      0.00       6.45       0.00       0.00       9.68       3.23
      9.68      38.71       6.45 |    100.00
-----------+------------------------------------------------------------------
-----------------------------------+----------
  BlueCol  |         1          3          1          7          4          6
         5         26          7 |        69
           |      1.45       4.35       1.45      10.14       5.80       8.70
      7.25      37.68      10.14 |    100.00
-----------+------------------------------------------------------------------
-----------------------------------+----------
    Craft  |         0          3          2          3          2          2
         7         39          7 |        84
           |      0.00       3.57       2.38       3.57       2.38       2.38
      8.33      46.43       8.33 |    100.00
-----------+------------------------------------------------------------------
-----------------------------------+----------
```

(output omitted)

I have often seen incorrect numbers taken from wrapped output. If your output wraps, fix it right away by changing the linesize to 80 and recycle the original output!

Limit your abbreviations

Variable abbreviations

In Stata, you can refer to a variable using the shortest abbreviation that is unique. As an extreme example, suppose you have a variable with the valid but unwieldy name age_at_1st_survey. If there is no other variable in your dataset that starts with a, you can abbreviate the long name simply as a. Although this is easy to type, your program will not work if you add a variable starting with a. For example, suppose you add a variable agesq that is the square of age_at_1st_survey. Now the abbreviation a generates the error:

```
a ambiguous abbreviation
r(111);
```

This error is received because Stata cannot tell if you are referring to age_at_1st_survey or agesq.

Abbreviations can lead to other, perplexing problems. Here is an example I recently encountered. The dataset has four binary variables bmi1_1019, bmi2_2024, bmi3_2530, and bmi4_31up indicating a person's body mass index (BMI). I got in the habit of using the abbreviations bmi1, bmi2, bmi3, and bmi4. Indeed, I had forgotten that these were abbreviations. Then I wanted to use svy: mean to test race differences in the mean of bm1:

```
svy: mean bmi1, over(black)
test [bmi1]black = [bmi1]white
```

The svy: mean command worked, but test gave the error:

```
equation [bmi1] not found
r(303);
```

Because I do not use svy commands regularly, I assumed that there must be another way to compute the test when using survey means. The problem could not be with the name bmi1 because I "knew" that was the right name. Eventually, I realized that the problem was the abbreviation. Although svy: mean allows abbreviations (e.g., bmi1 for bmi1_1019), the test command requires the full name:

```
test[bmi1_1019]black = [bmi1_1019]white
```

The time saved using the abbreviation was more than lost uncovering the problem caused by the abbreviation.

As tempting as it is to use abbreviations for variables, it is better not to use them. If you find that names are too long to type, consider changing the names (see section 5.11.2) or enter the variable names by clicking on the names in the Variables window. Then copy the names from the Command window to your do-file. To prevent Stata from allowing abbreviations for variable names, you can turn this feature on and off with the command:

```
set varabbrev {on|off}, permanently
```

Command abbreviations

Many commands and options can also be abbreviated, which can be confusing. For example, you can abbreviate the command name and variable name for

```
summarize education
```

as

```
su e
```

I find this to be too terse to be clear. A compromise is to use something like

```
sum educ
```

or

```
sum education
```

Consider a slight modification of a command I received in a do-file someone sent to me:

```
l a l in 1/3
```

I find it much clearer to write the command like this:

```
list age lwg in 1/3
```

Longer abbreviations are not necessarily better than shorter ones. For example, in a recent article that used Stata I saw the command:

```
nois sum mpg
```

I had not seen the `nois` command before so I checked the manual. Eventually, I realized the `nois` is an abbreviation for `noisily`. For me, `noi` is clearer than `nois`.

If you use abbreviations for commands, I suggest keeping them to three letters or more. In the rest of the book, I will abbreviate only a few commands where I find the abbreviations clear and convenient. Specifically, those in table 3.1.

Table 3.1. Stata command abbreviations used in the book

Full command name	Abbreviation
`generate`	`gen`
`label define`	`label def`
`label values`	`label val`
`label variable`	`label var`
`quietly`	`qui`
`summarize`	`sum`
`tabulate`	`tab`

As a general rule, command abbreviations make it harder for others to read your code. If you want your code to be completely legible to others, do not use command abbreviations.

Be consistent

All else being equal, you will make fewer errors and work faster if you find a standard way to do things. This applies to the style of your do-files (more on this below), how you format things, the order you run commands, and which commands you use. For example, when I create a variable with `generate`, I follow it with a variable label, a note, and a value label (if appropriate).

```
generate incomesqrt = sqrt(income)
label var incomesqrt "Square root of income"
notes incomesqrt:  sqrt of income \ dataclean01.do jsl 2006-07-18.
```

3.2.3 Templates for do-files

The more uniform your do-files are, the less likely you are to make errors and the easier it will be to read your output. Accordingly, I suggest that you create a template. You can load the template into your text editor, make changes, and save the files with a new name. This has several advantages. First, the template includes commands used in all do-files that you will not have to type (e.g., `capture log close`). Second, you will not forget to include commands that are in the template. Third, a standard structure makes it simpler to work uniformly across projects.

Commands that belong in every do-file

Before presenting two templates for do-files, I want to discuss commands that I suggest you include in every do-file. Here is a simple do-file named `wf3-example.do`, where the line numbers on the left are used to refer to a specific line but are not part of the file:

```
 1>  capture log close
 2>  log using wf3-example, replace text
 3>
 4>  //  wf3-example.do: compute descriptive statistics
 5>  //  scott long 03Apr2008
 6>
 7>  version 10
 8>  clear all
 9>  macro drop _all
10>  set linesize 80
11>
12>  * load the data and check descriptive statistics
13>  use wf-lfp, clear
14>  summarize
15>
16>  log close
17>  exit
```

Opening your log file

Line 1 can best be explained after I go through the rest of the program. Line 2 opens a log file to record the output. I recommend that you give the log file the same name as the do-file that created it (the prefix only, not the suffix `.do`). Because I have not specified a directory, the log is created in the current working directory. The `replace` option tells Stata that if `wf3-example.log` already exists, replace it. This is handy if you need to rerun the do-file while debugging it. If you do not add `replace`, the second time you run the program you get the error

```
. log using wf3-example, text
log file already open
r(604);
```

The `text` option specifies that the output is written in plain text rather than in Stata Markup and Control Language (SMCL). Although SMCL output looks nicer, only Stata can print it, so I do not use it. Line 16 closes the log file. This means that Stata will stop sending output to the file.

Blank lines

Lines 3, 6, 11, and 15 are blank to make the program easier to read. If you do not find blank lines to be useful, do not use them.

Comments about what the do-file does

Lines 4 and 5 explain what the do-file does so that this information will be included in your log file. I recommend including the name of the do-file, who wrote the do-file, the date it was written, and a summary of what the do-file does.

Controlling Stata

Lines 7–10 affect the way Stata runs. Line 7 indicates the version of Stata being used. Because `version 10` is in the file, if you run this do-file in later versions of Stata, you should get exactly the same output that you got today using Stata 10. Because the `version` command is located after the `log using` command, `version 10` will be included in the log that allows you to verify from printed output which version of Stata was used.[5] Lines 8 and 9 reset Stata so that your do-file will run as if it was the first thing done after starting Stata. This is important for making your do-file robust. Many commands leave information in memory that you do not want to affect your do-file. `clear all` removes from memory the data, value labels, matrices, scalars, saved results, and more. For a full description, see `help clear`. In Stata 9, you use the command `clear`, not `clear all`. Oddly, `clear all` clears everything but macros from memory. To do this, you use `macro drop _all`. Line 10 sets the line size for output to 80 columns. Even if the default line size for my copy of Stata is 80 (see appendix A for how to set the default line size), I want to explicitly set the line size in the do-file so that it will generate output that is formatted the same way if it is run with a copy of Stata with a different default line size. To see why this is important, you can try running `tabulate` for variables with a lot of categories using different line sizes.

Your commands

Your commands begin at line 12 and include comments to describe what you are doing.

Ending the do-file

Line 17 is critical. Stata only executes a command when it encounters a carriage return.[6] Without a carriage return at the end of line 16, the `log close` command does not run and your log file remains open. Although line 17 could be anything, including a blank, I prefer the `exit` command. This command tells Stata to terminate the do-file (i.e., do not run any more commands in the do-file). For example, I could include comments and commands after `exit`, such as

```
exit
1) Double check how the sample is selected.
2) Consider running these commands.
    describe
    summarize
    tab1 _all
```

The lines after `exit` are ignored by Stata.

5. I used to place the `version` command immediately after line 1, as suggested by Long and Freese (2006). When writing this book, a colleague showed me a problem that would have been simple to resolve if the `version` command had been part of the output that he had been trying to replicate. Instead, it took him two weeks to figure out why he could not replicate his earlier results.

6. The language of computers is filled with anachronisms. On a typewriter, the mechanism that holds the paper using a platten is called the carriage. When you type to the end of a line, you "return the carriage" to advance to the next line. Even though we no longer use a carriage to advance to a new line, we refer to the symbol that is created by pressing **Enter** as a carriage return.

capture log close

Now I can explain why line 1 is needed. Suppose that the first time I ran `wf3-example.do`, the program terminated with an error before executing `log close` in line 16. The log file would be left open, meaning that new results generated by Stata would continue to be sent to the log. When I rerun the program, assuming for the moment that line 1 is not in the do-file, line 2 would cause the error `r(604): log file already open` because I am trying to open a log file when a log file is already open. To avoid this error, I could add the command `log close` before the `log using` command. If I do this, the first time I run the do-file, the `log close` command will generate the error `r(606): no log file open` because I am trying to close a log file when no log file is open. The `capture` command in line 1 means "if this line generates an error, ignore it and continue to the next line". If you do not completely follow what I just explained, do not worry about it. Just get in the habit of beginning your do-files with the command `capture log close`.

A template for simple do-files

Based on the principles just considered, here is a template for simple do-files (file: `wf3-simple.do`).

```
capture log close
log using _name_, replace text
//  _name_.do:
//  scott long _date_
version 10
clear all
macro drop _all
set linesize 80
* my commands start here
log close
exit
```

I save this file in my working directory or to my computer's desktop, perhaps with the name `simple.do`. When I want to create a new do-file, I load `simple.do` into my editor, change `_name_` and `_date_`, and write my program. I save the file with its new name, say, `myprogram.do`, and then from the Command window, type `run myprogram`.

A more complex do-file template

For most of my work, I used a more elaborate template (file: `wf3-complex.do`):

```
capture log close
log using _name_, replace text
// program:    _name_.do
// task:
// project:
// author:     _who_ \ _date_
// #0
// program setup
version 10
clear all
set linesize 80
macro drop _all
// #1
// describe task 1
// #2
// describe task 2
log close
exit
```

This template makes it easier to document what the do-file is doing, especially by including numbered sections to the output for different steps of the analysis. By numbering sections, it is easier to find things within the file and to discuss the results with others (especially over email). When I send a log file to someone, I might write: "Do you think the results at #6 are consistent with our earlier findings?" If you start numbering parts of your do-files, I think you will find that it saves a lot of time and confusion.

There are many effective templates that can be used. The most important thing is to find a template that you like and use it consistently.

Aside on text editors

A full-featured text editor is probably the most valuable tool you can have for data analysis. A good editor speeds up your work, makes your do-files more uniform, and helps you debug programs. Although text editors have hundreds of valuable features, here are a few that are particularly useful. First, many editors can automatically insert text into a file. I have mine set up so that the keystroke Alt+0 inserts the simple do-file template (so I do not have to remember where I stored the template) and Alt+1 inserts the more complex template. Then the editor automatically inserts the date. Second, sophisticated text editors have a feature known as *syntax highlighting* that helps you find errors. These editors recognize predefined words and display them in different colors. For example, if you type the line `oloigit warm wc hc age k5`, the word `oloigit` will not be highlighted because it is not a Stata command. If you had typed `ologit`, the word would be highlighted because it is a valid command name. This is very handy for finding and fixing errors before they occur. The Workflow web site provides additional information.

3.3 Debugging do-files

In a perfect world, your do-files run the first time, every time. In practice, your do-files generate errors and probably lots of errors. Sometimes it is frustrating and time-consuming to determine the source of an error. While the principles for writing legible and robust do-files should make errors less likely and make it easier to resolve errors when they occur, you are still likely to spend more time than you like debugging your do-files. This section discusses how to debug do-files for both simple and complicated errors. I begin by reviewing a few simple strategies for finding problems. The section ends with two extended examples that illustrate how to fix more subtle bugs.[7]

3.3.1 Simple errors and how to fix them

To get started, I want to illustrate some very common errors.

Log file is open

If you have a log file open (for example, it might be left open because your last do-file ended with an error) and you try to open a log file, you get the message

```
. log using example1, replace
log file already open
r(604);
```

The simplest solution is to place `capture log close` at the top of your do-file.

Log file already exists

Because do-files are often run several times before they are debugged, you want to replace the log file that contains an error with the output from the corrected do-file. If your do-file contains the command

```
log using example2, text
```

and that log file already exists, you get the error

```
file e:\workflow\work\example2.log already exists
r(602);
```

The solution is the option `replace`:

```
log using example2, text replace
```

7. One theory of the origin of the term "bug" refers to a two-inch moth taped to Grace Murray Hopper's research log for September 9, 1947. This moth shorted a circuit in the Harvard University Mark II Aiken Relay Calculator (Kanare 1985).

Incorrect command name

The command

```
loget lfp k5 k618 age wc hc lwg inc
```

generates the error

```
unrecognized command:  loget
r(199);
```

The message makes it clear that something is wrong with the word `loget` and you are likely to quickly see that you mistyped `logit`. If you did not understand what `unrecognized command` meant, Stata can provide more information. In the Results window, `r(199)` appears in blue. Blue indicates that the highlighted word is linked to more information. If you click on `r(199)`, a Viewer window opens with the information:

```
[P]     error . . . . . . . . . . . . . . . . . . . . . . . . Return code 199
        unrecognized command;
        Stata failed to recognize the command, program, or ado-file name,
        probably because of a typographical or abbreviation error.
```

Sometimes, unrecognized commands will not be easy to see. For example,

```
. tabl lfp k5
unrecognized command:  tabl
r(199);
```

The problem is that I typed `tabl` instead of `tab1`, which can look very similar with some fonts. When I get an error related to the name of a command and everything looks fine, I often just retype the command and find that the second time I typed the command correctly.

Incorrect variable name

In the following `logit` command, the name of one of the variables is incorrect.

```
. logit lfp kO5 k618 age wc hc lwg inc
variable kO5 not found
r(111);
```

I meant to type `k05` (kay-zero-five), not `kO5` (kay-capital-oh-five). If you think a name is correct but you are getting an error, there are a few things to try. Suppose the error refers to a name beginning with "k". Type `describe k*` to describe all the variables that begin with `k`. Verify that the name in your do-file is listed. If it is and you still do not see the problem, you can click on the variable name in the Variables window. This will paste the name to the Command window. Copy the name from here to your do-file.

Stata reports only one incorrect name at a time. If you fixed the command above to

```
logit lfp k05 k618 age wc hc lwg inc
```

and k618 was the wrong name (e.g., it was supposed to be k0618), a new r(111) error message is generated.

Incorrect option

If you type an incorrect option, you get an error message like this:

```
. logit lfp k5 k618 age wc hc lwg inc, logoff
option logoff not allowed
r(198);
```

I wanted to turn off the iteration log for logit but incorrectly thought the option was logoff. To find the correct option, I could 1) try another name for the option, 2) type the help logit command from the Command window, 3) open the logit dialog box and find the option name, or 4) check the manual. Each would show you that the option I wanted was nolog.

Missing comma before options

This error confuses many people learning Stata:

```
. logit lfp wc nowc k5 k618 age hc lwg inc nocon
variable nocon not found
r(111);
```

The problem is that you need a comma before the nocon option:

```
logit lfp wc nowc k5 k618 age hc lwg inc, nocon
```

3.3.2 Steps for resolving errors

The errors above were easy to solve. In other cases, it can be very difficult to determine from the error message what is wrong. In later sections, I give examples of the multiple steps you might need to track down a problem. In this section, I provide some general strategies that you should consider if you do not see an obvious solution for the error you encountered.

Step 1: Update Stata and user-written programs

Before spending too much time debugging an error, make sure that your copy of Stata and any installed user-written ado-files are up to date. Your error might be caused by an error in the command that you are using, not by a mistake in your do-file. Updating Stata is simple, unless you are running Stata on a network. If you are on a network, you will have to talk to your network administrator (see appendix A for further information). While Stata is running and you are connected to the Internet, run the update all command and follow the instructions. This will update official Stata, including the executable, the help files, and the ado-files. If the do-file you are debugging uses commands written by others (e.g., `listcoef` in the SPost package), you should update those programs as well. The first thing to do is try the `adoupdate` command that was introduced in Stata 9.2. If you type the `adoupdate` command, it will check if your user-written ado-files are up to date. You can then either update the packages individually with `adoupdate` *package-name* or update all packages with `adoupdate, all`. To automatically update all your packages, try `adoupdate, update`. Unfortunately, this handy command only works with user-written packages where the author has made the package compatible with `adoupdate`. If some of your user-written commands are not checked with the `adoupdate` command (you will know this if they are not listed after the command is entered), you can run `findit` *command-or-package* and follow the instructions you receive.

Step 2: Start with a clean slate

When things do not work and your initial attempts fail to fix the problem, make sure that there is not information left in memory that is causing the problem (e.g., a matrix that should not be there). There are several ways to do this.

clear all and macro drop _all

From the command line, type `clear all` and `macro drop _all` or add them to your do-file. These commands tell Stata to forget everything that happened since you launched Stata. In Stata 9, use `clear` instead of `clear all`.

Restart Stata

If `clear all` and `macro drop _all` do not fix the problem, exit Stata, relaunch Stata, and try the program again.

Rebooting

Next reboot your computer and try the program again. After rebooting and before loading Stata, close all programs, including utilities such as macro programs, screen capture utilities, and so on. This might seem extreme, but if I had followed this advice three years ago, I would have saved myself and a very patient econometrician at StataCorp a great deal of trouble.

Use another computer

Still not working? You might try the program on another computer that is configured differently than your own. If it works there, the problem is caused by the way Stata is installed on your system.

Step 3: Try other data

Some errors are caused by problems in the dataset, such as perfect collinearity or zero variance for a variable. In other cases, the specific names or labels could be causing problems. The SPost command `mlogview` used to generate an error when certain characters were included in the value labels. If you get the same error using another dataset, you can be fairly sure that the problem is in your commands. If the error does not occur with the new data, focus on characteristics of your data.

Step 4: Assume everything could be wrong

It is easy to ignore parts of your program that you are "sure" are right. Most people who do a lot of programming have learned this lesson the hard way. As we will see, some error messages point to a part of the program that is actually correct. If the obvious solutions to an error do not work, review the entire program.

Step 5: Run the program in steps

I usually write a program a few commands at a time, rather than typing 100 lines at once. For example, I start with a do-file that only loads the data and runs descriptive statistics. If that works, I add the next set of commands. If that works, I add the next lines, and so on. This approach does not work as well if you have an extremely large sample or you are using a command that is computationally very demanding (e.g., `asmprobit`). In such cases, you can test you program using a small sample or block out parts of the program that have been tested.

Aside on selecting a random subsample

If you need a small sample for debugging your program, here is how you can take a random sample from your data (file: `wf3-subsample.do`):

```
. use wf-lfp, clear
(Workflow data on labor force participation \ 2008-04-02)
. set seed 11020
. generate isin = (runiform()>.8)
. label var isin "1 if in random sample (seed 11020)"
. label def isin 0 0_NoIn 1 1_InSample
. label val isin isin
. keep if isin
(601 observations deleted)
. tabulate isin, missing

   1 if in |
    random |
    sample |
     (seed |
    11020) |      Freq.     Percent        Cum.
-----------+-----------------------------------
1_InSample |        152      100.00      100.00
-----------+-----------------------------------
     Total |        152      100.00
. label data "20% subsample of wf-lfp."
. notes: wf3-subsample.do \ jsl 2008-04-03
. save x-wf3-subsample, replace
file x-wf3-subsample.dta saved
```

The command `set seed 11020` sets the seed for the random-number generator and is important if you want to create exactly the same sample later. You can pick any number for the seed. The command `generate isin = (runiform() > .8)` creates a binary variable equal to 1 if the random number is greater than .8. Because `runiform()` creates a uniform random variable with values from 0 to 1, `isin` will be 1 about 20% of the time. If you want a larger sample, replace .8 with a smaller number; for a smaller sample, replace .8 with a larger number. The last part of the program saves a dataset that contains roughly 20% of the original sample.

Note: The `runiform()` function was introduced in Stata 10.1. If you are using Stata 10, but have not updated to Stata 10.1 and you are connected to the Internet, run the `update all` command and follow the instructions. If you are running Stata 9, use the `uniform()` function instead of `runiform()`.

Step 6: Exclude parts of the do-file

If you have a long do-file that is generating an error, it is often useful to run only part of the file. This can be done using comments. You can add a * to the front of any line you do not want to run. For example,

```
* logit lfp wc hc
```

To comment out a series of lines, use /* and */. Everything between the opening /* and the closing */ is ignored when you run the do-file. This technique is used extensively with the extended examples presented later in this section.

Step 7: Starting over

Sometimes the fastest way to fix a problem is to start over. You checked the syntax of each command, you clicked on the blue error message to make sure you understand what the error means, you showed the problem to others who see no problems, yet the program keeps generating an error. This is a good time to start over. Hopefully, if you re-create the program without looking at the original version, you will not make the same mistake again. Of course, you might make the same error again. But, if you already tried everything you can think of, it is worth a try.

Why does this method sometimes work? Some errors are caused by subtle typing errors that you do not see even when looking at the code very carefully. Research on reading has shown that people construct much of what they read from what they think they should be reading. This is why it can be so hard to find typos. For example, you have written `tabl` rather than `tab1` or tried to analyze `varO1` or `var1` instead of `var01`. You can stare at this a long time and still not see it. If you start over, retyping all commands and variable names, there is a chance that you will not make the same typing error again. When starting over, here are some things to keep in mind.

Throw out all the original code

It is tempting to keep some of your original code that you "know" is correct. I once spent hours debugging a complex program until I discovered that the error was in a part of the program that was so simple and "obviously correct" that I skipped over it.

Use a new file

Start with a new file, rather than simply deleting everything in the original do-file. Why? It is possible to have a problem in a do-file that is caused by characters that are not visible and that your editor cannot delete. Your new program might look exactly like the old one, but a bit comparison of the two files will show that the files are different.

Try alternative approaches

When starting over, I often use a different approach rather than trying to do exactly what I did before. For example, if I think the command name is `tabl` and not `tab1`, I will intentionally enter the same incorrect command again. If instead I use a series of `tab` commands, the problem is resolved.

Step 8: Sometimes it is not your mistake

It is possible that there is an error in Stata or a user-written program that you are using. If you have tried everything you can think of to fix the problem, you might try posting the problem on Statalist (http://www.stata.com/statalist/), checking Stata's frequently asked questions (http://www.stata.com/support/faqs/), or contacting technical support at StataCorp (http://www.stata.com/support/). Before you do this, read section 3.4 about getting the most out of asking for help.

3.3.3 Example 1: Debugging a subtle syntax error

In this section, I go through the steps I would use to debug problems when the error message does not immediately point to a solution. I want to plot the prestige of a person's doctoral department against the prestige of the person's first academic job. These commands, which are so long they run off the page, were extracted from `wf3-debug-graph1.do`

```
use wf-acjob, clear
twoway (scatter job phd, msymbol(smcircle_hollow) msize(small)), ///
    ytitle(Where do you work?) yscale(range(1 5.)) ylabel(1(1)5, angle(ninety)) xtitle(Where did you
    xscale(range(1 5)) xlabel(1,5) caption(wf3-debug-graph1.do 2006-03-17, size(small)) scheme(s2man
```

The error message is

```
option 5 not allowed
r(198);
```

Because the message confuses me, I click on `r(198)` and obtain

```
[P]     error . . . . . . . . . . . . . . . . . . . . . . . . . Return code 198
        invalid syntax;
        __________ invalid;
        range invalid;
        __________ invalid obs no;
        invalid filename;
        __________ invalid varname;
        __________ invalid name;
        multiple by´s not allowed;
        __________ found where number expected;
        on or off required;
        All items in this list indicate invalid syntax.  These errors are
        often, but not always, due to typographical errors.  Stata attempts
        to provide you with as much information as it can.  Review the
        syntax diagram for the designated command.
        In giving the message "invalid syntax", Stata is not very helpful.
        Errors in specifying expressions often result in this message.
```

This message does not help much (even Stata warns me that the error message is not very helpful!), but it suggests that the problem might be related to an option that contains a 5.

Aside on why error messages can be misleading

Error messages do not always point to the real problem. The reason is that Stata knows how to parse the syntax of correct commands, not incorrect commands. Although Stata tries to make sense out of incorrect commands, it might not succeed. Think of error messages as suggestions that might point to the problem or that might be misleading.

The first thing I do to debug this program is to reformat the command so that it is easier to read (file: `wf3-debug-graph2.do`):

```
twoway (scatter job phd, msymbol(smcircle_hollow) msize(small)),    ///
    ytitle(Where do you work?) yscale(range(1 5.))                  ///
    ylabel(1(1)5, angle(ninety))                                    ///
    xtitle(Where did you graduate?) xscale(range(1 5)) xlabel(1,5) ///
    caption(wf3-debug-graph2.do 2006-03-17, size(small))            ///
    scheme(s2manual) aspectratio(1) by(fem)
```

The command is easier to read, but it generates the same error because I only changed the formatting. If you have sharp eyes and a good understanding of the `twoway` command, you might see the error, particularly because the error message suggests that the problem has something to do with a 5. Still, let us suppose that I do not know what is causing the problem.

Next I check that the variables are appropriate for this type of plot by creating a simple graph from the command line using the same variables (file: `wf3-debug-graph3.do`):

```
scatter job phd
```

This works, so I know that the problem is not with the data. Next I comment out part of the original command using the `/*` and `*/` delimiters. My strategy is to comment out most of the command and verify that the program runs. Then I gradually add back parts of the original code until I find exactly which part of the command is causing the problem. Often this makes it simple to see what is causing the error. The next time I try the program it looks like this (file: `wf3-debug-graph4.do`):

```
twoway (scatter job phd, msymbol(smcircle_hollow) msize(small)), /* ///
    ytitle(Where do you work?) yscale(range(1 5.))                  ///
    ylabel(1(1)5, angle(ninety))                                    ///
    xtitle(Where did you graduate?) xscale(range(1 5)) xlabel(1,5)  ///
    caption(wf3-debug-graph4.do 2008-04-03, size(small))            ///
    scheme(s2manual) aspectratio(1) by(fem) */
```

This works and adds symbols to the graph. Next I include options that refine the y axis (file: `wf3-debug-graph5.do`):

```
twoway (scatter job phd, msymbol(smcircle_hollow) msize(small)),      ///
    ytitle(Where do you work?) yscale(range(1 5.))                    ///
    ylabel(1(1)5, angle(ninety))                                   /* ///
    xtitle(Where did you graduate?) xscale(range(1 5)) xlabel(1,5)    ///
    caption(wf3-debug-graph5.do 2008-04-03, size(small))              ///
    scheme(s2manual) aspectratio(1) by(fem)                          */
```

This works too, so I decide that the error is not caused by the 5s in this part of my program. Next I uncomment the commands controlling the x axis (file: `wf3-debug-graph6.do`):

```
twoway (scatter job phd, msymbol(smcircle_hollow) msize(small)),        ///
    ytitle(Where do you work?) yscale(range(1 5.))                      ///
    ylabel(1(1)5, angle(ninety))                                        ///
    xtitle(Where did you graduate?) xscale(range(1 5)) xlabel(1,5)  /* ///
    caption(wf3-debug-graph6.do 2008-04-03, size(small))                ///
    scheme(s2manual) aspectratio(1) by(fem)                            */
```

This generates the original error, so I conclude that the problem is probably in this segment of code:

```
xtitle(Where did you graduate?) xscale(range(1 5)) xlabel(1,5)
```

The `xtitle()` option looks fine. I could verify this by rerunning the program after commenting out the `xscale()` and `xlabel()` commands. Because it is hard to make a mistake with a simple `xtitle()` option, I decide not to do this (yet). I assume that the problem is caused by the `xscale()` or `xlabel()` options. Looking closely, I see the error is with `xlabel(1,5)`. Although this looks like a reasonable way to indicate that labels should go from 1 to 5, the correct syntax is `xlabel(1(1)5)`. I change this and the program does just what I want it to do (file: `wf3-debug-graph7.do`).

If I did not see that the error was caused by `xlabel(1,5)`, I would run the command with only the `xtitle()` and `xscale()` options included (file: `wf3-debug-graph8.do`):

```
twoway (scatter job phd, msymbol(smcircle_hollow) msize(small)),       ///
    ytitle(Where do you work?) yscale(range(1 5.))                     ///
    ylabel(1(1)5, angle(ninety))                                       ///
    xtitle(Where did you graduate?) xscale(range(1 5)) /* xlabel(1,5) ///
    caption(wf3-debug-graph8.do 2008-04-03, size(small))               ///
    scheme(s2manual) aspectratio(1) by(fem) */
```

This also runs, so I would know that the problem is with the `xlabel()` option.

3.3.4 Example 2: Debugging unanticipated results

You might have a do-file that runs without error but produces strange or unanticipated results. To illustrate this type of problem, I use an example motivated by a question I received from a sophisticated Stata user.[8] I have nine binary indicators of functional limitations (e.g., Do you have problems standing? Walking? Reaching?). Before trying

8. Claudia Geist kindly allowed me to use this example. I have changed the data and variables, but the problem is the same one she encountered.

to scale these measures, I want to determine if there are certain combinations that occur commonly. For example, do troubles with walking tend to occur with other problems in lower-body function? Do some limitations tend to occur in pairs, but less often by themselves? And so on. I start by looking at the percentage of 1s for each variable (file: `wf3-debug-precision.do`). Because the variables are binary, I can simply compute the summary statistics:

```
. use wf-flims, clear
(Workflow data on functional limitations \ 2008-04-02)
. summarize hnd hvy lft rch sit std stp str wlk
    Variable |       Obs        Mean    Std. Dev.       Min        Max
-------------+--------------------------------------------------------
         hnd |      1644     .169708    .3754903          0          1
         hvy |      1644    .4288321    .4950598          0          1
         lft |      1644    .2475669    .4317301          0          1
         rch |      1644    .1703163    .3760248          0          1
         sit |      1644    .2104623     .407761          0          1
-------------+--------------------------------------------------------
         std |      1644    .3607056    .4803514          0          1
         stp |      1644    .3643552    .4813953          0          1
         str |      1644    .2974453    .4572732          0          1
         wlk |      1644    .2706813    .4444469          0          1
```

The distributions for the nine variables individually (or even 72 tabulations between pairs of variables) do not tell me all I want to know about how limitations cluster. A seemingly quick way to look at this is to create a new variable that combines the nine binary variables. For example, with the variables `str` and `wlk`, I create the variable `strwlk`:

```
generate strwlk = 10*str + wlk
```

`strwlk` is 0 if both `wlk` and `str` are 0, 1 if only `wlk` is 1, 10 if only `str` is 1, and 11 if both are 1.

```
. tabulate strwlk, missing
     strwlk |      Freq.     Percent        Cum.
------------+-----------------------------------
          0 |      1,091       66.36       66.36
          1 |         64        3.89       70.26
         10 |        108        6.57       76.82
         11 |        381       23.18      100.00
------------+-----------------------------------
      Total |      1,644      100.00
```

Seems easy, so I extend the idea to the nine variables:

```
generate flimall = hnd*100000000 + hvy*10000000 + lft*1000000 ///
    + rch*100000 + sit*10000 + std*1000 + stp*100 + str*10 + wlk
label var flimall "hnd-hvy-lft-rch-sit-stp-stp-str-wlk"
```

Next I tabulate `flimall` where the value 0 indicates no limitations in any function; 111,111,111 indicates limitations with all activities; and other combinations of 0s and 1s reflect other patterns of limitations. Here is the output:

```
. tabulate flimall, missing

hnd-hvy-lft |
-rch-sit-st |
d-stp-str-w |
         lk |      Freq.     Percent        Cum.
------------+-----------------------------------
          0 |        715       43.49       43.49
          1 |          5        0.30       43.80
         10 |          8        0.49       44.28
         11 |          2        0.12       44.40
  (output omitted)
    1100111 |          1        0.06       54.08
    1101100 |          1        0.06       54.14
   1.00e+07 |         86        5.23       59.37
  (output omitted)
   1.10e+08 |          7        0.43       88.56
   1.11e+08 |         15        0.91       91.42
  (output omitted)
------------+-----------------------------------
      Total |      1,644      100.00
```

Unfortunately, the large numbers are in scientific notation and I lose the information that I want. To fix this, I create a string variable:

```
generate sflimall = string(flimall, "%16.0f")
```

The `%16.0f` indicates that I want the string to correspond to a 16-digit number without decimal points (for details, see `help format` or [D] **format**; also see section 6.4.5, which discusses how data are stored in Stata). I add a label and tabulate the new variable:

```
label var sflimall "hnd-hvy-lft-rch-sit-std-stp-str-wlk"
tabulate sflimall, missing
```

(Continued on next page)

I see something very peculiar.

```
hnd-hvy-lft |
-rch-sit-st |
d-stp-str-w |
         lk |      Freq.     Percent        Cum.
------------+-----------------------------------
          0 |        715       43.49       43.49
          1 |          5        0.30       43.80
         10 |          8        0.49       44.28
        100 |         28        1.70       45.99
  (output omitted)
   10000000 |         86        5.23       53.83
  100000000 |         15        0.91       54.74
   10000001 |          4        0.24       54.99
  100000096 |          4        0.24       55.23
  (output omitted)
    1000001 |          1        0.06       55.29
   10000010 |          5        0.30       55.60
   10000011 |          5        0.30       55.90
  (output omitted)
------------+-----------------------------------
      Total |      1,644      100.00
```

The values are supposed to be all 0s and 1s, but I have a number 100000096. To figure out what went wrong, I run **tab1 hnd-wlk, missing** to verify that the variables only have values of 0 and 1. If I find four cases with 9s for **str** and 6 for **wlk**, I know that I have a problem with my original data, but the data look fine. Next I clean up the code to make it easier to find typos:

```
generate flimall = hnd*100000000 ///
                 + hvy*10000000 ///
                 + lft*1000000 ///
                 + rch*100000 ///
                 + sit*10000 ///
                 + std*1000 ///
                 + stp*100 ///
                 + str*10 ///
                 + wlk
```

The code looks fine, so I try the same approach but with only four variables. A good strategy when debugging is to see if you can get a similar but simpler program to work.

```
. generate flimall = std*1000 ///
>                  + stp*100 ///
>                  + str*10 ///
>                  + wlk
. generate  sflimall = string(flimall,"%9.0f")
. label var sflimall "std-stp-str-wlk"
```

```
. tabulate  sflimall, missing

std-stp-str |
       -wlk |      Freq.     Percent        Cum.
------------+-----------------------------------
          0 |        866       52.68       52.68
          1 |         16        0.97       53.65
         10 |         24        1.46       55.11
        100 |         80        4.87       59.98
       1000 |         73        4.44       64.42
       1001 |         13        0.79       65.21
        101 |          8        0.49       65.69
       1010 |         15        0.91       66.61
       1011 |         25        1.52       68.13
         11 |         13        0.79       68.92
        110 |         24        1.46       70.38
       1100 |         72        4.38       74.76
       1101 |         27        1.64       76.40
        111 |         20        1.22       77.62
       1110 |         45        2.74       80.35
       1111 |        323       19.65      100.00
------------+-----------------------------------
      Total |      1,644      100.00
```

Again this looks fine. I continue adding variables and things still work with eight variables. Further, it does not matter which eight I choose. I conclude that there is a problem going from eight to nine variables. The problem is because the nine-digit number I am creating with `flimall` is too large to be held accurately. Essentially, this means that 100,000,096 (the number above that seemed odd) is only an approximation to the correct result 100,000,100. Indeed, the number that raised suspicions is off by only 4 out of over 100 million. The solution is to store the information in double precision. With the addition of one word, the problem is fixed:

```
. generate double flimall = hnd*100000000 ///
>                          + hvy*10000000 ///
>                          + lft*1000000 ///
>                          + rch*100000 ///
>                          + sit*10000 ///
>                          + std*1000 ///
>                          + stp*100 ///
>                          + str*10 ///
>                          + wlk
```

See section 6.4.5 for more information on types of variables.

3.3.5 Advanced methods for debugging

If things are still not working, you can trace the error. Tracing refers to looking at each of the steps taken by Stata in executing your program (i.e., you trace the steps the program takes). This shows you what the program is doing behind the scenes, often revealing the specific step that causes the problem. To trace a program, type the command `set trace on`. Stata echoes every line of code it runs, both from your do-file and from your ado-files. To turn tracing off, type `set trace off`. For details on how to use this powerful feature, type `help trace` or see [P] **trace**.

3.4 How to get help

At some point, you will need to ask for help. Here are some things that make it easier for someone to help you and increase your chances of getting the help you need.

1. Try all the suggestions above to debug your program. Read the manual for the commands related to your error.
2. Make sure that your copy of Stata and user-written programs are up to date.
3. Write a brief description of the problem and the things you have done to resolve the problem (e.g., updated Stata, tried a different dataset). I often solve my own problems when I am composing a detailed email asking someone for help.
4. Create a do-file that generates the error using a small dataset. Do not send a huge dataset as an attachment. Make the do-file self-contained (e.g., it loads the dataset) and transportable (e.g., it does not hardcode the directory for the data).
5. Send the do-file, the log file in text format, and the dataset to the person you are asking for help.

When you ask for help, the clearer and more detailed the information you provide, the greater the chance that someone will be willing and able to help you.

3.5 Conclusions

Although this chapter contains many suggestions on using Stata, it only touches on the many features in the program. If you spend a lot of time with Stata, it is worth browsing the manuals. Often you will find a command or feature that solves a problem. I know that in writing this book I discovered many useful commands that I was unaware of. If you do not like reading manuals, consider a NetCourse (web course) from StataCorp (http://www.stata.com/netcourse/). The investment of time in learning the tools usually saves time in the long run.

4 Automating your work

A great deal of data management and statistical analysis involves doing the same task multiple times. You create and label many variables, fit a sequence of models, and run multiple tests. By automating these tasks, you can save time and prevent errors, which are fundamental to an effective workflow. In this chapter, I discuss six tools for automation in Stata.

> *Macros*: Macros are simply abbreviations for a string of characters or a number. These abbreviations are amazingly useful.
>
> *Saved results*: Many Stata commands save their results in memory. This information can be retrieved and used to automate your work.
>
> *Loops*: Loops are a way to repeat a group of commands. By combining macros with loops, you can speed up tasks ranging from creating variables to fitting models.
>
> *The include command*: `include` inserts text from one file into another, which is useful when the same commands are used multiple times in do-files.
>
> *Ado-files*: Ado-files let you write your own commands to customize Stata, automate your workflow, and speed up routine tasks.
>
> *Help files*: Although help files are primarily used to document ado-files, they can also be used to document your workflow.

Macros, saved results, and loops are essential for chapters 5–7. Although `include`, ado-files, and help files are very useful, they are not essential for later chapters. Still, I encourage you to read these sections.

4.1 Macros

Macros are the simplest tool for automating your work. A macro assigns a string of text or a number to an abbreviation. Here is a simple example. I want to fit the model

```
logit y var1 var2 var3
```

I can create the macro `rhs` with the names of the independent or right-hand-side variables:

```
local rhs "var1 var2 var3"
```

Then I can write the `logit` command as

```
logit y `rhs'
```

where the ` and ' indicate that I want to insert the contents of the macro **`rhs`**. The command `logit y `rhs'` works exactly the same as `logit y var1 var2 var3`. In the examples that follow, I show you many ways to use macros. For a more technical discussion, see [P] **macro**.

4.1.1 Local and global macros

Stata has two types of macros, local macros and global macros. Local macros can be used only within the do-file or ado-file in which they are defined. When that program ends, the local macro disappears. For example, if I create the local **`rhs`** in **`step1.do`**, that local disappears as soon as `step1.do` ends. By comparison, a global macro persists until you delete it or exit Stata. Although global macros can be useful, they can lead to do-files that unintentionally depend on a global macro created by another do-file or from the Command window. Such do-files are not robust and can lead to unpredictable results. Accordingly, I almost exclusively use local macros.

Local macros

Local macros that contain a string of characters are defined as

`local` *local-name* `"`*string*`"`

For example,

```
local rhs "var1 var2 var3 var4"
```

A local macro can also be set equal to a numerical expression:

`local` *local-name* `=` *expression*

For example,

```
local ncases = 198
```

The content of a macro is inserted into your do-file or ado-file by entering `*local-name*'. For example, to print the contents of the local **`rhs`**, type

```
. display "The local macro rhs contains: `rhs'"
The local macro rhs contains: var1 var2 var3 var4
```

or type

```
. display "The local ncases equals: `ncases'"
The local ncases equals: 198
```

The opening quote ` and closing quote ´ are different symbols that look similar with some fonts. To make sure you have the correct symbols, load the do-file `wf4-macros.do` from the Workflow package and compare the symbols it contains with those you can create from your keyboard.

Global macros

Global macros are defined much like local macros:

`global` *global-name* "*string*"

`global` *global-name* = *expression*

For example,

```
global rhs "var1 var2 var3 var4"
global ncases = 198
```

The content of a global macro is inserted by entering $*global-name*. For example,

```
. display "The local macro rhs contains: $rhs"
The local macro rhs contains: var1 var2 var3 var4
```

or

```
. display "The local ncases equals: $ncases"
The local ncases equals: 198
```

Using double quotes when defining macros

When defining a macro containing a string, you can include the string in quotes. For example,

```
local myvars "y x1 x2"
```

Or you can remove the quotation marks:

```
local myvars y x1 x2
```

I prefer using quotation marks because they clarify where the string begins and ends. Plus text editors with syntax highlighting can show everything that appears between quotation marks in a different color, which helps when debugging programs.

Creating long strings

You can create a macro that contains a long string in one step, such as

```
local demogvars "female black hispanic age agesq edhighschl edcollege edpostgrad"
```

The problem is that long commands are truncated or wrapped when viewed on screen or printed. As shown on page 58, this can make it harder to debug your program. To keep lines shorter than 80 columns (the `local` command above is 81 columns wide), I build long macros in steps. For example, I can create `demogvars` by starting with the first five variable names:

```
local demogvars "female black hispanic age agesq"
```

The next line takes the current content of `demogvars` and adds new names to the end. Remember, the content of `demogvars` is inserted by `` `demogvars´ ``:

```
local demogvars "`demogvars´ edhighschl edcollege edpostgrad"
```

Additional names can be added in the same way.

4.1.2 Specifying groups of variables and nested models

Macros can hold the names of variables that you are analyzing. Suppose that I want summary statistics and estimates for a logit of `lfp` on `k5`, `k618`, `age`, `wc`, `hc`, `lwg`, and `inc`. Without macros, I enter the commands like this (file: `wf4-macros.do`):

```
summarize lfp k5 k618 age wc hc lwg inc
logit     lfp k5 k618 age wc hc lwg inc
```

If I change the variables, say, deleting `hc` and adding `agesquared`, I need to change both commands:

```
summarize lfp k5 k618 age agesquared wc lwg inc
logit     lfp k5 k618 age agesquared wc lwg inc
```

Alternatively, I can define a macro with the variable names:

```
local myvars "lfp k5 k618 age wc hc lwg inc"
```

Then I compute the statistics and fit the model like this:

```
summarize `myvars´
logit     `myvars´
```

Stata replaces `` `myvars´ `` with the content of the macro. Thus the `` summarize `myvars´ `` command is exactly equivalent to `summarize lfp k5 k618 age wc hc lwg inc`.

Using a local macro to specify variables allows me to change the variables being analyzed by changing the local. For example, I can change the list of variables in the macro `myvars`:

```
local myvars "lfp k5 k618 age agesquared wc lwg inc"
```

Then I can use the same commands as before to analyze a different set of variables:

```
summarize `myvars´
logit     `myvars´
```

The idea of using a macro to hold variable names can be extended by using different macros for different groups of variables (e.g., demographic variables, health variables). These macros can be combined to specify a sequence of nested models. First, I create macros for four groups of independent variables:

```
local set1_age   "age agesquared"
local set2_educ  "wc hc"
local set3_kids  "k5 k618"
local set4_money "lwg inc"
```

To check that a local is correct, I `display` the content. For example,

```
. display "set3_kids: `set3_kids´"
set3_kids: k5 k618
```

Next I specify four nested models. The first model includes only the first set of variables and is specified as

```
local model_1 "`set1_age´"
```

The macro `model_2` combines the content of the local `model_1` with the variables in local `set2_educ`:

```
local model_2 "`model_1´ `set2_educ´"
```

The next two models are specified the same way:

```
local model_3 "`model_2´ `set3_kids´"
local model_4 "`model_3´ `set4_money´"
```

Next I check the variables in each model:

```
. display "model_1: `model_1´"
model_1: age agesquared
. display "model_2: `model_2´"
model_2: age agesquared wc hc
. display "model_3: `model_3´"
model_3: age agesquared wc hc k5 k618
. display "model_4: `model_4´"
model_4: age agesquared wc hc k5 k618 lwg inc
```

Using these locals, I estimate a series of logits:

```
logit lfp `model_1´
logit lfp `model_2´
logit lfp `model_3´
logit lfp `model_4´
```

There are several advantages to using locals to specify models. First, when specifying complex models, it is easy to make a mistake. For example, here are `logit` commands for a series of nested models from a project I am currently working on. Do you see the error?

```
logit y black
logit y black age10 age10sq edhs edcollege edpost incdollars childsqrt
logit y black age10 age10sq edhs edcollege edpost incdollars ///
    childsqrt bmi1 bmi3 bmi4 menoperi menopost mcs_12 pcs_12
logit y black age10 age10sq edhs edcollege edpost incdollars ///
    childsqrt bmi1 bmi3 bmi4 menoperi menopost mcs_12 ///
    pcs_12 sexactsqrt phys8_imp2 subj8_imp2
logit y black age10 age10sq edhs edcollege edpost incdollars ///
    childsqrt bmi1 bmi3 bmi4 menoperi menopost mcs_12 ///
    pcs_12 sexactsqrt phys8_imp2 subj8_imp2 selfattr partattr
```

Second, locals make it easy to revise model specifications. Even if I am successful in initially defining a set of models by typing each variable name for each model, errors creep in when I change the models. For example, suppose that I do not need a quadratic term for age. Using locals, I need to make only one change:

```
local set1_age "age"
```

This change is automatically applied to the specifications of all models:

```
local model_1 "`set1_age´"
local model_2 "`model_1´ `set2_educ´"
local model_3 "`model_2´ `set3_kids´"
local model_4 "`model_3´ `set4_money´"
```

In chapter 7, these ideas are combined with loops to simplify complex analyses.

4.1.3 Setting options with locals

I often use locals to specify the options for a command. This makes it easier to change options for multiple commands and helps organize the complex options sometimes needed for graphs.

Using locals with tabulate

Suppose that I want to compute several two-way tables using `tabulate`. This command has many options that control what is printed within the table and the summary statistics that are computed. For my first tables, I want cell percentages requiring the `cell` option, missing values requiring the `missing` option, numeric values rather than value labels for row and column labels requiring the `nolabel` option, and a chi-squared test of independence requiring the `chi2` option. I can put these options in a local:

```
local opt_tab "cell miss nolabel chi2"
```

I use this local to set the options for two `tabulate` commands:

```
tabulate wc hc, `opt_tab´
tabulate wc lfp, `opt_tab´
```

I could have dozens of `tabulate` commands that use the same options. If I later decide that I want to add row percentages and remove cell percentages, I need to change only one line:

```
local opt_tab "row miss nolabel chi2"
```

This change will be applied to all the tabulate commands that use opt_tab to set the options.

Using locals with graph

The options for the graph command can be very complicated. For example, here is a graph comparing the probability of tenure by the number of published articles for male and female biochemists:

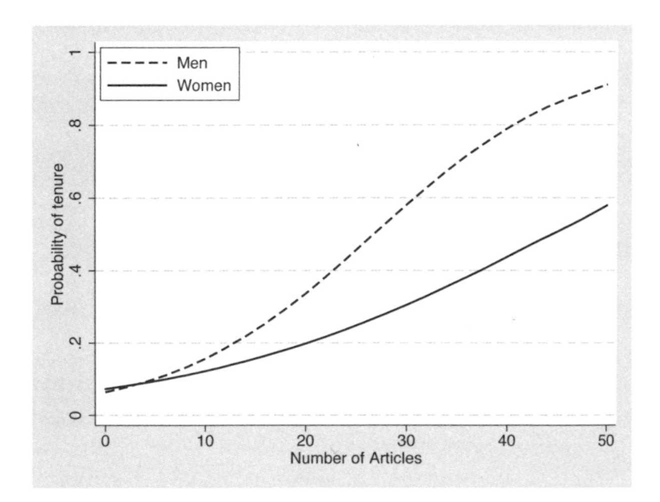

Even though this is a simple graph, the graph command is complex and hard to read (file: wf4-macros-graph.do):

```
graph twoway ///
  (connected pr_women articles, lpattern(solid) lwidth(medthick) ///
      lcolor(black) msymbol(i)) ///
  (connected pr_men articles,   lpattern(dash)  lwidth(medthick) ///
      lcolor(black) msymbol(i)) ///
  , ylabel(0(.2)1, grid glwidth(medium) glpattern(dash)) xlabel(0(10)50) ///
    ytitle("Probability of tenure") ///
    legend(pos(11) order(2 1) ring(0) cols(1))
```

Macros make it simpler to specify the options, to see which options are used, and to revise them. For this example, I can create macros that specify the line options for men and for women, the grid options, and the options for the legend:

```
local opt_linF   "lpattern(solid) lwidth(medthick) lcolor(black) msymbol(i)"
local opt_linM   "lpattern(dash)  lwidth(medthick) lcolor(black) msymbol(i)"
local opt_ygrid  "grid glwidth(medium) glpattern(dash)"
local opt_legend "pos(11) order(2 1) ring(0) cols(1)"
```

Using these macros, I create a **graph** command that I find easier to read:

```
graph twoway ///
  (connected pr_women articles,     `opt_linF')   ///
  (connected pr_men   articles,     `opt_linM')   ///
  , xlabel(0(10)50) ylabel(0(.2)1, `opt_ygrid')  ///
    ytitle("Probability of tenure")              ///
    legend(`opt_legend')
```

Moreover, if I have a series of similar graphs, I can use the same locals to specify options for all the graphs. If I want to change something, I only have to change the macros, not each **graph** command. For example, if I decide to use colored lines to distinguish between men and women, I change the macros containing line options:

```
local opt_linF   "lpattern(solid) lwidth(medthick) lcolor(red)  msymbol(i)"
local opt_linM   "lpattern(dash)  lwidth(medthick) lcolor(blue) msymbol(i)"
```

With these changes, I can use the same **graph twoway** command as before.

4.2 Information returned by Stata commands

Drukker's dictum: Never type anything that you can obtain from a saved result.

When writing do-files you never want to type a number if Stata can provide the number for you. Fortunately, Stata can provide just about any number you need. To understand what this means, consider a simple example where I mean center the variable **age**. I could do this by first computing the mean (file: **wf4-returned.do**):

```
. use wf-lfp, clear
(Workflow data on labor force participation \ 2008-04-02)
. summarize age
    Variable |       Obs        Mean    Std. Dev.       Min        Max
-------------+--------------------------------------------------------
         age |       753    42.53785    8.072574         30         60
```

Next I use the mean from **summarize** in the **generate** command:

```
. generate age_mean = age - 42.53785
```

The average of the new variable is very close to 0 as it should be (within .000001):

```
. summarize age_mean
    Variable |       Obs        Mean    Std. Dev.       Min        Max
-------------+--------------------------------------------------------
    age_mean |       753   -1.49e-06    8.072574  -12.53785   17.46215
```

I can do the same thing without typing the mean. The **summarize** command both sends output to the Results window and saves this information in memory. In Stata's terminology, **summarize** returns this information. To see the information returned by the last command, I use the **return list** command. For example,

```
. summarize age

    Variable |       Obs        Mean    Std. Dev.       Min        Max
-------------+--------------------------------------------------------
         age |       753    42.53785    8.072574         30         60

. return list

scalars:
                  r(N) =  753
              r(sum_w) =  753
               r(mean) =  42.53784860557769
                r(Var) =  65.16645121641095
                 r(sd) =  8.072574014303674
                r(min) =  30
                r(max) =  60
                r(sum) =  32031
```

The mean is returned to a scalar named `r(mean)`.[1] I use this value to subtract the mean from age:

```
. generate age_meanV2 = age - r(mean)
```

When I compare the two mean-centered variables, I find that the variable created using `r(mean)` is slightly closer to zero:

```
. summarize age_mean age_meanV2

    Variable |       Obs        Mean    Std. Dev.       Min        Max
-------------+--------------------------------------------------------
    age_mean |       753   -1.49e-06    8.072574  -12.53785   17.46215
  age_meanV2 |       753    6.29e-08    8.072574  -12.53785   17.46215
```

I could get even closer to zero by creating a variable using double precision:

```
. summarize age

    Variable |       Obs        Mean    Std. Dev.       Min        Max
-------------+--------------------------------------------------------
         age |       753    42.53785    8.072574         30         60

. generate double age_meanV3 = age - r(mean)

. label var age_meanV3 "age - mean(age) using double precision"

. summarize age_mean age_meanV2 age_meanV3

    Variable |       Obs        Mean    Std. Dev.       Min        Max
-------------+--------------------------------------------------------
    age_mean |       753   -1.49e-06    8.072574  -12.53785   17.46215
  age_meanV2 |       753    6.29e-08    8.072574  -12.53785   17.46215
  age_meanV3 |       753    3.14e-15    8.072574  -12.53785   17.46215
```

This example illustrates the first reason why you never want to enter a number by hand if the information is stored in memory. Values are returned with more numerical precision than shown in the output from the Results window. Second, using returned results prevents errors when typing a number. Finally, using a returned value is more robust. If you type the mean based on the output from `summarize` and later change the sample being analyzed, it is easy to forget to change the `generate` command where you typed the mean. Using `r(mean)` automatically inserts the correct quantity.

1. Scalar means a single numeric value.

Most Stata commands that compute numerical quantities return those quantities and often return additional information that is not in the output. To look at the returned results from commands that are not fitting a model, use **return list**. For estimation commands, use **ereturn list**. To find out what each return contains, enter **help** *command-name* and look at the section on saved results.

Using returned results with local macros

In the example above, I used the returned mean when generating a new variable. I can also place the returned information in a macro. For example, if I run **summarize age**, the mean and standard deviations are returned. I can assign these quantities to local macros:

```
. local mean_age = r(mean)
. local sd_age = r(sd)
```

I can now display this information:

```
. display "The mean of age `mean_age´ (sd=`sd_age´)"
The mean of age 42.53784860557769 (sd=8.072574014303674)
```

If you are using returned results to compute other quantities (e.g., to center a variable), you want to retain all 14 decimal digits. If you only want to display the quantity, you might want to round the result to fewer decimal digits. You can do this with the **string()** function. For example,

```
. local mean_agefmt = string(r(mean),"%8.3f")
. local sd_agefmt = string(r(sd),"%8.3f")
. display "The mean of age `mean_agefmt´ (sd=`sd_agefmt´)."
The mean of age 42.538 (sd=8.073).
```

The locals **mean_agefmt** and **sd_agefmt** have been printed with only three digits of precision and should not be used for computing other quantities.

Returned results are used in many ways in the later chapters. I encourage you to experiment with assigning returns to locals and using the **display** command. For more information, see **help display** and **help return**, or [P] **display**, [R] **saved results**, and [P] **return**.

4.3 Loops: foreach and forvalues

Loops let you execute a group of commands multiple times. Here is a simple example that illustrates the key features of loops. I have a four-category ordinal variable **y** with values from 1 to 4. I want to create the binary variables **y_lt2**, **y_lt3**, and **y_lt4** that equal 1 if **y** is less than the indicated value, else 0. I can create the variables with three **generate** commands (file: **wf4-loops.do**):

```
generate y_lt2 = y<2 if !missing(y)
generate y_lt3 = y<3 if !missing(y)
generate y_lt4 = y<4 if !missing(y)
```

where the `if` condition `!missing(y)` selects cases where `y` is not missing. I could create the same `generate` commands with a `foreach` loop:

```
1>  foreach cutpt in 2 3 4 {
2>      generate y_lt`cutpt´ = y<`cutpt´ if !missing(y)
3>  }
```

Let's look at each part of this loop. Line 1 starts the loop with the `foreach` command. `cutpt` is the name I chose for a macro to hold the cutpoint used to dichotomize `y`. Each time through the loop, the value of `cutpt` changes. `in` signals the start of a list of values that will be assigned in sequence to the local `cutpt`. The numbers `2 3 4` are the values to be assigned to `cutpt`. `{` indicates that the list has ended. Line 2 is the command that I want to run multiple times. Notice that it uses the macro `cutpt` that was created in line 1. Line 3 ends the `foreach` loop.

Here is what happens when the loop is executed. The first time through `foreach` the local `cutpt` is assigned the first value in the list. This is equivalent to the command `local cutpt "2"`. Next the `generate` command is run, where `` `cutpt´ `` is replaced by the value assigned to `cutpt`. The first time through the loop, line 2 is evaluated as

```
generate y_lt2 = y<2 if !missing(y)
```

Next the closing brace `}` is encountered, which sends us back to the `foreach` command in line 1. In the second pass, `foreach` assigns `cutpt` to the second value in the list, which means that the `generate` command is evaluated as

```
generate y_lt3 = y<3 if !missing(y)
```

This continues once more, assigning `cutpt` to 4. When the `foreach` loop ends, three variables have been generated.

Next I want to estimate binary logits on `y_lt2`, `y_lt3`, and `y_lt4`.[2] I assign my right-hand-side variables to the local `rhs`:

```
local rhs "yr89 male white age ed prst"
```

To run the logits, I could use the commands

```
logit y_lt2 `rhs´
logit y_lt3 `rhs´
logit y_lt4 `rhs´
```

2. I am using a series of binary logits to assess the parallel regression assumption in the ordinal logit model; see Long and Freese (2006) for details.

Or I could do the same thing with a loop:

```
foreach lhs in y_lt2 y_lt3 y_lt4 {
    logit `lhs´ `rhs´
}
```

Using foreach to fit three models is probably more trouble than it is worth. Suppose that I also want to compute the frequency distribution of the dependent variable and fit a probit model. I can add two lines to the loop:

```
foreach lhs in y_lt2 y_lt3 y_lt4 {
    tabulate `lhs´
    logit    `lhs´ `rhs´
    probit   `lhs´ `rhs´
}
```

If I want to change a command, say, adding the missing option to tabulate, I have to make the change in only one place and it applies to all three outcomes.

I hope this simple example gives you some ideas about how useful loops can be. In the next section, I present the syntax for the foreach and forvalues commands. The foreach command has options to loop through lists of existing variables, through lists of variables you want to create, or through numeric lists. The forvalues command is for looping though numbers. After going through the syntax, I present more complex examples of loops that illustrate techniques used in later chapters. For further information, use help or check [P] **foreach** and [P] **forvalues**.

The foreach command

The syntax is

foreach *local-name* { in | of *list-type* } *list* {

commands referring to `*local-name*´

}

where *local-name* is a local macro whose value is assigned by the loop. *list* contains the items to be assigned to *local-name*. With the in option, you provide a list of values or names and foreach goes through the list one at a time. For example,

```
foreach i in 1 2 3 4 5 {
```

will assign i the values 1, 2, 3, 4, and 5, or you can assign names to i:

```
foreach i in var1 var2 var3 var4 var5 {
```

The of option lets you specify the kind of list you are providing and Stata verifies that all the elements in the list are appropriate. The command foreach *local-name* of varlist *list* { is for lists of variables, where *list* is expanded according to standard variable abbreviation rules. For example,

```
foreach var of varlist lfp-inc {
```

expands `lfp-inc` to include all variables between `lfp` and `inc`. In `wf-lfp.dta`, this would be `lfp k5 k618 age wc hc lwg inc`. Stata verifies that each name in the list corresponds to a variable in the dataset in memory. If it does not, the loop ends with an error.

The command `foreach` *local-name* `of newlist` *newvarlist* is for a list of variables to be created. The names in *newvarlist* are not automatically created, but Stata verifies that the names are valid for generating new variables. The command `foreach` *local-name* `of numlist` *numlist* is used for numbered lists, where *numlist* uses standard number list notation. For details on the many ways to create sequences of numbers with *numlist*, type `help numlist` or see [U] **11.1.8 numlist**.

The forvalues command

The `forvalues` command loops through numbers. The syntax is

```
forvalues lname = range {
        commands referring to `local-name'
}
```

where range is specified as

Syntax	Meaning	Example	Generates
#1(#d)#2	From #1 to #2 in steps of #d	`1(2)10`	1, 3, 5, 7, 9
#1/#2	From #1 to #2 in steps of 1	`1/10`	1, 2, 3, ..., 10
#1 #t to #2	From #1 to #2 in steps of (#t−#1)	`1 4 to 15`	1, 4, 7, 10, 13

For example, to loop through ages 40 to 80 by 5s:

```
forvalues i = 40(5)80 {
```

Or to loop from 0 to 100 by .1:

```
forvalues i = 0(.1)100 {
```

4.3.1 Ways to use loops

Loops can be used in many ways that make your workflow faster and more accurate. In this section, I use loops for the following tasks:

- Listing variable and value labels
- Creating interaction variables
- Fitting models with alternative measures of education

- Recoding multiple variables the same way
- Creating a macro that holds accumulated information
- Retrieving information returned by Stata

The examples are simple, but illustrate features that are extended in later chapters. Hopefully, as you read these examples you will think of other ways in which loops can benefit your work. All the examples assume that `wf-loops.dta` has been loaded (file: `wf4-loops.do`).

Loop example 1: Listing variable and value labels

Surprisingly, Stata does not have a command to print a list of variable names followed only by their variable labels. The `describe` command lists more information than I often need, plus it contains details that often confuse people (e.g., what does `byte %9.0g warmlbl` mean?). To create a list of names and labels, I loop through a list of variables, retrieve each variable label, and print the information. To retrieve the variable label for a given variable, I use an extended macro function. Stata has dozens of extended macro functions that are used to create macros with information about variables, datasets, and other things. For example, to retrieve the variable label for `warm`, I use this command

```
local varlabel : variable label warm
```

To see the contents of `varlabel`, type

```
. display "Variable label for warm: `varlabel´"
Variable label for warm: Mom can have warm relations with child
```

To create a list for several variables, I loop through a list of variable names, extract each variable label, and print the results:

```
1>  foreach varname of varlist warm yr89 male white age ed prst {
2>      local varlabel : variable label `varname´
3>      display "`varname´" _col(12) "`varlabel´"
4>  }
```

Line 1 starts the loop through seven variable names. The first time through the loop, the local `varname` contains `warm`. Line 2 creates the local `varlabel` with the variable label for the variable in `varname`. Line 3 prints the results. Everything in this line should be familiar, except for `_col(12)`, which specifies that the label should start printing in the 12th column. Here is the list produced by the loop:

```
warm       Mom can have warm relations with child
yr89       Survey year: 1=1989 0=1977
male       Gender: 1=male 0=female
white      Race: 1=white 0=not white
age        Age in years
ed         Years of education
prst       Occupational prestige
```

If I want the labels to be closer to the names, I could change `_col(12)` to `_col(10)` or some other value. In section 4.5, I elaborate this simple loop to create a new Stata command that lists variable names with their labels.

Loop example 2: Creating interaction variables

Suppose that I need variables that are interactions between the binary variable **male** and a set of independent variables. I can do this quickly with a loop:

```
1>  foreach varname of varlist yr89 white age ed prst {
2>      generate  maleX`varname´ = male*`varname´
3>      label var maleX`varname´  "male*`varname´"
4>  }
```

Line 1 loops through the list of independent variables. Line 2 generates a new variable named `maleX`varname´`. For example, if `varname` is `yr89`, the new variable is `maleXyr89`. The variable label created in line 3 combines the names of the two variables used to create the interaction. For example, if `varname` is `yr89`, the variable label is `maleXyr89`. To examine the new variables and their labels, I use `codebook`:

```
. codebook maleX*, compact
```

Variable	Obs	Unique	Mean	Min	Max	Label
maleXyr89	2293	2	.1766245	0	1	male*yr89
maleXwhite	2293	2	.4147405	0	1	male*white
maleXage	2293	71	20.50807	0	89	male*age
maleXed	2293	21	5.735717	0	20	male*ed
maleXprst	2293	59	18.76625	0	82	male*prst

Although this variable label clearly indicates how the variable was generated, I prefer a label that includes the variable label from the source variable. I do this using the extended macro function introduced in *Loop example 1*:

```
1>  foreach varname of varlist yr89 white age ed prst {
2>      local varlabel : variable label `varname´
3>      generate  maleX`varname´ = male*`varname´
4>      label var maleX`varname´  "male*`varlabel´"
5>  }
```

Line 2 retrieves the variable label for `` `varname´ `` and line 4 uses this to create the new variable label. For `maleXage`, the label is `male*Age in years`. I could create an even more informative variable label by replacing line 4 with

```
label var maleX`varname´ "male*`varname´ (`varlabel´)"
```

For example, for `maleXprst`, the label would be `male*prst (Occupational prestige)`.

Loop example 3: Fitting models with alternative measures of education

Suppose I want to predict labor-force participation using education and five additional independent variables. My dataset has five measures of education (e.g., years of education, a binary indicator of attending high school), but I have no theoretical reason for choosing among them. I decide to try each measure in my model. First, I create a local containing the names of the education variables:

```
local edvars "edyrs edgths edgtcol edsqrtyrs edlths"
```

The other independent variables are

```
local rhs "male white age prst yr89"
```

I loop through the education variables and fit five ordinal logit models, each with a different measure of education:

```
foreach edvarname of varlist `edvars´ {
    display _newline "==> education variable: `edvarname´"
    ologit warm `edvarname´ `rhs´
}
```

This is equivalent to running these commands:

```
display _newline "==> education variable: edyrs"
ologit warm edyrs male white age prst yr89
display _newline "==> education variable: edgths"
ologit warm edgths male white age prst yr89
display _newline "==> education variable: edgtcol"
ologit warm edgtcol male white age prst yr89
display _newline "==> education variable: edsqrtyrs"
ologit warm edsqrtyrs male white age prst yr89
display _newline "==> education variable: edlths"
ologit warm edlths male white age prst yr89
```

I find the loop to be simpler and easier to debug than the repeated list of commands. In chapter 7, this idea is extended to collect information for selecting among the models using a Bayesian information criterion statistic (see page 306).

Loop example 4: Recoding multiple variables the same way

I often have multiple variables that I want to recode the same way. For example, I have six variables that measure social distance (e.g., would you be willing to have this person live next door to you?) using the same 4-point scale. The variables are

```
local sdvars "sdneighb sdsocial sdchild sdfriend sdwork sdmarry"
```

To dichotomize these variables, I use a loop:

```
1>  foreach varname of varlist `sdvars´ {
2>      generate  B`varname´ = `varname´
3>      label var B`varname´ "`varname´: (1,2)=0 (3,4)=1"
4>      replace   B`varname´ = 0 if `varname´==1 | `varname´==2
5>      replace   B`varname´ = 1 if `varname´==3 | `varname´==4
6>  }
```

Line 2 generates a new variable equal to the source variable. The new variable name adds B (for binary) to the source variable name (e.g., Bsdneighb from sdneighb). Line 3 adds a variable label. Line 4 assigns 0 to the new variable when the source variable is 1 or 2, where the | symbol means "or". Similarly, line 5 assigns 1 when the source variable is 3 or 4. The loop applies the same recoding to all the variables in the local sdvars.

Suppose that I have measures of income from five panels of data. The variables are named incp1 through incp5. I can transform each by adding .5 and taking the log:

```
foreach varname of varlist incp1 incp2 incp3 incp4 incp5 {
    generate  ln`varname´ = ln(`varname´+.5)
    label var ln`varname´ "Log(`varname´+.5)"
}
```

Loop example 5: Creating a macro that holds accumulated information

Typing lists is boring and often leads to mistakes. In the last example, typing the five income measures was simple, but if I had 20 panels it would be tedious. Instead, I can use a loop to create the list of names. First, I create a local varlist that contains nothing (known as a null string):

```
local varlist ""
```

I will use varlist to hold my list of names. Next I loop from 1 to 20 to build my list. Here I use forvalues because it automatically creates the sequence of numbers 1–20:

```
1>  forvalues panelnum = 1/20 {
2>      local varlist "`varlist´ incp`panelnum´"
3>  }
```

The local in line 2 can be confusing, so let me decode it from right to left (not left to right). The first time through the loop, incp`panelnum´ is evaluated as incp1 because `panelnum´ is 1. To the left, `varlist´ is a null string. Combining `varlist´ with incp`panelnum´ changes the local varlist from a null string to incp1. The second time through the loop, incp`panelnum´ is incp2. This is added to varlist, which now contains incp1 incp2. And so on.

Hopefully, my explanation of this loop was clear. Suppose that you are still confused (and macros can be confusing when you first use them). You could add display commands that print the contents of the local macros at each iteration of the loop:

```
local varlist ""
forvalues panelnum = 1/20 {
    local varlist "`varlist´ incp`panelnum´"
    display _newline "panelnum is: `panelnum´"
    display          "varlist  is: `varlist´"
}
```

The output looks like this:

```
panelnum is: 1
varlist  is: incp1

panelnum is: 2
varlist  is: incp1 incp2

panelnum is: 3
varlist  is: incp1 incp2 incp3

panelnum is: 4
varlist  is: incp1 incp2 incp3 incp4
  (output omitted)
```

Adding `display` to loops is a good way to verify that the loop is doing what you think it should. Once you have verified that the loop is working correctly, you can comment out the `display` commands (e.g., put a `*` in front of each line). As an exercise, particularly if any of the examples are confusing, add `display` commands to the loops in prior examples to verify how they work.

Loop example 6: Retrieving information returned by Stata

When Stata executes a command, it almost always leaves information in memory. You can use this information in many ways. For example, I start by computing summary statistics for one variable:

```
. summarize Bsdneighb

    Variable |       Obs        Mean    Std. Dev.       Min        Max
-------------+--------------------------------------------------------
   Bsdneighb |       490    .1938776    .3957381          0          1
```

After `summarize` runs, I type the command `return list` to see what information was left in memory. In Stata terminology, this information was "returned" by `summarize`:

```
. return list

scalars:
                  r(N) =  490
              r(sum_w) =  490
               r(mean) =  .1938775510204082
                r(Var) =  .1566086557322316
                 r(sd) =  .3957381150865198
                r(min) =  0
                r(max) =  1
                r(sum) =  95
```

This information can be moved into macros. For example, to retrieve the number of cases, type

```
local samplesize = r(N)
```

To compute the percentage of cases equal to one, I can multiply the mean in **r(mean)** by 100:

```
local pct1 = r(mean)*100
```

Next I use returned information in a loop to list the percentage of ones and the sample size for each measure of social distance:

```
1> foreach varname of varlist `sdvars´ {
2>     quietly summarize B`varname´
3>     local samplesize = r(N)
4>     local pct1 = r(mean)*100
5>     display "B`varname´:" _col(14) "Pct1s = " %5.2f `pct1´ ///
 >         _col(30) "N = `samplesize´"
6> }
```

Line 1 loops through the list of variables. Line 2 computes statistics for one variable at a time. After I was sure the loop worked correctly, I added **quietly** to suppress the output from **summarize**. Line 3 grabs the sample size from **r(N)** and puts it in the local **samplesize**. Similarly, line 4 grabs the mean and multiplies it by 100 to compute the percentage of ones. Line 5 prints the results using the format **%5.2f**, which specifies five columns and two decimal digits (type **help format** or see [D] **format** for further details). The output from the loop looks like this:

```
Bsdneighb:   Pct1s = 19.39   N = 490
Bsdsocial:   Pct1s = 27.46   N = 488
Bsdchild:    Pct1s = 71.73   N = 481
Bsdfriend:   Pct1s = 28.75   N = 487
Bsdwork:     Pct1s = 31.13   N = 485
Bsdmarry:    Pct1s = 52.75   N = 455
```

As a second example, I use the returns from **summarize** to compute the coefficient of variation (CV). The CV is a measure of inequality for ratio variables that equals the standard deviation divided by the mean. I compute the CV with these commands:

```
foreach varname of varlist incp1 incp2 incp3 incp4 {
    quietly summarize `varname´
    local cv = r(sd)/r(mean)
    display "CV for `varname´: " %8.3f `cv´
}
```

4.3.2 Counters in loops

In many applications using loops, you will need to count how many times you have gone through the loop. To do this, I create a local macro that will contain how often I have gone through the loop. Because I have not started the loop yet, I start by setting the counter to 0 (file: **wf4-loops.do**):

```
local counter = 0
```

Next I loop through the variables as I did above:

```
1> foreach varname of varlist warm yr89 male white age ed prst {
2>     local counter = `counter' + 1
3>     local varlabel : variable label `varname'
4>     display "`counter'. `varname'" _col(12) "`varlabel'"
5> }
```

Line 2 increments the counter. To understand how this works, start on the right and move left. I take `1` and add it to the current value of `counter`. I retrieve this value with `` `counter' ``. The first time through the loop, `` `counter' `` is 0, so `` `counter' + 1 `` is 1. Line 3 retrieves the variable label, and line 4 prints the results using the local `counter` to number each line. The results look like this:

```
1. warm    Mom can have warm relations with child
2. yr89    Survey year: 1=1989 0=1977
3. male    Gender: 1=male 0=female
4. white   Race: 1=white 0=not white
5. age     Age in years
6. ed      Years of education
7. prst    Occupational prestige
```

Counters are so useful that Stata has a simpler way to increment them. The command `local ++counter` is equivalent to `` local counter = `counter' + 1 ``. Using this, the loop becomes

```
local counter = 0
foreach varname in warm yr89 male white age ed prst {
    local ++counter
    local varlabel : variable label `varname'
    display "`counter'. `varname'" _col(12) "`varlabel'"
}
```

Using loops to save results to a matrix

Loops are critical for accumulating results from statistical analyses. To illustrate this application, I extend the example on page 101 so that instead of printing the percentage of ones and the sample size, I save this information in a matrix. I begin by creating a local with the names of the six binary measures:

```
local sdvars "Bsdneighb Bsdsocial Bsdchild Bsdfriend Bsdwork Bsdmarry"
```

I use an extended macro function to count the number of variables in the list:

```
local nvars : word count `sdvars'
```

By using this extended macro function, I can change the list of variables in `sdvars` and not worry about updating the count for the number of variables I want to analyze. You are always better off letting Stata compute a quantity than entering it by hand. For each variable, I need the percentage of ones and the number of nonmissing cases. I will save these in a matrix that will have one row for each variable and two columns. I use a `matrix` command ([P] **matrix**) to create a 6×2 matrix named `stats`:

```
matrix stats = J(`nvars´,2,.)
```

The J() function creates a matrix based on three arguments. The first is the number of rows, the second the number of columns, and the third is the value used to fill the matrix. Here I want the matrix to be initialized with missing values that are indicated by a period. The matrix looks like this:

```
. matrix list stats
stats[6,2]
    c1  c2
r1   .   .
r2   .   .
r3   .   .
r4   .   .
r5   .   .
r6   .   .
```

To document what is in the matrix, I add row and column labels:

```
matrix colnames stats = Pct1s N
matrix rownames stats = `sdvars´
```

The matrix now looks like this:

```
. matrix list stats
stats[6,2]
              Pct1s       N
Bsdneighb         .       .
Bsdsocial         .       .
 Bsdchild         .       .
Bsdfriend         .       .
  Bsdwork         .       .
 Bsdmarry         .       .
```

Next I loop through the variables in local `sdvars`, run `summarize` for each variable, and add the results to the matrix. I initialize a counter that will indicate the row where I want to put the information:

```
local irow = 0
```

Then I loop through the variables, compute what I need, and place the values in the matrix:

```
1>  foreach varname in `sdvars´ {
2>      local ++irow
3>      quietly sum `varname´
4>      local samplesize = r(N)
5>      local pct1 = r(mean)*100
6>      matrix stats[`irow´,1] = `pct1´
7>      matrix stats[`irow´,2] = `samplesize´
8>  }
```

Lines 1–5 are similar to the example on page 101. Line 6 places the value of `pct1` into row `irow` and column 1 of the matrix `stats`. Line 7 places the sample size in column 2.

After the loop has completed, I list the matrix using the option `format(%9.3f)`. This option specifies that I want to display each number in nine columns and show three decimal digits:

```
. matrix list stats, format(%9.3f)
stats[6,2]
              Pct1s         N
Bsdneighb    19.388   490.000
Bsdsocial    27.459   488.000
 Bsdchild    71.726   481.000
Bsdfriend    28.747   487.000
  Bsdwork    31.134   485.000
 Bsdmarry    52.747   455.000
```

This technique for accumulating results is used extensively in chapter 7.

4.3.3 Nested loops

You can nest loops by placing one loop inside of another loop. Consider the earlier example (page 93) of creating binary variables indicating if `y` was less than a given value. Suppose that I need to do this for variables `ya`, `yb`, `yc`, and `yd`. I could repeat the code used above four times, once for each variable. A better approach uses a `foreach` loop over the four variables (file: `wf4-loops.do`):

```
foreach yvar in ya yb yc yd { // loop 1 begins
      (content of loop goes here)
} // loop 1 ends
```

Within this loop, I insert a modification of the loop used before to dichotomize `y`. I refer to this as loop 2:

```
 1> foreach yvar in ya yb yc yd { // loop 1 begins
 2>     foreach cutpt in 2 3 4  { // loop 2 begins
 3>         * create binary variable
 4>         generate `yvar´_lt`cutpt´ = y<`cutpt´ if !missing(y)
 5>         * add labels
 6>         label var    `yvar´_lt`cutpt´ "y is less than `cutpt´?"
 7>         label define `yvar´_lt`cutpt´ 0 "Not<`cutpt´" 1 "Is<`cutpt´"
 8>         label values `yvar´_lt`cutpt´ `yvar´_lt`cutpt´
 9>    } // loop 2 ends
10> } // loop 1 ends
```

The first time through loop 1, the local `yvar` is assigned `ya`, so when `` `yvar´ `` appears in later lines, it is evaluated as `ya`. The second loop varies over the three values for `cutpt`. The locals from the two loops are combined in later lines. For example, in line 4 I create a variable named `` `yvar´_lt`cutpt´ ``. The local `yvar` is initially `ya` and the first value of `cutpt` is 2. Accordingly, the first variable created is `ya_lt2`. Then `ya_lt3` and `ya_lt4` are created. At this point, loop 2 ends and the value of `yvar` in loop 1 becomes `yb` and variables `yb_lt2`, `yb_lt3`, and `yb_lt4` are generated by loop 2.

4.3.4 Debugging loops

Loops can generate confusing errors. When this happens, I am often able to figure out what is wrong by using `display` to monitor the values of the local variables created in the loop. For example, this loop looks fine (file: `wf4-loops-error1.do`)

```
foreach varname in "sdneighb sdsocial sdchild sdfriend sdwork sdmarry" {
    generate B`varname' = `varname'
    replace  B`varname' = 0 if `varname'==1 | `varname'==2
    replace  B`varname' = 1 if `varname'==3 | `varname'==4
}
```

but it generates the following error:

```
sdsocial already defined
r(110);
```

To debug the loop, I start by removing `sdsocial` from the list to see if there was something specific to this variable that caused the error. When I do this, however, I get the same error for a different variable (file: `wf4-loops-error1a.do`):

```
sdchild already defined
r(110);
```

Because the second variable in the list causes the same error, I suspect that the problem is not with variables that I want to recode. Next I add a `display` command immediately after the `foreach` command (file: `wf4-loops-error1b.do`):

```
display "==> varname is: >`varname'<"
```

This command prints `==> varname is > ... <`, where ... is replaced by the contents of the local `varname`. I print `>` and `<` to make it easy to see if there are blanks at the beginning or end of the local. Here is the output:

```
==> varname is: >sdneighb sdsocial sdchild sdfriend sdwork sdmarry<
sdsocial already defined
r(110);
```

Now I see the problem. The first time through the loop, I wanted `varname` to contain `sdneighb` but instead it contains the entire list of variables `sdneighb sdchild sdfriend sdwork sdmarry`. This is because everything within quotes is considered to be a single item; the solution is to get rid of the quote marks:

```
foreach varname in sdneighb sdsocial sdchild sdfriend sdwork sdmarry {
```

Errors in loops are often caused by problems in the local variable created by the `foreach` or `forvalues` command. The specific error message you get depends on the commands used within the loop. Regardless of the error message, the first thing I do when I have a problem with a loop is to use `display` to show the value of the local created by `foreach` or `forvalues`. More times than not, this uncovers the problem.

Using trace to debug loops

Another approach to debugging loops is to trace the program execution (see page 81). Before the loop begins, type the command

```
set trace on
```

Then, as the loop is executed, you can see how each macro has been expanded. For example,

```
. foreach varname in "sdneighb sdsocial sdchild sdfriend sdwork sdmarry" {
  2.       gen B`varname´ = `varname´
  3.       replace  B`varname´ = 0 if `varname´==1 | `varname´==2
  4.       replace  B`varname´ = 1 if `varname´==3 | `varname´==4
  5. }
- foreach varname in "sdneighb sdsocial sdchild sdfriend sdwork sdmarry" {
- gen B`varname´ = `varname´
= gen Bsdneighb sdsocial sdchild sdfriend sdwork sdmarry = sdneighb sdsocial sdc
> hild sdfriend sdwork sdmarry
sdsocial already defined
  replace B`varname´ = 0 if `varname´==1 | `varname´==2
  replace B`varname´ = 1 if `varname´==3 | `varname´==4
  }
r(110);
```

With trace, lines that begin with = show the command after all the macros have been expanded. In this example, you can see right away what the problem is. To turn trace off, type the command `set trace off`.

4.4 The include command

Sometimes I repeat the same code multiple times within the same do-file or across multiple do-files. For example, when cleaning data, I might have many variables that use 97, 98, and 99 for missing values where I want to recode these values to the extended missing value codes `.a`, `.b`, and `.c`. Or I want to select my sample in the same way in multiple do-files. Of course, I can copy the same code into each file, but if I decide to change something, say, to use `.n` rather than `.c` for a missing value, I must change each do-file in each location where the recoding is done. Making such repetitious changes is time-consuming and error-prone. An alternative is to use the `include` command. The `include` command inserts code from a file into your do-file just as if you had typed it at the location of the `include` command. To give you an idea of how to use `include`, I provide two examples. The first example uses an include file to select the sample in multiple do-files. The second example uses include files to recode data. The section ends with some warnings about things that can go wrong. The `include` command was added in Stata 9.1, where `help include` is the only documentation; in Stata 10, also see [P] **include**.

4.4.1 Specifying the analysis sample with an include file

I have a series of do-files that analyze `mydata.dta`.[3] For these analyses I want to use the same cases selected with the following commands:

```
use mydata, clear
keep if panel==1  // only use 1st panel
drop if male==1   // restrict analysis to males
drop if inc>=.    // drop if missing on income
```

I could type these commands at the beginning of each do-file. Instead, I prefer to use the `include` command. I create a file called `mydata-sample.doi`, where I chose the suffix `doi` to indicate that the file is an include file. You can use any suffix you want, but I suggest you always use the same suffix to make it easier to find your include files. My analysis program uses the include file like this:

```
* load data and select sample
include mydata-sample.doi
* get descriptive statistics
summarize
* run base model
logit y x1 x2 x3
```

This is exactly equivalent to the program:

```
* load data and select sample
use mydata, clear
keep if panel==1 // only use 1st panel
drop if male==1  // restrict analysis to males
drop if inc>=.   // drop if missing on income
* get descriptive statistics
summarize
* run base model
logit y x1 x2 x3
```

If I use different analysis samples for different purposes, I can create a series of include files, say,

```
mydata-males-p1.doi
mydata-males-allpanels.doi
mydata-females-p1.doi
mydata-females-allpanels.doi
```

By selecting one of these to include in a do-file, I can quickly select the sample I want to use.

4.4.2 Recoding data using include files

I also use include files for data cleaning when I have a lot of variables that need to be changed in similar ways. Here is a simple example. Suppose that variable `inneighb`

3. Recall that if a filename does not start with `wf`, it is not part of the Workflow package that you can download.

uses 97, 98, and 99 as missing values. I want to recode these values to be extended missing values. For example (file: `wf4-include.do`),

```
* inneighb: recode 97, 98 & 99
clonevar inneighbR = inneighb
replace  inneighbR = .a if inneighbR==97
replace  inneighbR = .b if inneighbR==98
replace  inneighbR = .c if inneighbR==99
tabulate inneighb inneighbR, miss nolabel
```

Because I want to recode `insocial`, `inchild`, `infriend`, `inmarry`, and `inwork` the same way, I use similar commands for each variable:

```
* insocial: recode 97, 98 & 99
clonevar insocialR = insocial
replace  insocialR = .a if insocialR==97
replace  insocialR = .b if insocialR==98
replace  insocialR = .c if insocialR==99
tabulate insocial insocialR, miss nolabel
* inchild:  recode 97, 98 & 99
clonevar inchildR = inchild
replace  inchildR = .a if inchildR==97
replace  inchildR = .b if inchildR==98
replace  inchildR = .c if inchildR==99
tabulate inchild inchildR, miss nolabel
```

(*and so on for* `infriend`, `inmarry`, *and* `inwork`)

Or I can use a loop:

```
foreach varname in inneighb insocial inchild infriend {
    clonevar `varname´R = `varname´
    replace  `varname´R = .a if `varname´R==97
    replace  `varname´R = .b if `varname´R==98
    replace  `varname´R = .c if `varname´R==99
    tabulate `varname´ `varname´R, miss nolabel
}
```

I can do the same thing with an include file. I create the file `wf4-include-2digit-recode.doi` that contains:

```
clonevar `varname´R = `varname´
replace  `varname´R = .a if `varname´R==97
replace  `varname´R = .b if `varname´R==98
replace  `varname´R = .c if `varname´R==99
tabulate `varname´ `varname´R, miss nolabel
```

As in the `foreach` loop, these commands assume that the local `varname` contains the name of the variable being cloned and recoded. For example,

```
local varname inneighb
include wf4-include-2digit-recode.doi
```

For the next variable,

```
local varname insocial
include wf4-include-2digit-recode.doi
```

and so on. I create other include files for other types of recoding. For example, wf4-include-3digit-recode.doi has the commands

```
clonevar `varname´R = `varname´
replace  `varname´R = .a if `varname´R==997
replace  `varname´R = .b if `varname´R==998
replace  `varname´R = .c if `varname´R==999
tabulate `varname´ `varname´R, miss nolabel
```

My program to recode all variables looks like this:

```
// recode two-digit missing values
local varname inneighb
    include wf4-include-2digit-recode.doi
local varname insocial
    include wf4-include-2digit-recode.doi
local varname inchild
    include wf4-include-2digit-recode.doi
local varname infriend
    include wf4-include-2digit-recode.doi
// recode three-digit missing values
local varname inmarry
    include wf4-include-3digit-recode.doi
local varname inwork
    include wf4-include-3digit-recode.doi
```

Or I could use loops:

```
// recode two-digit missing values
foreach varname in inneighb insocial inchild infriend {
    include wf4-include-2digit-recode.doi
}
// recode three-digit missing values
foreach varname in inmarry inwork {
    include wf4-include-3digit-recode.doi
}
```

I can create a different include file for each type of recoding that needs to be done. I find this to be very helpful in large data-cleaning projects as shown on page 236.

4.4.3 Caution when using include files

Although include files can be very useful, you need to be careful about preserving, documenting, and changing them. When backing up your work, it is easy to forget the include files. If you cannot find the include file that is used by a do-file, the do-file will not work correctly. Accordingly, you should carefully name and document your include files. I give include files the suffix .doi so that I can easily recognize them when looking at a list of files. I use a prefix that links them to the do-files that call them. For example, if mypgm.do uses an include file and no other do-files use this include file, I name the include file mypgm.doi. If I have an include file that is used by many do-files, I start the name of the include file with the same starting letters of the do-file. For example, cwh-men-sample.doi might be included in cwh-01desc.do and cwh-02logit.do. I document include files both in my research log and within the file itself. For example, the include file might contain

```
// include:    cwh-men-sample.doi
// used by:    cwh*.do analysis files
// task:       select cases for the male sample
// author:     scott long \ 2007-08-05
```

The advantage of include files is that they let you easily use the same code in multiple do-files or multiple times in the same do-file. If you change an include file, you must be certain that the change is appropriate for all do-files that use the include file. For example, suppose that `cwh-sample.doi` selects the sample for my analysis in the CWH project. The do-files `cwh-01desc.do`, `cwh-02table.do`, `cwh-03logit.do`, and `cwh-04graph.do` all include `cwh-sample.doi`. When reviewing the results for `cwh-01desc.do`, I decide that I want to include cases that I had initially dropped. If I change `cwh-sample.doi`, this will affect the other do-files. The best approach is to always follow the rule that once you have finished your work on a do-file or include file, if you change it, you should give it a new name. For example, the do-file becomes `cwh-01descv2.do` and includes `cwh-samplev2.doi`. For details on the importance of renaming changed files, see section 5.1.

The `include` command should not be used when other methods will produce clearer code. For example, the `foreach` version of the code fragment on page 108 is easier to understand than the corresponding code using `include` that follows because the `include` version hides what is being done in `wf4-include-2digit-recode.doi`. But, as the block of code contained in `wf4-include-2digit-recode.doi` grows, the `include` version becomes more attractive.

4.5 Ado-files

This section provides a basic introduction to writing ado-files.[4] Ado-files are like do-files, except that they are automatically run. Indeed, `.ado` stands for automatically loaded do-file. To understand how these files work, it helps to know something about the inner workings of Stata (see appendix A for further details). The Stata for Windows executable is a file named `wstata.exe` or `mpstata.exe` that contains the compiled program that is the core of Stata. When you click the Stata icon, this file is launched by the operating system. Some commands are contained in the executable, such as `generate` and `summarize`. Many other commands are not part of the executable but instead are ado-files. Ado-files are programs written using features from the executable to complete other tasks. For example, the executable does not have a program to fit the negative binomial regression model. Instead, this model is fitted by the ado-file `nbreg.ado`. Stata 10 has nearly 2,000 ado-files. A clever and powerful feature of Stata is that when you run a command, you cannot tell whether it is part of the executable or is an ado-file. This means that Stata users can write new commands and use them just like official Stata commands.

4. When you install the Workflow package, the ado-files and help files from this section are placed in your working directory. Because Stata automatically installs user-written ado-files and help files to the `PLUS` directory (see page 350), I have named these files with the suffixes `_ado` and `_hlp` (e.g., `wf._ado`, `wf._hlp`) so they will be downloaded to your working directory. Using your file manager, you should rename the files to remove the underscores.

Suppose that I have written the ado-file `listcoef.ado` and type `listcoef` in the Command window. Because `listcoef` is not an internal command, Stata automatically looks for the file `listcoef.ado`. If the file is found, it is run. This happens very quickly, so you will not be able to tell if `listcoef` is part of the executable, an ado-file that is part of official Stata, or an ado-file written by someone else. This is a very powerful feature of Stata.

Although ado-files can be extremely complex (for example, from the Command window, run `viewsource mfx.ado` to see an ado-file from official Stata), it is possible to write your own ado-files that are simple yet very useful.

4.5.1 A simple program to change directories

The `cd` command changes your working directory. For example, my work for this book is located in `e:\workflow\work`. To make this my working directory, I type the command

```
cd e:\workflow\work
```

Because I work on other projects and each project has its own directory, I change directories frequently. To make this easier, I can write an ado-file called `wf.ado` that automatically changes my working directory to `e:\workflow\work`. The ado-file is

```
program define wf
    version 10
    cd e:\workflow\work
end
```

The first line names the new command and the last line indicates that the code for the command has ended. The second line indicates that the program assumes you are running Stata 10 or later. Line 3 changes the working directory. I save `wf.ado` in my `PERSONAL` directory (type `adopath` to find where your `PERSONAL` directory is located). To change the working directory, I simply type `wf`.

I can create ado-files for each project. For example, my work on SPost is located in `e:\spost\work\`. So I create `spost.ado`:

```
program define spost
    version 10
    cd e:\spost\work
end
```

For scratch work, I use the `d:\scratch` directory. So the ado-file is

```
program define scratch
    version 10
    cd d:\scratch
end
```

In Windows, I often download files to the desktop. To quickly check these files, I might want to try them in Stata before moving them to their permanent location. To change to the desktop, I need to type the awkward command

```
cd "c:\Documents and Settings\Scott Long\Desktop"
```

It is much easier to create a command called `desk`:

```
program define desk
    version 10
    cd "c:\Documents and Settings\Scott Long\Desktop"
end
```

Now I can move around directories for different projects easily:

```
. wf
e:\workflow\work
. desk
c:\Documents and Settings\Scott Long\Desktop
. spost
e:\spost\work
. wf
e:\workflow\work
. scratch
d:\scratch
```

If you have not written an ado-file before, this is a good time to try writing a few that change to your favorite working directories.

4.5.2 Loading and deleting ado-files

Before proceeding to a more complex example, I need to further explain what happens to ado-files once they are loaded into memory and what happens if you need to change an ado-file that is already loaded. Suppose that you have the file `wf.ado` in your working directory when you start Stata. If you enter the `wf` command, Stata will look for `wf.ado` and run the file automatically. This loads the `wf` command into memory. Stata will try to keep this command in memory as long as possible. This means that if you enter the `wf` command again, Stata will use the command that is already in memory rather than running `wf.ado` again. If you change `wf.ado`, say, to fix an error or add a feature, and try to run it again, you get an error:

```
. run wf.ado
wf already defined
r(110);
```

Stata will not create a new version of the `wf` command because there is already a version in memory. The solution is to drop the command stored in memory. For example,

```
. program drop wf
. run wf.ado
```

When debugging an ado-file, I start the file with the `capture program drop` *command-name* command. If the command is in memory, it will be dropped. If it is not in memory, `capture` prevents an error that occurs if you try to drop a command that is not in memory.

4.5.3 Listing variable names and labels

As a more complex example, I will automate the loop used on page 96 to list variable names and labels. I start by creating the nmlabel command that works very simply. Then I add options to introduce new programming features.[5] For a command to run automatically, you need to give the file the same name as the command. For example, nmlabel.ado should define the nmlabel command. In the examples that follow, I create several versions of the nmlabel command. When you download the Workflow package, these are named to reflect their version and have suffixes ._ado rather than .ado (e.g., nmlabel-v1._ado). The suffix ._ado is necessary to download the files into your working directory; if the suffix was .ado, the file would be placed in your PLUS directory. Before working with these files, change the suffixes to .ado. For example, change nmlabel-v1._ado to nmlabel-v1.ado. If you want a particular version of the command to run automatically, you need to rename the file, such as renaming nmlabel-v1.ado to nmlabel.ado. After renaming, it will run automatically if you enter nmlabel.

Version 1

My first version of nmlabel lists the names and labels with no options. It looks like this (file: nmlabel-v1.ado)

```
 1>  *! version 1.0.0 \ jsl 2007-08-05
 2>  capture program drop nmlabel
 3>  program define nmlabel
 4>      version 10
 5>      syntax varlist
 6>      foreach varname in `varlist´ {
 7>          local varlabel :  variable label `varname´
 8>          display in yellow "`varname´" _col(10) "`varlabel´"
 9>      }
10>  end
```

and is saved as nmlabel.ado. Line 1 is a special type of comment. If a comment begins with *!, I can list the comment using the which command:

```
. which nmlabel
.\nmlabel.ado
*! version 1.0.0 \ jsl 2007-08-05
```

The output .\nmlabel.ado tells me that the file is located in my working directory, indicated by .\. Next the comment is echoed. If the file was in my PERSONAL directory, which would produce the following output:

```
. which nmlabel
.c:\ado\personal\nmlabel.ado.
*! version 1.0.0 \ jsl 2007-08-05
```

5. After writing nmlabel.ado as an example, I found it so useful that I created a similar command called nmlab to be part of my personal collection of ado-files. This file is installed as part of the Workflow package.

When writing an ado-file, you can initially save it in your working directory. When it works the way you want it to, move it to your PERSONAL directory so that Stata can find the file regardless of what your current working directory is.

Returning to the ado-file, the third line names the command. Line 4 says that the program is written for version 10 and later of Stata. If I run the command in version 9 or earlier, I will get an error. Line 5 is an example of the powerful syntax command, which controls how and what information you can provide your program and generates warnings and errors if you provide incorrect information (see help syntax or [P] **syntax** for more information). The syntax element varlist means that I am going to provide the program with a list of variable names from the dataset that is currently in memory. If I enter a name that is not a variable in my dataset, syntax reports an error. Lines 6–9 are the loop used in section 4.3. In line 10, end indicates that the program has ended. Here is how the command works:

```
. nmlabel lfp-wc
lfp      Paid Labor Force: 1=yes 0=no
k5       # kids < 6
k618     # kids 6-18
age      Wife´s age in years
wc       Wife College: 1=yes 0=no
```

I typed the abbreviation lfp-wc rather than lfp k5 k618 age wc. The syntax command automatically changed the abbreviation into a list of variables.

Version 2

Reviewing the output, I think it might look better if there was a blank line between the echoing of the command and the list of variables. To do this, I add an option skip that will determine whether to skip a line. Although this option is not terribly useful, it shows you how to add options using the powerful syntax command. The new version of the program looks like this (file: nmlabel-v2.ado):[6]

```
 1>  *! version 2.0.0 \ jsl 2007-08-05
 2>  capture program drop nmlabel
 3>  program define nmlabel
 4>     version 10
 5>     syntax varlist [, skip]
 6>     if "`skip´"=="skip" {
 7>         display
 8>     }
 9>     foreach varname in `varlist´ {
10>         local varlabel :  variable label `varname´
11>         display in yellow "`varname´" _col(10) "`varlabel´"
12>     }
13>  end
```

The syntax command in line 5 adds [, skip]. The , indicates that what follows is an option (in Stata options are placed after a comma). The word skip is the name I

6. If you have already run nmlabel-v1.ado, you need to drop the program nmlabel before running nmlabel-v2.ado. To do this, enter program drop nmlabel.

chose for the option. The [ ]'s indicate that the option is optional—that is, you can specify `skip` as an option but you do not have to. If I enter the command with the `skip` option, say, `nmlabel lfp wc hc, skip`, the `syntax` command in line 5 creates a local named `skip`. Think of this as if I ran the command

```
local skip "skip"
```

This can be confusing, so I want to discuss it in more detail. When I specify the `skip` option, the `syntax` command creates a macro named `skip` that contains the string `skip`. If I do not specify the `skip` option, `syntax` creates the local `skip` as a null string:

```
local skip ""
```

Line 6 checks whether the contents of the macro `skip` (the contents is indicated by `` `skip´ ``) are equal to the string `skip`. If they are, the `display` command in line 7 is run, creating a blank line. If not, the `display` command is not run.

To see how this works, I trace the execution of the ado-file by typing `set trace on`. Here is the output, where I have added the line numbers:

```
1>  . nmlabel lfp k5, skip
2>  ----------------------------------------------------------- begin nmlabel ---
3>  - version 10
4>  - syntax varlist [, skip]
5>  - if "`skip´"=="skip" {
6>  = if "skip"=="skip" {
7>  - display
8>
9>  - }
```
(output omitted)

Line 1 is the command I typed in the Command window. Line 2 indicates that this is a trace for a command named `nmlabel`. Line 3 reports that the `version 10` command was executed. The `-` in front of the command is how `trace` indicates that what follows echoes the code exactly as it appears in the ado-file. Line 4 echoes the `syntax` command, and line 5 echoes the `if` statement. Line 6 begins with `=` to indicate that what follows expands the code from the ado-file to insert values for things like macros. Here `` `skip´ `` has been replaced by its value, which is `skip`.

Returning to the code for version 2 of `nmlabel` on page 114, lines 9–12 loop through the variables being listed by `nmlabel`. To see what happens, I can look at the output from the trace:

(Continued on next page)

```
- foreach varname in `varlist´ {
= foreach varname in lfp k5 {
- local varlabel : variable label `varname´
= local varlabel : variable label lfp
- display in yellow "`varname´" _col(10) "`varlabel´"
= display in yellow "lfp" _col(10) "In paid labor force? 1=yes 0=no"
lfp      In paid labor force? 1=yes 0=no
- }
- local varlabel : variable label `varname´
= local varlabel : variable label k5
- display in yellow "`varname´" _col(10) "`varlabel´"
= display in yellow "k5" _col(10) "# kids < 6"
k5       # kids < 6
- }
```
end nmlabel

Not only is `set trace on` a good way to see how your ado-file works, but it is invaluable when debugging your program. To turn trace off, type `set trace off`.

Version 3

Next I want to add line numbers to my list. To do this, I need a new option and a counter as illustrated in section 4.3.2. Here is my new program (file: `nmlabel-v3.ado`):

```
 1>  *! version 3.0.0 \ jsl 2007-08-05
 2>  capture program drop nmlabel
 3>  program define nmlabel
 4>      version 10
 5>      syntax varlist [, skip NUMber ]
 6>      if "`skip´"=="skip" {
 7>          display
 8>      }
 9>      local varnumber = 0
10>      foreach varname in `varlist´ {
11>          local ++varnumber
12>          local varlabel :  variable label `varname´
13>          if "`number´"=="" {                         // do not number lines
14>              display in yellow "`varname´" _col(10) "`varlabel´"
15>          }
16>          else {                                      // number lines
17>              display in green "#`varnumber´: " ///
18>                  in yellow "`varname´" _col(13) "`varlabel´"
19>          }
20>      }
21>  end
```

The `syntax` command in line 5 adds `NUMber`, which means that there is an option named `number` that can be abbreviated as `num` (the capital letters indicate the shortest abbreviation that is allowed). Line 9 creates a counter, and line 11 increments its value. Lines 13–15 say that if the option `number` is not selected (i.e., `"`number´"` is not `"number"`), then print things just as before. Line 16 starts the portion of the program that runs when the `if` condition in line 13 is not true. Lines 17 and 18 print the information I want including a line number. Line 19 ends the `else` condition from line 16. The new version of the command produces output like this:

```
. nmlabel lfp k5 k618 inc, num
#1: lfp     In paid labor force? 1=yes 0=no
#2: k5      # kids < 6
#3: k618    # kids 6-18
#4: inc     Family income excluding wife´s
```

Version 4

Version 3 looks good, except that long variable names will get in the way of the labels. I could change `_col(13)` to `_col(18)`, but why not add an option instead? In this version of **nmlabel**, I add `COLnum(integer 10)` to **syntax** to create an option named **colnum()** that can be abbreviated as **col()**. **integer** 10 means that if I do not specify the **colnum()** option, the local **column** will automatically be set equal to 10. If I do not want to begin the labels in column 10, I use the **colnum()** option such as `nmlabel lfp, col(25)` and the labels begin in column 25. Here is the new ado-file (file: nmlabel-v4.ado):

```
capture program drop nmlabel
program define nmlabel
    version 10
    syntax varlist [, skip NUMber COLnum(integer 10)]
    if "`skip´"=="skip" {
        display
    }
    local varnumber = 0
    foreach varname in `varlist´ {
        local ++varnumber
        local varlabel :  variable label `varname´
        if "`number´"=="" {                                   // do not number lines
            display in yellow "`varname´" _col(`colnum´) "`varlabel´"
        }
        else {                                                // number lines
            local colnumplus2 = `colnum´ + 2
            display in green "#`varnumber´: " ///
                in yellow "`varname´" _col(`colnumplus2´) "`varlabel´"
        }
    }
end
```

I encourage you to study the changes. Although some of the changes might not be obvious, you should be able to figure them out using the tools from chapters 3 and 4.

4.5.4 A general program to change your working directory

We now have enough tools to write a more general program for changing your working directory.[7] Instead of having a separate ado-file for each directory, I want a command

7. This example was suggested by David Drukker. When you install the Workflow package, the **wd** command will be downloaded to your working directory with the name wd._ado. The suffix ._ado is necessary to download the file to your working directory; if the suffix was .ado, the file would be placed in your PLUS directory. Before working with this file, you should rename it wd.ado.

wd, where wd wf changes to my working directory for the workflow project, wd spost changes to my working directory for SPost, and so on. Here is the program:

```
 1> *! version 1.0.0 \ scott long 2007-08-05
 2> capture program drop wd
 3> program define wd
 4>     version 10
 5>     args dir
 6>     if "`dir´"=="wf" {
 7>         cd e:\workflow\work
 8>     }
 9>     else if "`dir´"=="spost" {
10>         cd e:\spost\work
11>     }
12>     else if "`dir´"=="scratch" {
13>         cd d:\scratch
14>     }
15>     else if "`dir´"=="" { // list current working directory
16>         cd
17>     }
18>     else {
19>         display as error "Working directory `dir´ is unknown."
20>     }
21> end
```

The args command in line 5 retrieves a single argument from the command line. If I type wd wf, then args will take the argument wf and assign it to the local macro dir. Line 6 checks if dir is wf. If so, line 7 changes to the directory e:\workflow\work. Similarly, lines 9–11 check if the argument is spost and then changes to the appropriate working directory. If wf is run without an argument, lines 15–17 display the current working directory. If any other argument is given, lines 18–20 display an error. You can customize this program with your own else if conditions that use the abbreviations you want for making changes to working directories that you specify. Then, after you put wd.ado in your PERSONAL directory, you can easily change your working directories. For example,

```
. wd wf
e:\workflow\work

. wd
e:\workflow\work\

. wd scratch
d:\scratch

. wd spost
e:\spost\work
```

4.5.5 Words of caution

If you write ado-files, be careful of two things. First, you must archive these files. If you have do-files that depend on ado-files you have written, your do-files will not work if you lose the ado-files. Second, if you change your ado-files, you must verify that your old do-files continue to work. For example, if I decide that I do not like the name number for the option that numbers the list of variables and change the option name to `addnumbers`, do-files that use the command `nmlabel` with the `number` option will no longer work. With ado-files, you must be careful that improvements do not break programs that used to work.

4.6 Help files

When you type `help` *command*, Stata searches the ado-path for the file *command*`.sthlp` or *command*`.hlp`.[8] If the file is found, it is shown in the Viewer window. In this section, I start by showing you how to write a simple help file for the `nmlabel` command written in section 4.5.3. Then I show you how I use help files to remind me of options and commands that I frequently use.

4.6.1 nmlabel.hlp

To document the `nmlabel` command, I create a text file called `nmlabel.hlp`. When I type `help nmlabel` a Viewer window displays the file; see figure 4.1.

8. The advantage of using the suffix `.sthlp` rather than `.hlp` is that many email systems refuse to accept attachments that have the `.hlp` suffix because they might contain a virus.

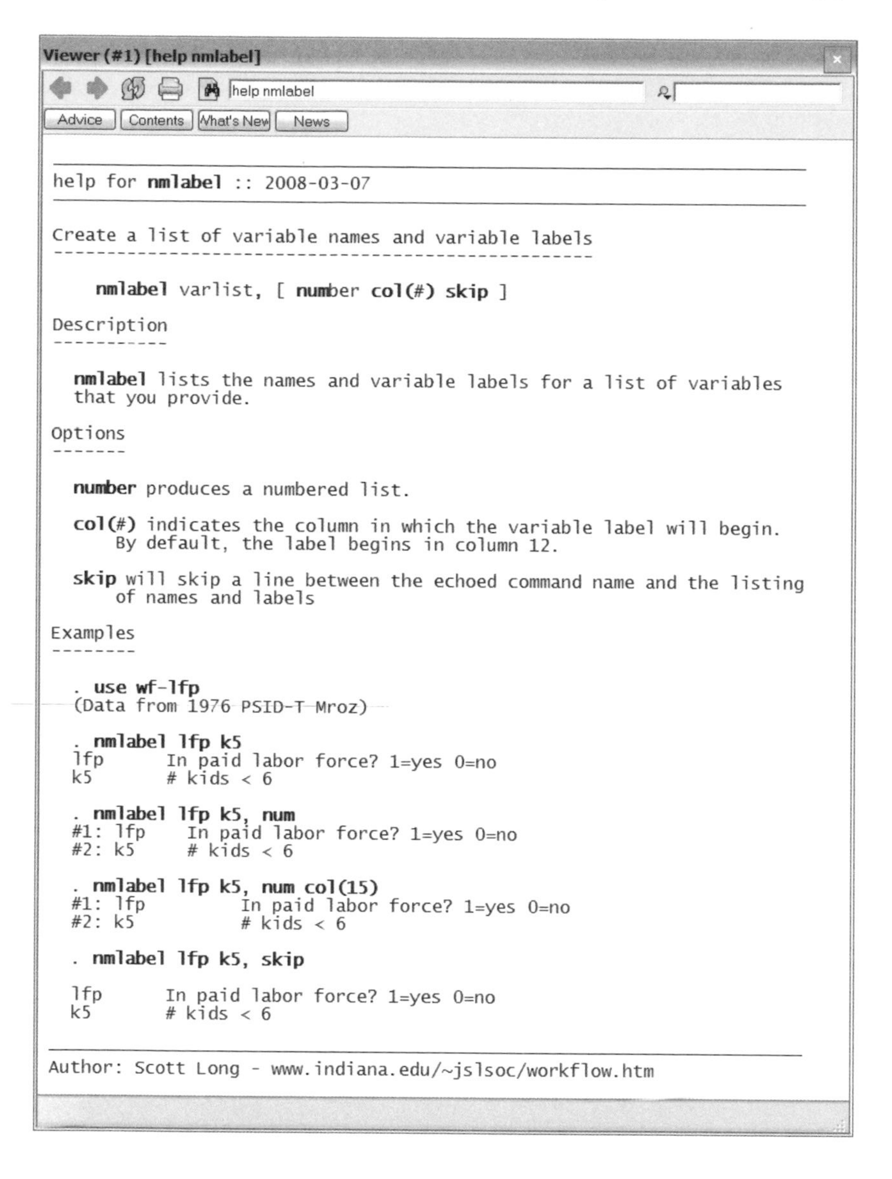

Viewer (#1) [help nmlabel]

help nmlabel

Advice | Contents | What's New | News

```
help for nmlabel :: 2008-03-07

Create a list of variable names and variable labels
---------------------------------------------------

    nmlabel varlist, [ number col(#) skip ]

Description
-----------

  nmlabel lists the names and variable labels for a list of variables
  that you provide.

Options
-------

  number produces a numbered list.

  col(#) indicates the column in which the variable label will begin.
      By default, the label begins in column 12.

  skip will skip a line between the echoed command name and the listing
      of names and labels

Examples
--------

  . use wf-lfp
  (Data from 1976 PSID-T Mroz)

  . nmlabel lfp k5
  lfp        In paid labor force? 1=yes 0=no
  k5         # kids < 6

  . nmlabel lfp k5, num
  #1: lfp    In paid labor force? 1=yes 0=no
  #2: k5     # kids < 6

  . nmlabel lfp k5, num col(15)
  #1: lfp         In paid labor force? 1=yes 0=no
  #2: k5          # kids < 6

  . nmlabel lfp k5, skip

  lfp        In paid labor force? 1=yes 0=no
  k5         # kids < 6

Author: Scott Long - www.indiana.edu/~jslsoc/workflow.htm
```

Figure 4.1. Viewer window displaying `help nmlabel`

The file **nmlabel.hlp** is a text file that looks like this:

```
.-
help for ^nmlabel^ :: 2008-03-07
.-

Create a list of variable names and variable labels
---------------------------------------------------

    ^nmlabel^ varlist^,^[ ^num^ber ^col(^#^)^^skip^]

Description
-----------

  ^nmlabel^ lists the names and variable labels for a list of variables
  that you provide.

Options
-------

  ^number^ produces a numbered list.

  ^col(^#^)^ indicates the column in which the variable label will begin.
      By default, the label begins in column 12.

  ^skip^ will skip a line between the echoed command name and the listing
      of names and labels

Examples
--------

  . ^use wf-lfp^
  (Data from 1976 PSID-T Mroz)

  . ^nmlabel lfp k5^
  lfp       In paid labor force? 1=yes 0=no
  k5        # kids < 6

  . ^nmlabel lfp k5, num^
  #1: lfp    In paid labor force? 1=yes 0=no
  #2: k5     # kids < 6

  . ^nmlabel lfp k5, num col(15)^
  #1: lfp         In paid labor force? 1=yes 0=no
  #2: k5          # kids < 6

  . ^nmlabel lfp k5, skip^

  lfp       In paid labor force? 1=yes 0=no
  k5        # kids < 6

.-
Author: Scott Long - www.indiana.edu/~jslsoc/workflow.htm
```

The file includes two shortcuts for making the file easier to write and easier to read in the Viewer windows. In the first line, `.-` is interpreted by the Viewer window as a solid line from the left border to the right. The carets ^ are used to toggle bold text on and off. For example, at the top of the Viewer window, the word "nmlabel" is in bold because the text file contains `^nmlabel^`, or consider the sequence `^col(^#^)^`. The first six characters "`^col(^`" makes `col(` bold, then `#` is not bold, and `^)^` makes `)` bold. If I wanted to make the help file fancier, with links to other files, automatic indentation, italics, and many other features, I could use the Stata Markup and Control Language (SMCL). See [U] **18.11.6 Writing online help** and [R] **help** for further information.

4.6.2 help me

I use a help file named `me.hlp` to give me quick access to information that I frequently use. This includes both summaries of options and fragments of code that I can copy-and-paste into a do-file. I put `me.hlp` in the `PERSONAL` directory. Then, when I type `help me`, a Viewer window opens (see figure 4.2) and I can quickly find this information (file: `me.hlp`):

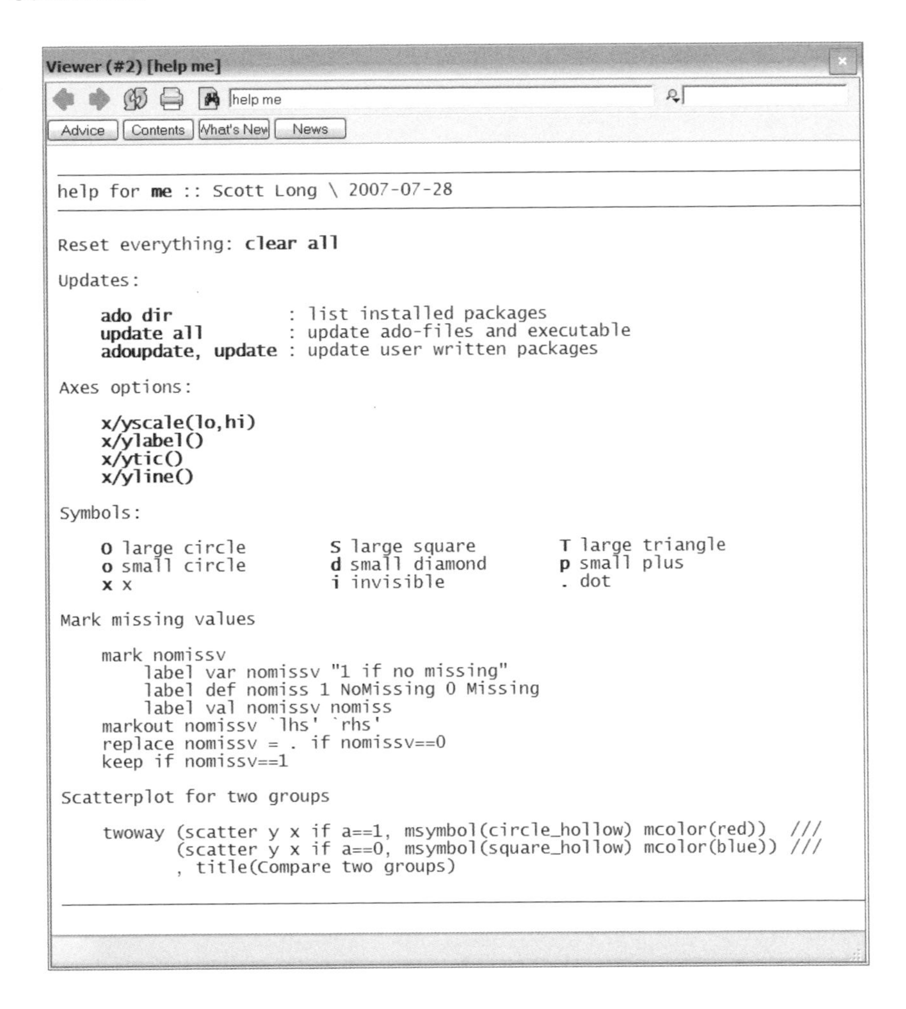

```
Viewer (#2) [help me]
help me
Advice  Contents  What's New  News

help for me :: Scott Long \ 2007-07-28

Reset everything: clear all

Updates:

    ado dir            : list installed packages
    update all         : update ado-files and executable
    adoupdate, update  : update user written packages

Axes options:

    x/yscale(lo,hi)
    x/ylabel()
    x/ytic()
    x/yline()

Symbols:

    O large circle        S large square        T large triangle
    o small circle        d small diamond       p small plus
    x x                   i invisible           . dot

Mark missing values

    mark nomissv
        label var nomissv "1 if no missing"
        label def nomiss 1 NoMissing 0 Missing
        label val nomissv nomiss
    markout nomissv `lhs' `rhs'
    replace nomissv = . if nomissv==0
    keep if nomissv==1

Scatterplot for two groups

    twoway (scatter y x if a==1, msymbol(circle_hollow) mcolor(red))   ///
           (scatter y x if a==0, msymbol(square_hollow) mcolor(blue))  ///
           , title(Compare two groups)
```

Figure 4.2. Viewer window displaying `help me`

4.7 Conclusions

Automation is fundamental to an effective workflow, and Stata provides many tools for automating your work. Although this chapter provides a lot of useful information, it is only a first step in learning to program in Stata. If you want to learn more, consider taking a NetCourse from StataCorp (http://www.stata.com/netcourse/). NetCourse 151—*Introduction to Stata Programming* is a great way to learn to use Stata more effectively even if you do not plan to do advanced programming. NetCourse 152—*Advanced Stata Programming* teaches you how to write sophisticated commands in Stata. If you spend a lot of time using Stata, learning how to automate your work will make your work easier and more reliable, plus it will save you time.

5 Names, notes, and labels

This chapter marks the transition from discussions of broad strategy in chapter 2 and general tools in chapters 3 and 4 to discussions of the specific tasks that you encounter as you move from an initial dataset to published findings. Chapter 5 discusses names, notes, and labels for variables, datasets, and do-files; these topics are essential for effective organization and documentation. Chapter 6 discusses cleaning data, constructing variables, and other common tasks in data management. For most projects, the vast majority of your time will be spent getting your data ready for statistical analysis. Finally, chapter 7 discusses the workflow of statistical analysis and presentation. Topics include organizing your analyses, extracting results for presentation, and documenting where the results you present came from. These three chapters incorporate two ideas that I find indispensable for an effective workflow. First, the concept of posting a file refers to deciding that a file is final and can no longer be changed. Posting files is critical because otherwise you risk inconsistent results that cannot be replicated. The second idea is that data analysis should be divided between data management and statistical analysis. Data management includes cleaning your data, constructing variables, and creating datasets. Statistical analysis involves examining the structure of your data using descriptive statistics, model estimates, hypothesis tests, graphical summaries, and other methods. Creating a dual workflow for data management and statistical analysis simplifies writing documentation, makes it easier to fix problems, and facilitates replication.

5.1 Posting files

Posting a file is a simple idea that is essential for data analysis. At some point when writing a do-file, you decide that the program is working correctly. When this happens, you should post your work. Posting means that the do-file and log file, along with datasets that were created, are placed in the directory where you save completed work (e.g., `c:\cwh\Posted\`). The fundamental principles for posted files is simple but absolute:

> *Posting principle*: Once a file is posted, it should never be changed.

If you change a posted file, you risk producing inconsistent results based on different variables that have the same name or two datasets with the same name but different content. I have seen this problem repeatedly and the only practical way that I know to avoid it is to have a strict policy that once a file is posted, it cannot be changed.

An implication of this rule is that only posted files should be shared with others or incorporated into papers or presentations.

The posting principle does not mean that you cannot change a do-file during the process of debugging. As you debug a do-file, you create the same dataset each time you run the program and might change the way a variable is created. That is not a problem because the files have not been posted, but once the files are posted, you must not change them.

Nor does posting a file mean that you cannot correct errors in do-files that have been posted. Rather it means that to fix the errors you need to create new files and possibly new variables. For example, suppose that `mypgm01.do` creates `mydata01.dta` with variables `var01`–`var99`. After posting these files, I discover a mistake in how `var49` was created. To fix this, I create a revised `mypgm01V2.do` that correctly generates the variable that I now name `var49V2` and saves the new dataset `mydata01V2.dta`. I can keep the original `var49` in the new dataset or I can delete it, but I must not change `var49`. I can delete `mydata01.dta` or I can keep it, but I must not change it. Because posted files are never changed, I can never have results for `var49` where the meaning of `var49` has changed. Nor is it possible for two people to analyze datasets with the same name but different content.

Finally, the practice of posting files does not mean that you must post each file immediately after you decide that it is complete and verified. I often work on a dozen related do-files at a time until I get things the way I want them. For me, this is the most efficient way to work. Something I learn while debugging one do-file might lead me to change another do-file. At some point, I decide that all the do-files and datasets are the way I want. Then the iterative process of debugging and program development ends. When this happens, I move the do-files, log files, and datasets from my working directory into a directory with completed work. That is, I post the files. After the files are posted, and only after they are posted, I can include the results in a paper, make the datasets available to collaborators, or share the log files with colleagues.

Although I find that most people agree in theory with the idea of posting, in practice the rule is violated frequently. I have been part of a project where a researcher posted a dataset, quickly realized a mistake, and ten minutes later replaced the posted file with a different file that had the same name. During those ten minutes, I had downloaded the file. It took us a lot of time to figure out why we were getting different results from the "same" dataset. I recently received a dataset that had the same name as an earlier one but was a different size. When I asked if the dataset was the same, I was told, "Exactly the same except that I changed the married variable".

The simplest thing is to make no exceptions to the rule for posting files. Once you allow exceptions, you start down a slippery slope that is bound to lead to problems. When a dataset is posted, if anything is changed, the dataset gets a new name. If a posted do-file is changed, it gets a new name. And so on. If you do not make an absolute distinction between files that are in process and those that are complete and posted, you risk producing inconsistent results and undermining your ability to replicate findings.

5.2 The dual workflow of data management and statistical analysis

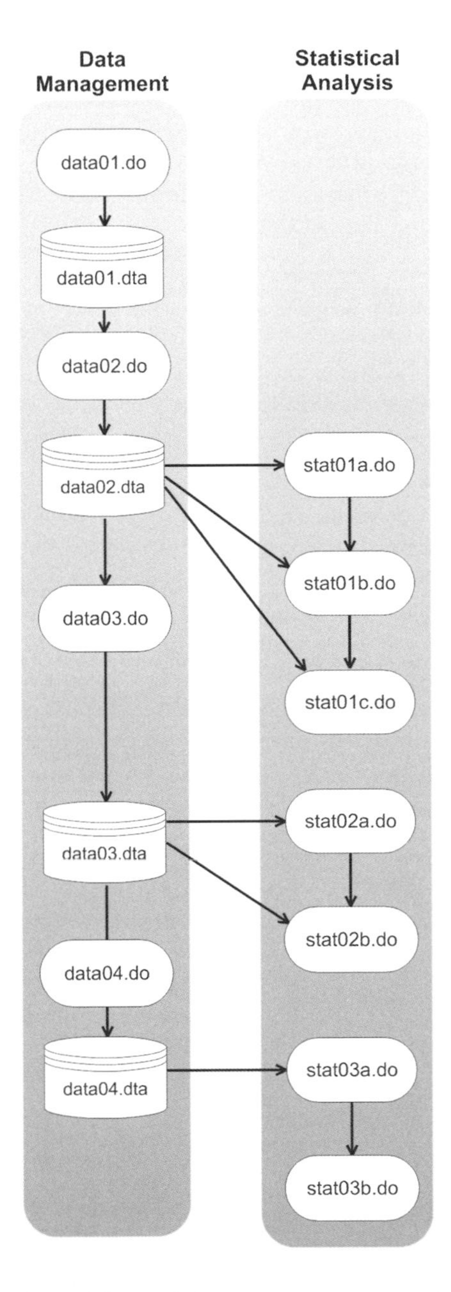

Figure 5.1. The dual workflow of data management and statistical analysis

I distinguish between programs for data management and programs for statistical analysis. I refer to this as a dual workflow as illustrated in figure 5.1. The two sets of do-files are distinct in the sense that programs for data management do not depend on programs for statistical analysis. Operationally, this means that I can run the data-management programs in sequence without running any of the programs for statistical analysis. This is possible because programs for statistical analysis never change the dataset (they might tell you how you want to change the dataset, but they do not make the change). Programs for statistical analysis do, however, depend on the datasets created by data-management programs. For example, `stat03a.do` will not work unless `data04.dta` has been created by `data04.do`.

A dual workflow makes it easier to correct errors when they occur. For example, if I find an error in `var15` in `data02.dta`, I only have to look for the problem in the data-management programs because the statistical analysis programs never create variables that are saved. If I find a problem in `data02.do`, I create the corrected do-file `data02V2.do`, which saves `data02V2.dta`, and corrects the problem in `data02.dta`. Then I revise, rename, and rerun any of the `stat*.do` do-files that depend on the changed data.

This workflow implies that you do not create and save new variables in your analysis do-files. For example, if I have a variable named `gender` coded 1 for men and 2 for women and decide that I want a variable `female` coded 1 for female and 0 for male, I would create a new dataset that added `female`, rather than creating `female` in the do-files for statistical analyses. I prefer this approach because I rarely create a variable that I use only once. I might think I will use it only once, but in practice I often need it for other, unanticipated analyses. Searching earlier do-files to find how a variable was created is time consuming and error prone. Also I might forget that I created a variable with the same name and later create a variable with the same name but a different meaning. Saving the variable in a dataset is easier and safer.

The distinction between data management and statistical analysis is not always clear. For example, I might use factor analysis to create a scale that I want to include in a dataset. The task of specifying, fitting, and testing a factor model is part of statistical analysis. But, constructing a scale to save is part of data management. In such situations, I might violate the principle of a dual workflow and create a dataset with a program that is part of the statistical analysis workflow. More likely, I would keep programs to fit, test, and perfect the factor model as part of the statistical analysis workflow. Once I have decided on the model I want for creating factor scores, I would incorporate that model into a program for data management. The dual workflow is not a Procrustean bed but rather is a principle that generally makes your work more efficient and facilitates replication.

5.3 Names, notes, and labels

With the principles of posting and a dual workflow in mind, we are ready to consider the primary topics of this chapter: names, notes, and labels for variables, datasets, and do-files. Is it worth your time to read an entire chapter about something as seemingly simple as picking names and labeling things? I think so. Many problems in data analysis occur because of misleading names and incomplete labels. An unclear name can lead to the wrong variable in a model or to incorrect interpretations of results. Less drastically, inconsistent names and ineffective labels make it harder to find the variables that you want and more difficult to interpret your output. On the other hand, clear, consistent, and thoughtful names and labels speed things up and prevent errors. Planning names and labels is one of the simplest things you can do to increase the ease and accuracy of your data analysis. Because choosing better names and adding full labels does not take much time, relative to the time lost by not doing this, the investment is well worth it.

Section 5.4 describes naming do-files in a way that keeps them organized and facilitates replication. Section 5.5 describes changing the filename of a dataset and adding an internal note that documents how the dataset was changed every time you change a dataset, no matter how small the change. The next five sections focus on variables. Section 5.6 is about naming variables, with topics ranging from systems for organizing names to how names appear in the Variables window. Section 5.7 describes variable labels. These short descriptions are included in the output of many commands and are essential for an effective workflow. Section 5.8 introduces the `notes` command for documenting variables. This command is incredibly useful, yet I find that many people are unaware of it. Section 5.9 describes labels for values and tools for keeping track of these labels. Section 5.10 is about a unique feature of Stata, the ability to create labels in multiple languages within one dataset. This is most obviously valuable with languages such as French, English, and German but is also a handy way to include long and short labels in the same language. Although you have no choice about the names and labels in data collected by others, you can change those names and create new labels that work better. A workflow for changing variable names and labels is presented in section 5.11 that includes an extended example using programming tools from chapter 4. Even if you are already familiar with commands such as `label variable`, `label define`, and `label values`, I think this section will help you work faster and more accurately.

5.4 Naming do-files

A single project can require hundreds of do-files. How you name these files affects how easily you can find results, document work, fix errors, and revise analyses. Most importantly, carefully named do-files make it easier to replicate your work. My recommendation for naming do-files is simple:

> *The run order rule*: Do-files should be named so that when run in alphabetical order they exactly re-create your datasets and replicate your statistical analyses.

For simplicity, I refer to the order in which a group of do-files needs to be run as the run order. The reasons you want names that reflect the run order differs slightly depending on whether the do-files are used to create datasets or to compute statistical analyses.

5.4.1 Naming do-files to re-create datasets

Creating a dataset often requires that several do-files are run in a specific order. If you run them in the wrong order, they will not work correctly. For example, suppose that I need two do-files to create a dataset. The first do-file merges the variable `hlthexpend` from `medical.dta` and the variable `popsize` from `census.dta` to create `health01.dta`. The second do-file creates a variable with `generate hlthpercap = hlthexpend/popsize` and then saves the dataset `health02.dta`. If I name the do-files `merge.do` and `addvar.do`, the names do not reflect the run order needed to create `health02.dta`. However, if I name them `data01-merge.do` and `data02-addvar.do`, the order is clear. Of course, in such a simple example, I could easily determine the sequence in which the do-files need to be run no matter how I name them. With dozens of do-files written over months or years, names that indicate the sequence in which the programs need to be run are extremely helpful.

Naming do-files to indicate the run order also makes it simpler to correct mistakes. Suppose that I need ten do-files to create `mydata01.dta` and that the programs need to run in the order `data01.do`, `data02.do`, through `data10.do`. After running the ten do-files and posting `mydata01.dta`, I realize that `data06.do` incorrectly deleted several observations. To fix the error, I create the corrected do-file `data06V2.do` and run the sequence of programs `data06V2.do` through `data10V2.do`. Because of the way I named the files, I know exactly which do-files need to be run and in what order to create a corrected dataset named `mydata01V2.dta`.

5.4.2 Naming do-files to reproduce statistical analysis

If you write robust do-files, as discussed in chapter 3 (see page 51), results should not depend on the order in which the programs are run. Still, I recommend that you sequentially name your analysis do-files so that the last do-file in the sequence produces the latest analyses. Suppose that you are computing descriptive statistics and fitting logit models. You might need a half dozen do-files as you refine your choice of variables and decide on the descriptive statistics that you want. Similarly, you might write several do-files as you explore the specification of your model. I suggest naming the do-files to correspond to the run order for each task. For example, you might have `desc01.do`–`desc06.do` and `logit01.do`–`logit05.do`, where you know that `desc06.log` and `logit05.log` have the latest results. This naming scheme prevents the problem of thinking that you are looking at the latest analyses when you are not.

5.4.3 Using master do-files

Sometimes you will need to rerun a sequence of do-files to reproduce all the work related to some part of your project. For example, when I complete the do-files to create a dataset, I want to verify that all the programs work correctly before posting the files. Or after discovering an error in one program in a sequence of related jobs, I want to fix the error and verify that all the programs continue to work correctly. A master do-file makes this simple. A master do-file is simply a do-file that runs other do-files. For example, I can create the master do-file `dual-dm.do` to run all the programs from the left column of figure 5.1:

```
//  dual-dm.do: do-file for data management
//  scott long \ 2008-03-14
do data01.do
do data02.do
do data03.do
do data04.do
exit
```

To rerun the four do-files in sequence, I type the command

```
do dual-dm.do
```

Similarly, for the statistical analysis, I can create `dual-sa.do`

```
//  dual-sa.do: do-file for statistical analysis
//  scott long \ 2008-03-14
* descriptive statistics
do stat01a.do
do stat01b.do
do stat01c.do
* logit models
do stat02a.do
do stat02b.do
* graphs of predictions
do stat03a.do
do stat03b.do
exit
```

which can be run by typing

```
do dual-sa.do
```

Suppose that I find a problem in `data03.do` that affects the creation of `data03.dta` and consequently the creation of `data04.dta`. This would also affect the statistical analyses based on these datasets. I need to create `V2` versions of several do-files for data management and statistical analysis as shown in figure 5.2.

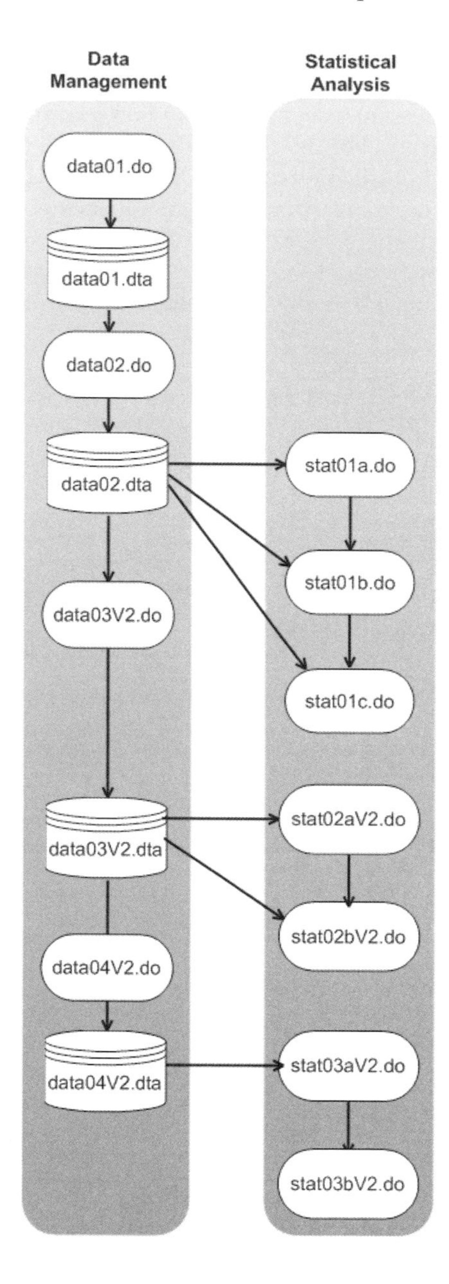

Figure 5.2. The dual workflow of data management and statistical analysis after fixing an error in `data03.do`

After revising the do-files, my master do-files become

```
//  dual-dm.do: do-file for data management
//  scott long \ 2008-03-14; revised 2008-03-17
do data01.do
do data02.do
do data03V2.do
do data04V2.do
exit
```

and

```
//  dual-sa.do: do-file for statistical analysis
//  scott long \ 2008-03-14; revised 2008-03-17
* descriptive statistics
do stat01a.do
do stat01b.do
do stat01c.do
* logit models
do stat02aV2.do
do stat02bV2.do
* graphs of predictions
do stat03aV2.do
do stat03bV2.do
exit
```

By running the following commands, all my work will be corrected:

```
do dual-dm.do
do dual-sa.do
```

Master log files

Stata allows you to have more than one log file open at the same time. This provides a convenient way to combine all the log files generated by a master do-file into one log. For example (file: `wf5-master.do`),

```
 1> capture log close master
 2> log using wf5-master, name(master) replace text
 3> //  program:    wf5-master.do
 4> //  task:       Creating a master log file
 5> //  project:    workflow chapter 5
 6> //  author:     jsl \ 2008-04-03
 7> do wf5-master01-desc.do
 8> do wf5-master02-logit.do
 9> do wf5-master03-tabulate.do
10> log close master
11> exit
```

Line 2 opens `wf5-master.log`. The `name(master)` option assigns the log a nickname referred to as the "logname". When you have more than one log file active, you need a logname for all but one of the logs. Line 1 closes `master` if it is already open, with `capture` meaning that if it is not open, ignore the error generated by the `log close` command. Lines 3–6 are recorded in `wf5-master.log`. In addition, the output from the do-files run in lines 7–9 are sent to `wf5-master.log`. Line 10 closes the master log file. When `wf5-master.do` was run, four log files were created:

```
wf5-master.log
wf5-master01-desc.log
wf5-master02-logit.log
wf5-master03-tabulate.log
```

The file `wf5-master.log` contains all the information from the three other log files. Instead of printing three files (or dozens in a complex set of analyses), I can print one file. If I am including results on the web, I need to post only one log file.

5.4.4 A template for naming do-files

Although my primary consideration in naming do-files is that the alphabetized names indicate the run order, there are other factors to consider:

- Use names that remind you of what is in the file and that help you find the file later. For example, `logit01.do` is better than `pgm01.do`.
- Anticipate revising your do-files and adding new do-files. If you find an error in a do-file, what will you name the corrected file and will the new name retain the sort order? If you need to add a step between two do-files, will your system allow you to add the do-file with a name that retains the run order?
- Choose names that are easy to type. Names that are too long or that have special characters should be avoided.

With these considerations in mind, I suggest the following template for naming do-files, where no spaces are included in the filename:

project [-*task*] *step* [*letter*] [V*version*] [-*description*].do

For example, `fl-clean01a-CheckLabels.do` or `fl-logit01aV2-BaseModel.do`. Here are the details:

project-task The *project* is a short mnemonic such as `cwh` for a study of cohort, work, and health; `fl` for a study of functional limitations; and `sgc` for the Stigma in a Global Context project. As needed, I divide the project into *task*s. For example, I might have `cwh-clean` for jobs related to cleaning data for the `cwh` project.

step and *letter* Within a project and task, the two-digit *step* indicates the order in which the do-files are run. For example, `fl-desc01.do`, `fl-desc02.do`, etc. If the project is complex, I might also use a letter, such as `fl-desc01a.do` and `fl-desc01b.do`.

version A *version* number is added if there is a revision to a do-file that has been posted. For example, if `fl-desc01a.do` was posted before an error was discovered, the replacement file is named `fl-desc01aV2.do`. I have never needed ten revisions so only one digit is used.

description The *description* is used to make it easier to remember what a do-file is for. The description does not affect the sort order and is not required to make the name unique. For example, I am not likely to remember what `fl-desc01a.do` does, but `fl-desc01a-health.do` reminds me that the program is computing descriptive statistics for health variables. When I refer to do-files, say in my research log, I do not need to include the description as part of the name. That is, I could refer to `fl-desc01a.do` rather than `fl-desc01a-health.do`.

Expanding the template

What happens if you have a very large project with complicated programs that require lots of modifications and additions? The proposed template scales easily. For example, between `fl-pgm01a.do` and `fl-pgm01b.do` I can insert `fl-pgm01a1.do` and `fl-pgm01a2.do`. Between these jobs I can insert `fl-pgm01a1a.do` and `fl-pgm01a1b.do`.

Collaborative projects

In a collaborative project, I often add the author's initials to the front of the job name. For example, I could use `jsl-fl-desc01a.do` rather than `fl-desc01a.do`.

Using subdirectories for complex analyses

As discussed in chapter 2, I use subdirectories to organize the do-files from large projects. This can best be explained by an example. Eliza Pavalko and I (Long and Pavalko 2004) wrote a paper examining how using different measures of functional limitations affected substantive conclusions. These measures were created from questions that ask if a person has trouble with physical activities such as standing, walking, stooping, or lifting. Using questions on nine activities, we constructed hundreds of scales to measures a person's overall limitations, where the scales were based on alternative measures used in the research literature. When the paper was finished, we had nearly 500 do-files to construct the scales and run analyses. Here is how we kept track of them.

The mnemonic for the project is `fl` standing for functional limitations. All project files had names that started with `fl` and were saved in the project directory `\flalt`. Posted files are placed in `\flalt\Posted` within these subdirectories:

(*Continued on next page*)

Directory	Task
`\fl00-data`	Datasets
`\fl01-extr`	Extract data from source files
`\fl02-scal`	Construct scales of functional limitations
`\fl03-out`	Construct outcome measures
`\fl04-desc`	Descriptive statistics for source variables
`\fl05-lca`	Fit latent-class models
`\fl06-reg`	Fit regression models

The first directory holds datasets, the next three directories are for data management and scale construction, and the last three directories are used for statistical analyses. If I need an additional step, say, verifying the scales, I can add a subdirectory that is numbered so that it sorts to the proper location in the sequence (e.g., `\fl03-1-verify`). Each subdirectory holds the do-files and log files for that task with datasets kept in `\fl00-data`. The do-files within a subdirectory are named so that if they are run in alphabetical order, they reproduce the work for that task. Even though it is unlikely that I will finish the work in task `\fl01-extr` before I start `\fl04-desc` (e.g., while looking at the descriptive statistics, I am likely to decide that I need to extract other variables), my goal is to organize the files so that I could correctly reproduce everything by running the jobs in order: all the jobs in `\fl01-extr`, followed by all the jobs in `\fl02-scal`, and so on. This is very helpful when trying to replicate your work or when you need to make revisions.

5.5 Naming and internally documenting datasets

The objective when naming datasets is to be certain that you never have two datasets with the same name but different content. Because datasets are often revised as you add variables, I suggest a simple convention that makes it easy to indicate the version of the dataset:

dataset-name##`.dta`

For example, if the initial dataset is `mydata01.dta`, the next one is `mydata02.dta`, and so on. Every time I change the current version, no matter how small the change, I increment the number by one. The most common objections I get to creating a new dataset every time a change is made are "I'm getting too many datasets!" and "These datasets take up too much space!" Storage is cheap so you can easily keep many versions of your data, or you can delete earlier versions of a dataset because you can reproduce them with your do-files (assuming you have an effective workflow). Alternatively, you can compress the datasets before archiving them. For example, the dataset `attr04.dta` has information on attrition from the National Longitudinal Survey. The file is 2,065,040 bytes long but when compressed (see page 264) is reduced to 184,552 bytes. When I compress a dataset, I like to combine the dataset with a do-file and log file that describes the data. The do-file might simply contain

```
log using attr04-dta, replace
use attr04, clear
describe
summarize
notes
log close
```

When I unzip the dataset I can quickly verify the content of the dataset without having to load the dataset or check my research log.

Never name it final!

Although it is tempting to name a dataset as final, this usually leads to confusion. For example, after a small error is found in `mydata-final.dta`, the next version is called `mydata-final2.dta`, and then later `mydata-reallyfinal.dta`. If `final` is in the name, you run the risk that you and others might believe that the dataset is final when there is an updated version. Recently, I was copied on a message that asked, "Does `final2` have a date attached so I know it is the most recent version?"

5.5.1 One time only and temporary datasets

If I create a dataset that I expect to use only once, I give it the name of the do-file that created it. For example, suppose that `demogcheck01.do` merges data from two datasets to verify that the demographic data from the two sources are consistent. Because I do not anticipate further analyses using this dataset, but I want to keep it if I have questions later, I would name it `demogcheck01.dta`. Then the name of the dataset documents its origin.

I often create temporary datasets when building a dataset for analysis (see section 5.11 for an example). I keep these datasets until the project is completed, but they are not posted. To remind me that these files are not critical, I name them beginning with `x-`. Accordingly, if I find a dataset that starts with `x-`, I know that I can delete it. For example, suppose that I am merging demographic information from `demog05.dta` and data on functional limitations from `flim06.dta`. My program `fl-mrg01.do` extracts the demographic data and stores it in `x-fl-mrg01.dta`; `fl-mrg02.do` extracts the limitation data and stores it in `x-fl-mrg02.dta`. Then `fl-mrg03.do` creates `fl-paper01.dta` by merging `x-fl-mrg01.dta` and `x-fl-mrg02.dta`. I delete the `x-` files when I finish the project.

I also find that prefacing a file with `x-` can prevent a problem when collaborating. Suppose that I am constructing a dataset that my collaborator and I both plan to analyze. I write a series of do-files to create the dataset that I will eventually name `fl-paper01.dta`. Initially, I am not sure if I have extracted all the variables that we need or created all the scales we planned. Rather than distributing a dataset named `fl-paper01.dta`, I create `x-fl-paper01.dta`. Because the name begins with `x-`, my colleague and I know that this is not a final dataset so there is no chance of accidentally running serious analyses. When we agree that the dataset is correct, I create `fl-paper01.dta` and post the dataset.

5.5.2 Datasets for larger projects

When working on projects using lots of variables, I prefer a separate dataset for each type of variable rather than one dataset for all variables. For example, in a project using the National Longitudinal Survey, we grouped variables by content and created these datasets:

Dataset	Content
attd##.dta	Attitude variables
attr##.dta	Attrition information
cntl##.dta	Control variables such as age and education
emps##.dta	Employment status
fami##.dta	Characteristics of the family
flim##.dta	Health and functional limitations

By dividing the variables, each member of the project could work on a different part of the data without risk of interfering with the work done by other team members. This was important because each set of variables took dozens of do-files and weeks to complete. When a segment of the data was completed, the new dataset was posted along with the associated do-files and log files. To run substantive analyses, we extracted variables from the multiple, source datasets and merged them into one analysis dataset.

5.5.3 Labels and notes for datasets

When you save a dataset, you should add internal documentation with a dataset label, a note, and a data signature. These are all forms of what is referred to as *metadata*—data about data. The advantage of metadata is that it is internal to the dataset so that when you have the dataset you have the documentation. To add a dataset label, use the command

`label data "`*label*`"`

For example,

```
label data "CWH analysis file \ 2006-12-07"
save cwh01, replace
```

The data label is echoed when you use the data:

```
. use cwh01, clear
(CWH analysis file \ 2006-12-07)
```

I use `notes` to add further details:

`notes:` *note*

Because no variable name is specified, the note applies to a dataset rather than to a variable (see section 5.8). In the note, I include the name of the dataset, a brief description, and details on who created the dataset with what do-file on what date. For example,

```
notes: cwh01.dta \ initial CWH analysis dataset \ cwh-dta01a.do jsl 2006-12-07.
label data "CWH analysis file \ 2006-12-07"
save cwh01, replace
```

After I load the dataset, I can easily determine how the dataset was created. For example,

```
. notes _dta

_dta:
  1.  cwh01.dta \ initial CWH analysis dataset \ cwh-dta01a.do jsl 2006-12-07.
```

Each time I update the data (e.g., create `cwh02.dta` from `cwh01.dta`), I add a note. Listing the notes provides a quick summary of the do-files used to create the dataset:

```
. use cwh05, clear
(CWH analysis file \ 2006-12-22)

. notes _dta

_dta:
  1.  cwh01.dta \ initial CWH analysis dataset \ cwh-dta01a.do jsl 2006-12-07.
  2.  cwh02.dta \ add attrition \ cwh-dta02a.do jsl 2006-12-07.
  3.  cwh03.dta \ add demographics \ cwh-dta03c.do jsl 2006-12-09.
  4.  cwh04.dta \ add panel 5 data \ cwh-dta04a.do jsl 2006-12-19.
  5.  cwh05.dta \ exclude youngest cohort \ cwh-dta05a.do jsl 2006-12-22.
```

As an example of how useful this is, while writing this book I lost the do-file that created a dataset used in an example. I had the dataset, but needed to modify the do-file that created it so I could add another variable. To find the file, I loaded the dataset, checked the notes to find the name of the do-file that created it, and searched my hard drive for the missing do-file. A good workflow makes up for lots of mistakes!

5.5.4 The datasignature command

The `datasignature` command protects the integrity of your data and should be used every time you save a dataset.[1] `datasignature` creates a string of numbers and symbols, referred to as the *data signature* or simply *signature*, which is based on five characteristics of the data. For example (file: `wf5-datasignature.do`),

```
. use wf-datasig01, clear
(Workflow data for illustrating datasignature #1 \ 2008-04-02)

. datasignature
  753:8(54146):1899015902:1680634677
```

1. The `datasignature` command in Stata 10 is not the same as `datasignature` in Stata 9. The newer command is much easier to use.

The string `753:8(54146):1899015902:1680634677` is the signature for `wf-datasig01.dta` (below I explain where this string comes from). If I load a dataset that does not have this signature, whether it is named `wf-datasig01.dta` or something else, I am certain that the datasets differ. On the other hand, if I load a dataset that has this signature, I am almost certain that I have the right dataset. (The reason that I am not completely certain is discussed below.) This can be useful in many ways. You and a colleague can verify whether you are analyzing the same dataset. If you are revising labels, as discussed later in this chapter, you can check if you mistakenly changed the data itself, not just the labels. If you store datasets on a LAN where others have read and write privileges, you can determine if someone changed the dataset but forgot to save it with a different name. `datasignature` is an easy way to prevent many problems.

The signature consists of five numbers, known as *checksums*, that describe the dataset. Anyone with the same dataset using the same rules for computing the checksums will obtain the same values. The first checksum is the number of cases (753 in the example above). If I load a dataset with more or less observations, this number will not match and I will know I have the wrong data. The second is the number of variables (8 in our example). If I load a dataset that does not have 8 variables, this part of the signature will not match. The third part of the signature is based on the names of the variables. To give you a simplified idea of how this works, consider the variables in `wf-datasig01.dta`:

```
. describe, simple
lfp   k5    k618  age   wc    hc    lwg   inc
```

These names are 22 $(= 3+2+4+3+2+2+3+3)$ characters long. If I load a dataset where the length of the names is not 22, I know that I have the wrong dataset. The fourth and fifth numbers are checksums that characterize the values of variables.

The idea behind a data signature is that if the signature of a dataset that you use matches the signature of a dataset you saved, it is very likely that the two datasets are the same. The signature is not perfect, however. If you have a lot of computing power, you could probably find two datasets with the same signature but different content (Mackenzie 2008). In practice, this is extremely unlikely so you can reasonably assume that if the data signatures from two datasets match, the data are the same. For full details on how the signature is computed, type `help datasignature` or see [D] **datasignature**.

A workflow using the datasignature command

I suggest that you always compute a data signature and save it with your dataset. When you use a dataset, you should confirm that the embedded signature matches the signature of the data in memory. The `datasignature set` command computes the signature. For example,

```
. datasignature set
  753:8(54146):1899015902:1680634677       (data signature set)
```

Once the signature is set, it is automatically saved when you save the dataset. For example,

```
. notes: wf-datasig02.dta \ add signature \ wf5-datasignature.do jsl 2008-04-03.
. label data "Workflow dataset for illustrating datasignature \ 2008-04-03"
. save wf-datasig02, replace
file wf-datasig02.dta saved
```

When I load the dataset, I can confirm that the dataset in memory generates the same signature as the one that was saved:

```
. use wf-datasig02, clear
(Workflow dataset for illustrating datasignature \ 2008-04-03)
. datasignature confirm
  (data unchanged since 03apr2008 09:58)
```

Because the signature matches, I am confident that I have the right data.

Why would a signature fail to match? Suppose that my colleague used `wf-datasig02.dta` that I created on 3 April 2008. He renamed a variable, changed the dataset label, and violated good workflow by saving the changed data with the same name:

```
. use wf-datasig02, clear
(Workflow dataset for illustrating datasignature \ 2008-04-03)
. rename k5 kids5
. save wf-datasig02, replace
file wf-datasig02.dta saved
```

He did not run the **datasignature set** command before saving the dataset. When I load the dataset and check the signature, I am told that the dataset has changed:

```
. use wf-datasig02, clear
(Workflow data for illustrating datasignature \ 2008-04-03)
. datasignature confirm
  data have changed since 03apr2008 09:58
r(9);
```

I know immediately that there is a problem.

Changes datasignature does not detect

The **datasignature confirm** command does not detect every change in a dataset. First, the signature does not change if you only change labels. For example,

```
. use wf-datasig02, clear
(Workflow dataset for illustrating datasignature \ 2008-04-03)
. label var k5  "Number of children less than six years of age"
. datasignature confirm
  (data unchanged since 03apr2008 09:58)
```

The signature does not change because it does not contain checksums based on variable or value labels. Because changed labels can cause a great deal of confusion, I hope this information is added to a later version of the command.

Second, `datasignature confirm` does not detect changes if the person saving the dataset embeds a new signature. For example, I load a dataset that includes a signature:

```
. use wf-datasig02, clear
(Workflow dataset for illustrating datasignature \ 2008-04-03)
. datasignature confirm
  (data unchanged since 03apr2008 09:58)
```

Next I rename variables `k5` and `k618`:

```
. rename k5 kids5
. rename k618 kids618
```

Now I reset the signature and change the data label:

```
. datasignature set, reset
  753:8(61387):1899015902:1680634677       (data signature reset)
. notes: Rename kids variables \ datasig02.do jsl 2008-04-04.
. label data "Workflow data for illustrating datasignature \ 2008-04-04"
```

By mistake, I save the dataset with the same name:

```
. save wf-datasig02, replace
file wf-datasig02.dta saved
```

The next time I load `wf-datasig02.dta`, I check the signature:

```
. use wf-datasig02, clear
(Workflow data for illustrating datasignature \ 2008-04-04)
. datasignature confirm
  (data unchanged since 04apr2008 11:23)
```

Appropriately, `datasignature confirm` finds that the embedded signature matches the dataset in memory. The problem is that I should not have saved the dataset with the same name `wf-datasig02.dta`. Because I used `label data` and `notes:`, the dataset contains information that points to the problem. First, the data label has the date 2008-04-04, whereas the original dataset was saved on 2008-04-03. The notes also show a problem:

```
. notes
_dta:
  1.  wf-datasig01.dta \ no signature \ wf-datasig01-supportV2.do jsl
      2008-03-09.
  2.  wf-datasig02.dta \ add signature \ wf5-datasig01.do jsl 2008-04-03.
  3.  wf-datasig02.dta \ rename kids variables \ datasig02.do jsl 2008-04-04.
```

Given my workflow for saving datasets, there should not be two notes indicating that the same dataset was saved by different do-files on different dates.

5.6 Naming variables

Variable names are fundamental to everything you do in data management and statistical analysis. You want names that are clear, informative, and easy to use. Choosing effective names takes planning. Unfortunately, planning names is an uninspiring job, is harder than it first appears, and seems thankless because the payoff generally comes much later. Everyone should think about how variables are named before they begin their analysis. Even if you use data collected by others, you need to choose names for the variables that you want to add. You might also want to revise the original names (discussed in section 5.11). In this section, I consider issues ranging from general approaches for organizing names to practical considerations that affect your choice of names.

5.6.1 The fundamental principle for creating and naming variables

The most basic principle for naming variables is simple:

Never change a variable unless you give it a new name.

Replication is nearly impossible if you have two versions of a dataset that contain variable `var27`, but where the content of the variable has changed. Suppose that you want to recode `var27` to truncate values above 100. You should not replace the values in the existing variable `var27` (file: `wf5-varnames.do`):

```
replace var27 = 100 if var27>100  // do NOT do this!
```

Instead, you should use either `generate` or `clonevar` to create copies of the original variable and then change the copy. The syntax for these commands is

`generate` *newvar* = *sourcevar* [*if*] [*in*]

`clonevar` *newvar* = *sourcevar* [*if*] [*in*]

The `generate` command creates a new variable but does not transfer labels and other characteristics. The `clonevar` command creates a variable that is an exact duplicate of an existing variable including variable and value labels; only the name is different. For example, I can create two copies of the variable `lfp`:

```
. use wf-names, clear
(Workflow data to illustrate names \ 2008-04-03)
. generate lfp_gen = lfp
(327 missing values generated)
. clonevar lfp_clone = lfp
(327 missing values generated)
```

The original `lfp` and the generated `lfp_gen` have the same descriptive statistics but `lfp_gen` does not have value or variable labels. `lfp_clone`, however, is identical to `lfp`:

```
. codebook lfp*, compact
Variable    Obs Unique       Mean  Min  Max  Label
--------------------------------------------------------------------
lfp         753      2  .5683931    0    1  Paid labor force?
lfp_gen     753      2  .5683931    0    1
lfp_clone   753      2  .5683931    0    1  Paid labor force?
--------------------------------------------------------------------

. describe lfp*
              storage  display     value
variable name   type   format      label      variable label
--------------------------------------------------------------------
lfp             byte   %9.0g       lfp        Paid labor force?
lfp_gen         float  %9.0g
lfp_clone       byte   %9.0g       lfp        Paid labor force?
```

Returning to our earlier example, after you generate or clone `var27`, you can change the copy. With `generate`, type

```
generate var27trunc = var27
replace  var27trunc = 100 if var27trunc>100 & !missing(var27trunc)
```

Or with `clonevar` type

```
clonevar var27trunc = var27
replace  var27trunc = 100 if var27trunc>100 & !missing(var27trunc)
```

Because truncating a variable can substantially affect later results, you probably agree that I should create a new variable with a different name. Suppose that I am not "really" changing the values. Imagine that `educ` uses 99 to indicate missing values and I decide to recode these values to `.`, the sysmiss. Do I really need to create a new variable for this? In one sense, I have not changed the data—missing are still missing. However, you never want to risk that a changed variable will be confused with the original. The easiest thing to do is to always create a new variable no matter how small the change. Here I would use these commands:

```
clonevar educV2 = educ
replace  educV2 = . if educV2==99
```

If you violate this rule, you can end up with results that are difficult or impossible to replicate and findings that are unclear or wrong.

5.6.2 Systems for naming variables

There are three basic systems for naming variables: sequential naming, source naming, and mnemonic naming.[2] Each has its advantages and in practice you might use a combination of all three.

2. This discussion is based in part on ICPSR (2005).

Sequential naming systems

Sequential names use a stub followed by sequential digits. For example, the 2002 International Social Survey Program (http://www.issp.org) uses the names v1, v2, v3, ..., v362. The National Longitudinal Survey uses names that start with R and end with seven digits such as R0000100, R0002203, and R0081000. Some sequential names use padded numbers (e.g.,v007, v011, v121), while others do not (e.g., v7, v11, v121). Stata's aorder command (see page 155) alphabetizes sequential names as if they are padded with zeros, even if they are not padded.

The numbers used in sequential names might correspond to the order in which the questions were asked, to some other aspect of the data, or be meaningless. Although sequential naming is often necessary with large datasets, these names do not work well for data analysis. Because the names do not reflect content, it is easy to use the wrong variable, it is hard to remember the name of the variable you need, and it is difficult to interpret output. For example, was the command supposed to be this?

```
logit R0051400 R0000100 R0002203 R0081000
```

Or this?

```
logit R00541400 R1000100 R0002208 R0081000
```

Because of the risk of using the wrong variable when using sequential names, I often refer to a printed list of variable names, descriptive statistics, and variable labels, such as produced by codebook, compact.

Source naming systems

Source names use information about where a variable came from as part of the name. The first three questions from a survey might be named q1, q2, and q3. If a question had multiple parts, the variables might be named q4a, q4b, and q4c. In older datasets, names might index the card and column where a variable is located (e.g., c1c15). With source names, you are likely to have some variables that do not fit into the scheme, which requires using some names that are not based on the source. For example, there might be variables with information about the site of data collection or from debriefing questions that are not numbered as part of the survey instrument. If you are creating a dataset using source names, be sure to plan how you will name all the variables that will be needed.

Names based on the source question are more useful than purely sequential names because they refer to the questionnaire. Still, it is hard to look at a model specification using source names and be certain that you have selected the correct variables.

Mnemonic naming systems

Mnemonic names use abbreviations that convey content (e.g., `id`, `female`, `educ`). I much prefer this system because the names partially document your commands and the output. A command like this

```
logit lfp age educ kids
```

is easier to use than this

```
logit R0051400 R0000100 R0002203 R0081000
```

or this

```
logit q17 q31 q19 q02
```

Although mnemonic names have many advantages, you need to choose the names carefully because finding names that are short, unambiguous, and informative is hard. Mnemonic names created "on the fly" can be misleading and difficult to use.

5.6.3 Planning names

If you are collecting your own data, you should plan names before the dataset is created. If you are extracting variables from an existing dataset, you should plan which variables you need and how you want to rename them before data extraction begins. Large datasets such as the National Longitudinal Survey (NLS, http://www.bls.gov/nls) or the National Longitudinal Study of Adolescent Health (http://www.cpc.unc.edu/addhealth) have thousands of variables, and you might want to extract hundreds of them. For example, Eliza Pavalko and I (Long and Pavalko 2004) used data from the NLS on functional limitations. We extracted variables measuring limitations for nine activities in each of four panels for two cohorts and created over 200 scales. It took several iterations to come up with names that were clear and consistent.

When planning names, think about how you will use the data. The more complex the project, the more detailed your plan needs to be. Will the project last a few weeks or several years? Do you anticipate a small number of analyses or will the analyses be detailed and complex? Are you the only one using the data or will it be shared with others? Will you be adding a new wave of data or another country? The answers to these and similar questions need to be anticipated as you plan your names.

After you make general decisions on how to name variables, I suggest that you create a spreadsheet to help you plan. For example, in a study of stigma (http://www.indiana.edu/~sgcmhs/), we received datasets from survey centers in 17 countries. Each center used source names for most variables. To create mnemonic names, we began by listing the original name and question. We then classified variables into categories (e.g., questions about treatment, demographics, measures of social distance). One member of the research team then proposed a set of mnemonic names that was circulated for comments. After several iterations, we came up with names that we agreed upon. Figure 5.3 is a portion of the large spreadsheet that we used (file: `wf5-names-plan.xls`):

	A	B	C	D
1	Question	Question ID	Proposed name	Variable Category
14	Question stem: What should NAME do about this situation:...			
15	...Talk to family	q2-1	tofam	treatment_option
16	...Talk to friends	q2-2	tofriend	treatment_option
17	...Talk to a religious leader	q2-3	torel	treatment_option
18	...Go to a medical doctor	q2-4	todoc	treatment_option
19	...Go to a psychiatrist	q2-5	topsy	treatment_option
20	...Go to a counselor or another mental health professional	q2-6	tocou	treatment_option
21	...Go to a spiritual or traditional healer	q2-7	tospi	treatment_option
22	...Take non-prescription medication	q2-8	tonpm	treatment_option
23	...Take prescription medication	q2-9	topme	treatment_option
24	...Check into a hospital	q2-10	tohos	treatment_option
25	...Pray	q2-11	topray	treatment_option
26	...Change lifestyle	q2-12	tolifest	treatment_option
27	...Take herbs	q2-13	toherb	treatment_option
28	...Try to forget about it	q2-14	toforg	treatment_option
29	...Get involved in other activities	q2-15	toothact	treatment_option
30	...Get involved in a group	q2-16	togroup	treatment_option

Figure 5.3. Sample spreadsheet for planning variable names

5.6.4 Principles for selecting names

Although choosing a system for naming variables is the first step, there are additional factors to consider when selecting names (file: `wf5-varnames.do`).

Anticipate looking for variables

Before you decide on names (and labels, which are discussed in section 5.7), think about how you will find variables during your analysis. This is particularly important with large datasets. There are two aspects of finding a variable to consider. First, how will the names work with Stata's `lookfor` command? Second, how will the names appear in a sorted list?

The `lookfor` *string* command lists all variables that have *string* in their names or variable labels. Of course, `lookfor` is only useful if you use names and labels that include the strings that you are likely to search for. For example, if I name three indicators of race `raceblck`, `racewhite`, and `raceasian`, then `lookfor race` will find these variables. For example,

```
. lookfor race

                storage  display     value
variable name     type   format      label      variable label
-------------------------------------------------------------------------
racewhite         byte   %9.0g       Lyn        Is white?
raceblack         byte   %9.0g       Lyn        Is black?
raceasian         byte   %9.0g       Lyn        Is asian?
```

If I use the names `black`, `white`, and `asian`, then `lookfor race` will not find them unless "race" is part of their variable labels. There is a trade-off between short names and being able to find things. For example, if I abbreviate `race` as `rce` to create shorter names, I must remember to use `lookfor rce` to find these variables because `lookfor race` will not find them.

You can sort variables so that they appear in alphabetical order in the Variables window (see the discussion of `order` and `aorder` on page 155). This is handy for finding variables, especially if you like to click on a name in the Variables window to insert the name into a command. When choosing names, think about how the names will appear when sorted. For example, suppose I have several variables that measure a person's preference for social distance from someone with mental illness. These questions deal with different types of contact, such as having the person as a friend, having the person marry a relative, working with the person, having her as a neighbor, and so on. I could choose names such as `friendsd`, `marrysd`, `worksd`, and `neighbsd`. If I sorted the names, the variables will not be next to one another. If I name the variables `sdfriend`, `sdmarry`, `sdwork`, and `sdneighb`, they appear together in an alphabetized list. Similarly, the names `raceblck`, `racewhite`, and `raceasian` work better than `blckrace`, `whiterace`, and `asianrace`. If I have binary indicators of educational attainment (e.g., completing high school, completing college), the names `edhs`, `edcol`, and `edphd` work better than `hsed`, `coled`, and `phded`.

Use simple, unambiguous names

There is a trade-off between the length of a name and its clarity. Although the name `lQ_23v` is short, it is hard to remember and hard to type. A name like `socialdistancescale2` is descriptive but too long for typing and is likely to be truncated in your output or when converting your data to another format. In a large dataset, it is impossible to find names that meet all your goals for being clear and easy to use. Keeping names short often conflicts with making names clear and being able to find them with `lookfor`. With planning, however, you can select names that are much more useful than if you create names without anticipating their later use. Here are some things to consider when looking for simple, effective names.

Use shorter names Stata allows names of up to 32 characters but often truncates long names when listing results. You need to consider not only how clear a name is but also how clear it is when truncated in the output. For example, I generate three variables with names that are 32 characters long and use the `runiform()` function to assign uniform random numbers to the variables (file: `wf5-varnames.do`):

```
generate a2345678901234567890123456789012 = runiform()
generate a23456789012345678901234567890_1 = runiform()
generate a23456789012345678901234567890_2 = runiform()
```

When analyzed, the names are truncated in a way that is confusing:

```
. summarize

    Variable |       Obs        Mean    Std. Dev.       Min        Max
-------------+--------------------------------------------------------
a23456789~12 |       100    .4718318    .2695077   .0118152   .9889972
a234567890~1 |       100    .4994476    .2749245   .0068972   .9929506
a23456789~_2 |       100    .4973259    .3026792   .0075843   .9889733
```

Because most Stata commands show at least 12 characters for the name, I suggest the following guideline:

Use names that are at most 12 characters long.

For the original variables in a dataset, limit names to 10 characters so that you have two characters to indicate version if the variable is revised. For example,

```
generate socialdistV2 = socialdist if socialdist>=0 & !missing(socialdist)
```

Some statistics packages do not allow long variables names. For example, when I converted the variables above to a LIMDEP dataset (http://www.limdep.com), the names were changed to `a2345678`, `a2345670`, and `a2345671`. The only way to verify how the converted names mapped to the source names was by looking at the raw data. If I plan to use software that limits names to eight characters, I either restrict variable names to eight characters in Stata, or I create a new Stata dataset in which I explicitly shorten the names. After I rename a variable, I revise the variable label to document the original name. For example,

```
rename socialdistance socdist
label var socdist "socialdistance \ social distance from person with MI"
```

or

```
rename socialdistance socdist
label var socdist "social distance from person with MI (socialdistance)"
```

Now when I convert the dataset I have control over the names that are used.

Use clear and consistent abbreviations Because long names are harder to type and might be truncated, I often use abbreviations as part of the variable names. For example, I might use `ed` as an abbreviation for `education` and create variables such as `ed_lths` and `ed_hs` rather than `educationlths` and `educationhs`. Abbreviations, however, by their nature are ambiguous. To make them as clear as possible, plan your abbreviations and get feedback from a colleague before you finalize them. Then use those abbreviations consistently and keep the list of abbreviations as part of the project documentation. A convenient way to do this is with the `notes` command as discussed in section 5.8.

Use names that convey content All else being equal, names that convey content are easier to use than those that do not. Names such as `educ` or `socdist` are easier to use and less likely to cause errors than names such as `q32part2` or `R003197`. There are other ways to make names more informative. For binary variables, I suggest names that indicate the category that is coded as 1. For example, if 0 is male and 1 is female, I would name the variable `female`, not `gender`. (When you see a regression coefficient for `gender`, is it the effect of being male or being female?) If you have multiple scales coded in different directions (i.e., `scale1` is coded 1 = disagree, 2 = neutral, and 3 = agree, whereas `scale2` is coded 1 = agree, 2 = neutral, and 3 = disagree), I suggest names that indicate the direction of the scale. For example, I might use the names `sdist1P`, `sdist2N`, and `sdist3N`, where `N` and `P` indicate negative and positive coding.

Be careful with capitalization Stata distinguishes between names with the same letters but different capitalization. For example, `educ`, `Educ`, and `EDUC` are three different variables. Although such names are valid and distinct in Stata, they are likely to cause confusion. Further, some statistical packages do not differentiate between uppercase and lowercase. Worse, programs that convert between formats might simply drop the "extra" variables. When I converted a Stata dataset containing `educ`, `Educ`, and `EDUC` to a format that is case insensitive, the conversion program dropped two of the variables without warning and without indicating which variable was kept! I do, however, use capitalization to highlight information. For example, I use `N` to indicate negatively coded scales and `P` for positively coded scales. Capitalization emphasizes this so I prefer `scale1N` and `scale2P` to `scale1n` and `scale2p`. I would never create a pair of variables called `scale1n` and `scale1N`. I use the capitals in table 5.1 as standard abbreviations within variable names:

Table 5.1. Recommendations for capital letters used when naming variables

Letter	Meaning	Example
B	Binary variable	`highschlB`
I	Indicator variable	`edIhs`, `edIgths`, `edIcol`
L	Value labels used by multiple variables	`Lyesno`
M	Indicator of data being missing[a]	`educM`
N	A negatively coded scale	`sdworkN`
O	Too close to the number 0, so I do not use it	
P	A positively coded scale.	`sdkidsP`
S	The unchanged, source variable	`educS`; `Seduc`
V	Version number for modified variables	`marstatV2`
X	A temporary variable	`Xtemp`

[a] These are binary variables equal to 1 if the source variable is missing, and 0 otherwise. For example, `educM` would be 1 if `educ` is missing, and 0 otherwise.

Try names before you decide

Selecting effective names and labels is an iterative process. After you make initial selections, check how well the names work with the Stata commands you anticipate using. If the names are truncated or confusing in the output from `logit` and you plan to run a lot of logit models, consider different names. Continue revising and trying names until you are satisfied.

5.7 Labeling variables

Variable labels are text strings of up to 80 characters that are associated with a variable. These labels are listed in the output of many commands to document the variables being analyzed. Variable labels are easy to create, and they can save a great deal of confusion. My recommendation for variable labels is simple:

Every variable should have a variable label.

If you receive a dataset that does not include labels, add them. When you create a new variable, always add a variable label. It is tempting to forgo labeling a variable that you are "sure" you will not need later. Too often, such variables find their way into a saved dataset (e.g., you create a temporary variable while constructing a variable but forget to delete the temporary variable).[3] When you later encounter these unlabeled variables, you might forget what they are for and be reluctant to delete them. A quick label such as

```
label var checkvar "Scott´s temp var; can be dropped"
```

avoids this problem. The accumulation of stray variables is a bigger problem in collaborative projects when several people can add variables, and you do not want to delete a variable someone else needs. In the long run, the time you spend adding labels is less than the time you lose trying to figure out what a variable is.

5.7.1 Listing variable labels and other information

Before considering how to add variable labels and principles for choosing labels, I want to review the ways you can examine variable labels. There are many reasons why you might want a list of variables with their labels—to construct tables of descriptive statistics in a paper, to remind you of the names of variables as you plan your analyses, or to help you clean your data (file: `wf5-varlabels.do`).

---

3. One way to avoid the problem of saving temporary variables is to use the `tempvar` command. For details, see `help tempvar` or [P] **macro**.

codebook, compact

The `codebook, compact` command lists variable names, labels, and some descriptive statistics. The syntax is

`codebook` [*varlist*] [*if*] [*in*], `compact`

The `if` and `in` qualifiers allow you to select the cases for computing descriptive statistics. Here is an example of the output:

```
. codebook id tc1fam tc2fam tc3fam vignum, compact
Variable      Obs Unique      Mean  Min   Max  Label
-------------------------------------------------------------------------
id           1080   1080     540.5    1  1080  Identification number
tc1fam       1074     10  8.755121    1    10  Q43 How important is it to turn ...
tc2fam       1074     10  8.755121    1    10  Q43 How Impt: Turn to family for...
tc3fam       1074     10  8.755121    1    10  Q43 Family help important
vignum       1080     12  6.187963    1    12  Vignette number
-------------------------------------------------------------------------
```

If your variable labels are truncated on the right, you can increase the line size. For example, `set linesize 120`. Unfortunately, `codebook` does not give you a choice of which statistics are shown and there is no measure of variance.

describe

The `describe` command lists variable names, variable labels, and characteristics of the variables. The syntax is

`describe` [*varlist*] [*if*] [*in*] [, `simple fullnames numbers`]

If *varlist* is not given, all variables are listed. If you have long variable names, by default they are truncated in the list. With the `fullnames` option, the entire name is listed. The `numbers` option numbers the variables. For other options, use `help describe`. Here is an example of the default output:

```
. describe id tc1fam tc2fam tc3fam vignum

              storage  display     value
variable name   type   format      label      variable label
-------------------------------------------------------------------------
id              int    %9.0g                  Identification number
tc1fam          byte   %21.0g      Ltenpt     Q43 How important is it to turn
                                                to family for help
tc2fam          byte   %21.0g      Ltenpt     Q43 How Impt: Turn to family for
                                                help
tc3fam          byte   %21.0g      Ltenpt     Q43 Family help important
vignum          byte   %35.0g      vignum   * Vignette number
```

Storage type tells you the numerical precision used for storing that variable (see the `compress` command on page 264 for further details). Display format, reasonably enough, describes the way a variable is displayed. I have never had to worry about this because

Stata seems to figure out how to display things just fine. However, if you are curious, see [U] **15.5 Formats: controlling how data are displayed** for details. The value label column lists the name of the value label associated with each variable (see section 5.9 for information on value labels). The *'s indicate that there is a note associated with that variable (see section 5.8 for further details). If you only want a list of names, add the `simple` option. For example, to create a list of all variables in your dataset, type

```
. describe, simple
id            tc1fam        tc2mhprof     Ed            var14
vignum        tc2fam        tc3mhprof     ED            var15
  (output omitted)
tcfam         tc1mhprof     ed            var13
```

Or, to quickly find the variables included in a varlist shorthand notation, say, `id-opdoc`, type

```
. describe id-opdoc, simple
id        female    opnoth    opfriend  opdoc
vignum    serious   opfam     oprelig
```

nmlab

Stata does not have a command that lists only variable names and labels. Because I find such lists to be useful, I adapted the code used as an example in chapter 4 to create the command `nmlab`. Most simply, type

```
. nmlab id tc1fam tc2fam tc3fam vignum
id      Identification number
tc1fam  Q43 How important is it to turn to family for help
tc2fam  Q43 How Impt: Turn to family for help
tc3fam  Q43 Family help important
vignum  Vignette number
```

The `number` option numbers the list, whereas `column(#)` changes the start column for the variable labels. The `vl` option adds the name of the value label, as discussed below. Just typing `nmlab` lists all the variables in the dataset.

tabulate

This command shows you the variable label and the value labels (see section 5.9.3):

```
. tabulate tcfam, missing
  Q43 How Impt: |
 Turn to family |
       for help |      Freq.     Percent        Cum.
----------------+-----------------------------------
1Not at all Impt |          9        0.83        0.83
              2 |          4        0.37        1.20
              3 |         11        1.02        2.22
  (output omitted)
```

Although `tabulate` does not truncate long labels, longer labels are often more difficult to understand than shorter ones:

```
. tabulate tcfamV2, missing
Question 43: How |
 important is it |
  to you to turn |
   to the family |
    for support? |      Freq.     Percent        Cum.
-----------------+-----------------------------------
1Not at all Impt |          9        0.83        0.83
               2 |          4        0.37        1.20
               3 |         11        1.02        2.22
  (output omitted)
```

The Variables window

Because variable labels are shown in the Variables window, I also make sure that the labels work well here. For example,

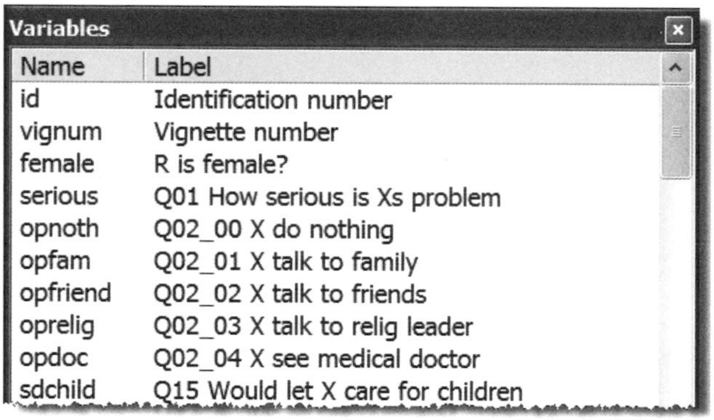

Variables

Name	Label
id	Identification number
vignum	Vignette number
female	R is female?
serious	Q01 How serious is Xs problem
opnoth	Q02_00 X do nothing
opfam	Q02_01 X talk to family
opfriend	Q02_02 X talk to friends
oprelig	Q02_03 X talk to relig leader
opdoc	Q02_04 X see medical doctor
sdchild	Q15 Would let X care for children

If your variable labels do not appear in the window or if there is a large gap between the names and the start of the label, you need to change the column in which the labels begin. By default, this is column 32, which means you need a wide Variables window or you will not see the labels. In Windows and Macintosh for Stata 10, you can use the mouse to resize the columns. In Unix, you can change the space allotted for variable names with the command

```
set varlabelpos #
```

where # is the maximum number of characters to display for variable names. Once you change this setting, it persists in later sessions. Because I typically limit names to 12 characters or less, I use set the variable label position to 12.

Changing the order of variables in your dataset

Commands such as `codebook`, `describe`, `nmlab`, and `summarize` list variables in the order they are arranged in the dataset. You can see how variables are ordered by looking at the Variables window or by browsing your data (type `browse` to open a spreadsheet view of your data). When a new variable is created, it is placed at the end of the list. You can change the order of variables with the `order`, `aorder`, and `move` commands. Changing the order lets you put frequently used variables first to make them easier to click on in the Variables window. You can alphabetize names to make them easier to find, place related variables together, and do other similar things. The `aorder` command arranges the variables in *varlist* alphabetically. The syntax is

`aorder` [*varlist*]

If no *varlist* is given, all variables are alphabetized. The `order` command allows you to move a group of variables to the front of the dataset:

`order` *varlist*

To move one variable, use the command

`move` *variable-to-move target-variable*

where *variable-to-move* is placed in front of the *target-variable*. For many datasets, I run this pair of commands:

`aorder`

`order` *id*

where *id* is the name of the variable with the ID number. This arranges variables alphabetically, except that the ID variable appears first. The best way to learn how these commands work is to open a dataset, try the commands, and watch how the list of variables in the Variables window changes.

5.7.2 Syntax for label variable

Now that we know how to look at variable labels, we can create them. The `label variable` command assigns a text label of up to 80 characters to a variable. The syntax is

`label variable` *varname* "*label*"

Although I generally do not abbreviate commands, I often use the abbreviation `label var`, which is shorter, yet still clear. For example,

```
label var artsqrt "Square root of # of articles"
```

To remove a label, you use the command

```
label variable varname
```

For example,

```
label var artsqrt
```

5.7.3 Principles for variable labels

Just like names, you can create more-useful labels by planning. Here are some things to think about as you plan your labels.

Beware of truncation

A variable label should be long enough to provide the essential information but short enough that the content can be grasped quickly. Although variable labels can be 80 characters long, many commands truncate labels that are longer than about 30 characters. Accordingly, I recommend

Put the most important information in the first 30 columns of a variable label.

Here is an example of what can happen if you use the labels typically found in secondary data. The data we received used labels that were slightly condensed versions of the questions from the survey. For example, one group of questions asked a person who they would turn to if they needed care:

```
tc1fam     Q43 How important is it to turn to family for help
tc1friend  Q44 How important is it to turn to friends for help
tc1relig   Q45 How important is it to turn to a minister, priest, rabbi or other religious
tc1doc     Q46 How important is it to go to a general medical doctor for help
tc1psy     Q47 How important is it to go to a psychiatrist for help
tc1mhprof  Q48 How important is it to go to a mental health professional
```

The labels are so long that they are useless for commands that truncate the labels at column 30. For example,

```
. codebook tc1*, compact
Variable     Obs Unique      Mean  Min  Max  Label
-------------------------------------------------------------------------------
tc1doc      1074     10  8.714153    1   10  Q46 How important is it to go to ...
tc1fam      1074     10  8.755121    1   10  Q43 How important is it to turn t...
tc1friend   1073     10  7.799627    1   10  Q44 How important is it to turn t...
tc1mhprof   1045     10   7.58756    1   10  Q48 How important is it to go to ...
tc1psy      1050     10  7.567619    1   10  Q47 How important is it to go to ...
tc1relig    1039     10   5.66025    1   10  Q45 How important is it to turn t...
-------------------------------------------------------------------------------
```

A better set of labels looks like this:

```
. codebook tc2*, compact
Variable      Obs Unique       Mean  Min   Max  Label
-------------------------------------------------------------------------
tc2doc       1074     10   8.714153    1    10  Q46 How Impt: Go to a gen med doc...
tc2fam       1074     10   8.755121    1    10  Q43 How Impt: Turn to family for ...
tc2friend    1073     10   7.799627    1    10  Q44 How Impt: Turn to friends for...
tc2mhprof    1045     10    7.58756    1    10  Q48 How Impt: Go to a mental heal...
tc2psy       1050     10   7.567619    1    10  Q47 How Impt: Go to a psych for Help
tc2relig     1039     10    5.66025    1    10  Q45 How Impt: Turn to a religious...
-------------------------------------------------------------------------
```

We eventually chose even shorter labels:

```
. codebook tc3*, compact
Variable      Obs Unique       Mean  Min   Max  Label
-------------------------------------------------------------------------
tc3doc       1074     10   8.714153    1    10  Q46 Med doctor help important
tc3fam       1074     10   8.755121    1    10  Q43 Family help important
tc3friend    1073     10   7.799627    1    10  Q44 Friends help important
tc3mhprof    1045     10    7.58756    1    10  Q48 MH prof help important
tc3psy       1050     10   7.567619    1    10  Q47 Psychiatric help important
tc3relig     1039     10    5.66025    1    10  Q45 Relig leader help important
-------------------------------------------------------------------------
```

Given our familiarity with the survey instrument, these labels tell us everything we need to know.

Although I find short variable labels work best for analysis, I sometimes want to see the original labels. For example, I might want to verify the exact wording of a question or know exactly how the categories are labeled. Stata's `language` command allows you to have both long, detailed labels for documenting your variables and shorter labels that work better in your output. This is discussed in section 5.10.

Test labels before you post the file

After creating a set of labels, you should check how they work with commands such as `codebook, compact` and `tabulate`. If you do not like how the labels appear in the output, try different labels. Rerun the test commands and repeat the cycle until you are satisfied.

5.7.4 Temporarily changing variable labels

Sometimes I need to temporarily change or eliminate a variable label. For example, `tabulate` does not list the name of a variable if it has a variable label. Yet, when cleaning data, I often want to know the variable name. To see the variable name in the `tabulate` output, you need to remove the variable label by assigning a null string as the label:

```
label variable varname ""
```

I can do this for a group of variables using a loop (file: `wf5-varlabels.do`):

```
. foreach varname in pub1 pub3 pub6 pub9 {
  2.       label var `varname´ ""
  3.       tabulate `varname´, missing
  4. }
       pub1 |      Freq.     Percent        Cum.
------------+-----------------------------------
          0 |         77       25.00       25.00
          1 |         75       24.35       49.35
          2 |         36       11.69       61.04
  (output omitted)
```

Another reason to change the variable label temporarily is to revise labels in graphs. By default, the variable label is used to label the axes.

5.7.5 Creating variable labels that include the variable name

Recently, I was asked, "Do you know of a Stata command that will add the variable name to the beginning of the variable label?" Although there is not a Stata command to do this, it is easy to do this using a loop and a local macro (file: `wf5-varname-to-label.do`).[4] Here are the current labels:

```
. use wf-lfp, clear
(Workflow data on labor force participation \ 2008-04-02)
. nmlab
lfp   In paid labor force? 1=yes 0=no
k5    # kids < 6
k618  # kids 6-18
age   Wife´s age in years
wc    Wife attended college? 1=yes 0=no
hc    Husband attended college? 1=yes 0=no
lwg   Log of wife´s estimated wages
inc   Family income excluding wife´s
```

To see why I want to add the name of the variable to the label, consider the output from `tabulate`:

```
. tabulate wc hc, missing
       Wife |
   attended |   Husband attended
   college? |  college? 1=yes 0=no
 1=yes 0=no |   0_NoCol  1_College |     Total
------------+----------------------+----------
    0_NoCol |       417        124 |       541
  1_College |        41        171 |       212
------------+----------------------+----------
      Total |       458        295 |       753
```

4. If you want to try creating your own command with an ado-file, I suggest you write a command that adds a variable's name to the front of its label.

It would be convenient to know the names of the variables in this table. This can be done by adding the variable name to the front of the variable label. I start by using **unab** to create a list of the variables in the dataset, where **_all** is Stata shorthand for "all the variables in memory":

```
. unab varlist : _all
. display "varlist is: `varlist´"
varlist is: lfp k5 k618 age wc hc lwg inc
```

Next I loop through the variables:

```
1>  foreach varname in `varlist´ {
2>      local varlabel : variable label `varname´
3>      label var `varname´ "`varname´: `varlabel´"
4>  }
```

Line 2 is an extended macro function that creates the local **varlabel** with the variable label for the variable named in local **varname**. Extended macro functions, which are used extensively in section 5.11, retrieve information about variables, datasets, labels, and other things and place the information in a macro. The command begins with **local varlabel** to indicate that you want to create a local macro named **varlabel**. The **:** is like an equal-sign, saying that the local equals the content described on the right. For example, **local varlabel : variable label lfp** assigns local **varlabel** to the variable label for **lfp**. Line 3 creates a new variable label that begins with the variable name (i.e., **`varname´**), adds a colon, and inserts the current label (i.e., **`varlabel´**). Here are the new variable labels:

```
. nmlab
lfp   lfp: In paid labor force? 1=yes 0=no
k5    k5: # kids < 6
k618  k618: # kids 6-18
age   age: Wife´s age in years
wc    wc: Wife attended college? 1=yes 0=no
hc    hc: Husband attended college? 1=yes 0=no
lwg   lwg: Log of wife´s estimated wages
inc   inc: Family income excluding wife´s
```

Now when I use **tabulate**, I see both the variable name and its label:

```
. tabulate wc hc, missing
   wc: Wife |
   attended | hc: Husband attended
   college? |  college? 1=yes 0=no
 1=yes 0=no |   0_NoCol  1_College |     Total
------------+----------------------+----------
    0_NoCol |       417        124 |       541
  1_College |        41        171 |       212
------------+----------------------+----------
      Total |       458        295 |       753
```

I changed the variable labels without changing the names of the variables. In general, I think this is fine. If I wanted to keep the new labels, I would save these in a new dataset.

5.8 Adding notes to variables

The **notes** command attaches information to a variable that is saved in the dataset as metadata. **notes** is incredibly useful for documenting your work, and I highly recommend that you add a note when creating new variables. The syntax for **notes** is

`notes` [*varname*]: *text*

Here is how I routinely use **notes** when generating new variables. I start by creating **pub9trunc** from **pub9** and adding a variable label (file: **wf5-varnotes.do**):

```
. generate pub9trunc = pub9
(772 missing values generated)
. replace pub9trunc = 20 if pub9trunc>20 & !missing(pub9trunc)
(8 real changes made)
. label variable pub9trunc "Pub 9 truncated at 20: PhD yr 7 to 9"
```

I use **notes** to record how the variable was created, by what program, by whom, and when:

```
. notes pub9trunc: pubs>20 recoded to 20 \ wf5-varnotes.do jsl 2008-04-03.
```

The note is saved when I save the dataset. Later, if I want details on how the variable was created, I run the command:

```
. notes pub9trunc
pub9trunc:
  1.  pubs>20 recoded to 20 \ wf5-varnotes.do jsl 2008-04-03.
```

I can also add longer notes (up to 8,681 characters in Small Stata and 67,784 characters in other versions). For example,

```
. notes pub9trunc: Earlier analyses (pubreg04a.do 2006-09-20) showed
> that cases with a large number of articles were outliers. Program
> pubreg04b.do 2006-09-21 examined different transformations of pub9
> and found that truncation at 20 was most effective at removing
> the outliers. \ jsl 2008-04-03.
```

Now, when I check the notes for **pub9trunc**, I see both notes:

```
. notes pub9trunc
pub9trunc:
  1.  pubs>20 recoded to 20 \ wf5-varnotes.do jsl 2008-04-03.
  2.  Earlier analyses (pubreg04a.do 2006-09-20) showed that cases with a large
      number of articles were outliers. Program pubreg04b.do 2006-09-21 examined
      different transformations of pub9 and found that truncation at 20 was most
      effective at removing the outliers. \ jsl 2008-04-03.
```

With this information and my research log, I can easily reconstruct how and why I created the variable.

The **notes** command has an option to add a time stamp. In the text of the note, the letters TS (for time stamp) surrounded by blanks are replaced by the date and time. For example,

```
. notes pub9trunc: pub9 truncated at 20 \ wf5-varnotes.do jsl TS
. notes pub9trunc in 3
pub9trunc:
  3.  pub9 truncated at 20 \ wf5-varnotes.do jsl 3 Apr 2008 11:28
```

5.8.1 Commands for working with notes

Listing notes

To list all notes in a dataset, type

```
notes
```

To list the notes for selected variables, use the command

notes list *variable-list*

If you have multiple notes for a variable, they are numbered. To list notes from *start-#* to *end-#*:

notes list *variable-list* in *start-#*[/*end-#*]

For example, if **vignum** has many notes, I can look at just the second and third:

```
. notes list vignum in 2/3
vignum:
  2.  BGR - majority vs. minority = bulgarian vs. turk
  3.  ESP - majority vs. minority = spaniard vs. gypsy
```

You can also list notes with **codebook** using the **notes** option. For example,

```
. codebook pub1trunc, notes

-------------------------------------------------------------------------------
pub1trunc                                                           (unlabeled)
-------------------------------------------------------------------------------

                  type:  numeric (float)

                 range:  [0,20]                       units:  1
         unique values:  17                       missing .:  772/1080

                  mean:   2.53247
              std. dev:   3.00958

           percentiles:        10%       25%       50%       75%       90%
                                 0        .5         2         4         6

pub1trunc:
  1.  pubs# truncated at 20 \ wf5-varnotes.do jsl 2008-04-03.
```

Removing notes

To remove notes for a given variable, use the command

notes drop *variable-name* [in #[/#]]

where in #/# specifies which notes to drop. For example, notes drop vignum in 2/3.

Searching notes

Although there currently is no Stata command to search notes, this feature is planned for future versions of Stata. For now, the only way do this is to open a log file and run

```
notes
```

Then close the log and use a text editor to search the log file.

5.8.2 Using macros and loops with notes

You can use macros when creating notes. For example, to create similar notes for several variables, I use a local that I call tag with information for "tagging" each variable:

```
local tag "pub# truncated at 20 \ wf5-varnotes.do jsl 2008-04-09."
notes pub1trunc: `tag'
notes pub3trunc: `tag'
notes pub6trunc: `tag'
notes pub9trunc: `tag'
```

Then

```
. notes pub*
pub1trunc:
  1.  pub# truncated at 20 \ wf5-varnotes.do jsl 2008-04-09.
pub3trunc:
  1.  pub# truncated at 20 \ wf5-varnotes.do jsl 2008-04-09.
  (output omitted)
```

The advantage of using macros is that exactly the same information is added to each variable. You can also create notes within a loop. For example,

```
local tag "wf5-varnotes.do jsl 2008-04-09."
foreach varname in pub1 pub3 pub6 pub9 {
    clonevar  `varname'trunc = `varname'
    replace   `varname'trunc = 20 if `varname'trunc>20 ///
        & !missing(`varname'trunc)
    label var `varname'trunc "`varname' truncated at 20"
    notes     `varname'trunc: `varname' truncated at 20 \ `tag'
}
```

5.9 Value labels

Value labels assign text labels to the numeric values of a variable. The rule for value labels is

> *Categorical variables should have value labels unless the variable has an inherent metric.*

Although there is little benefit from having value labels for something like the number of young children in the family, a variable indicating attending college should be labeled. To see why labels are important, consider `k5`, which is the number of young children in the family, and `wc`, indicating whether the wife attended at least some college coded as 0 and 1. Without value labels, the tabulation of `wc` and `k5` looks like this (file: `wf5-vallabels.do`):

```
. tabulate wc_v1 k5
  Did wife |
    attend |          # of children younger than 6
  college? |         0          1          2          3 |     Total
-----------+--------------------------------------------+----------
         0 |       444         85         12          0 |       541
         1 |       162         33         14          3 |       212
-----------+--------------------------------------------+----------
     Total |       606        118         26          3 |       753
```

Although it is reasonable to assume that 1 stands for yes and 0 stands for no, what would you decide if the output looked like this?

```
. tabulate wc_v2 k5
  Did wife |
    attend |          # of children younger than 6
  college? |         0          1          2          3 |     Total
-----------+--------------------------------------------+----------
         1 |       444         85         12          0 |       541
         2 |       162         33         14          3 |       212
-----------+--------------------------------------------+----------
     Total |       606        118         26          3 |       753
```

A value label attaches a label to each value. Here I use a label that includes both the value and a description of the category:

```
. tabulate wc_v3 k5
  Did wife |
    attend |          # of children younger than 6
  college? |         0          1          2          3 |     Total
-----------+--------------------------------------------+----------
      0_No |       444         85         12          0 |       541
     1_Yes |       162         33         14          3 |       212
-----------+--------------------------------------------+----------
     Total |       606        118         26          3 |       753
```

5.9.1 Creating value labels is a two-step process

Stata assigns labels in two steps. In the first step, `label define` associates labels with values; that is, the labels are defined. In the second step, `label values` assigns a defined label to one or more variables.

Step 1: Defining labels

In the first step, I define a set of labels to be associated with values without indicating which variables use these labels. For yes/no questions with yes coded as 1 and no coded as 0, I could define the label as

```
label define yesno 1 yes 0 no
```

For a five-point scale with low values indicating negative responses, I could define

```
label define lowneg5 1 StDisagree 2 Disagree 3 Neutral 4 Agree 5 StAgree
```

For scales where low values are positive, I could define

```
label define lowpos5 1 StAgree 2 Agree 3 Neutral 4 Disagree 5 StDisagree
```

Step 2: Assigning labels

After labels are defined, `label values` assigns the defined labels to one or more variables. For example, because `wc` and `hc` are yes/no questions, I can use the label definition `yesno` for both variables:

```
label values wc yesno
label values hc yesno
```

Or, in the latest version of Stata 10, I can assign labels to both variables in one command:

```
label values wc hc yesno
```

Why a two-step system?

The primary advantage of a two-step system for creating value labels is that it facilitates having consistent labels across variables and simplifies making changes to labels used by multiple variables. For example, surveys often have many yes/no variables and many positively or negatively ordered five-point scales. For these three types of variables, I need three label definitions:

```
label define yesno 0 No 1 Yes
label define neg5  1 StDisagree 2 Disagree 3 Neutral 4 Agree    5 StAgree
label define pos5  1 StAgree    2 Agree    3 Neutral 4 Disagree 5 StDisagree
```

If I assign the `yesno` label to all yes/no questions, I know that these questions have exactly the same labels. The same holds for assigning `neg5` and `pos5` to variables that are negative or positive five-point scales. Defining labels only once makes it more likely that labels are assigned correctly.

This system also has advantages when changing value labels. Suppose that I want to shorten the labels and begin each label with its value. All I need to do is change the existing definitions using the `modify` option:

```
label define yesno 0 0No 1 1Yes, modify
label define neg5  1 1StDis    2 2Disagree 3 3Neutral ///
                   4 4Agree    5 5StAgree, modify
label define pos5  1 1StAgree  2 2Agree    3 3Neutral ///
                   4 4Disagree 5 5StDis, modify
```

The revised labels are automatically applied to all variables for which these definitions have been assigned.

Removing labels

To remove an assigned value label, use `label values` without specifying the label. For example, to remove the `yesno` label assigned to `wc`, type

```
label values wc
```

In the latest version of Stata 10, you can use a new syntax where a period indicates that the label is being removed:

```
label values wc .
```

Although I have removed the `yesno` label from `wc`, the label definition has not been deleted and can be used by other variables.

5.9.2 Principles for constructing value labels

You will save time and have clearer output if you plan value labels before you create them. Your plan should determine which variables can share labels, how missing values will be labeled, and what the content of your labels will be. As you plan your labels, here are some things to consider.

1) Keep labels short

Because value labels are truncated by some commands, notably `tabulate` and `tab1`, I recommend

Value labels should be eight or fewer characters in length.

Here's an example of what can happen if you use longer labels. I have created two label definitions that could be used to label variables measuring social distance (file: `wf5-vallabels.do`):

```
. labelbook sd_v1 sd_v2
------------------------------------------------------------------------------

value label sd_v1
  (output omitted)
  definition
           1   Definitely Willing
           2   Probably Willing
           3   Probably Unwilling
           4   Definitely Unwilling

   variables:  sdchild_v1
------------------------------------------------------------------------------

    value label sd_v2
  (output omitted)
  definition
           1   1Definite
           2   2Probably
           3   3ProbNot
           4   4DefNot

   variables:  sdchild_v2
```

The sd_v1 definitions use labels that are identical to the wording on the questionnaire. These labels were assigned to sdchild_v1. The sd_v2 labels are shorter and add the category number to the label; these were assigned to schild_v2. With tabulate, the original definitions are worthless:

```
. tabulate female sdchild_v1
      R is |         Q15 Would let X care for children
   female? | Definitel   Probably    Probably   Definitel |     Total
-----------+--------------------------------------------+----------
     0Male |        41         99         155         197 |       492
   1Female |        73         98         156         215 |       542
-----------+--------------------------------------------+----------
     Total |       114        197         311         412 |     1,034
```

The sd_v2 definitions are much better:

```
. tabulate female sdchild_v2
      R is |         Q15 Would let X care for children
   female? | 1Definite  2Probably    3ProbNot     4DefNot |     Total
-----------+--------------------------------------------+----------
     0Male |        41         99         155         197 |       492
   1Female |        73         98         156         215 |       542
-----------+--------------------------------------------+----------
     Total |       114        197         311         412 |     1,034
```

2) Include the category number

When looking at tabulated results, I often want to know the numeric value assigned to a category. You can see the values associated with labels by using the nolabel option of tabulate, but with this option, you no longer see the labels. For example,

```
. tabulate sdchild_v1, nolabel

  Q15 Would |
 let X care |
        for |
   children |      Freq.     Percent        Cum.
------------+-----------------------------------
          1 |        114       11.03       11.03
          2 |        197       19.05       30.08
          3 |        311       30.08       60.15
          4 |        412       39.85      100.00
------------+-----------------------------------
      Total |      1,034      100.00
```

A better solution is to use value labels that include both a label and the value for each category as illustrated with the label sd_v2.

Adding values to value labels

One way to include numeric values in value labels is to add them when you define the labels (file: wf5-vallabels.do):

```
label define defnot 1 1Definite 2 2Probably 3 3ProbNot 4 4DefNot
```

If you already have label definitions that do not include the values, you can use the numlabel command to add them. Suppose that I start with these labels:

```
label define defnot 1 Definite 2 Probably 3 ProbNot 4 DefNot
```

To add values to the front of the label, I use the command:

```
numlabel defnot, mask(#) add
```

Before explaining the command, let us look at the new labels:

```
. label val sdchild defnot

. tabulate sdchild

  Q15 Would |
 let X care |
        for |
   children |      Freq.     Percent        Cum.
------------+-----------------------------------
  1Definite |        114       11.03       11.03
  2Probably |        197       19.05       30.08
   3ProbNot |        311       30.08       60.15
    4DefNot |        412       39.85      100.00
------------+-----------------------------------
      Total |      1,034      100.00
```

The mask() option for numlabel controls how the values are added. The mask(#) option adds only numbers (e.g., 1Definite); mask(#_) adds numbers followed by an underscore (e.g., 1_Definite); and mask(#.) adds the values followed by a period and a space (e.g., 1. Definite).

You can remove values from labels with the `remove` option. For example, `numlabel defnot, mask(#_) remove` removes values that are followed by an underscore.

Creating new labels before adding numbers

The `numlabel` command changes existing labels. Once they are changed, the original labels are no longer in the dataset. This can be a problem if you want to replicate prior results. With the `label copy` command, added in the February 25, 2008 update of Stata 10, you can solve this problem by making copies of the original labels. For example, I can create a new value label definition named `defnotNew` that is an exact copy of `defnot`:

```
label copy defnot defnotNew
```

Then I revise the copy, leaving the original label intact:

```
. numlabel defnotNew, mask(#_) add

. label val sdchild defnotNew

. tabulate sdchild

  Q15 Would
 let X care
        for
   children |      Freq.     Percent        Cum.
------------+-----------------------------------
 1_Definite |        114       11.03       11.03
 2_Probably |        197       19.05       30.08
  3_ProbNot |        311       30.08       60.15
   4_DefNot |        412       39.85      100.00
------------+-----------------------------------
      Total |      1,034      100.00
```

To reassign the original labels,

```
. label val sdchild defnot

. tabulate sdchild

  Q15 Would
 let X care
        for
   children |      Freq.     Percent        Cum.
------------+-----------------------------------
   Definite |        114       11.03       11.03
   Probably |        197       19.05       30.08
    ProbNot |        311       30.08       60.15
     DefNot |        412       39.85      100.00
------------+-----------------------------------
      Total |      1,034      100.00
```

3) Avoid special characters

Adding spaces and characters such as `.`, `:`, `=`, `%`, `@`, `{`, and `}` to labels can cause problems with some commands (e.g., `hausman`), even though `label define` allows you to use these characters in your labels. To avoid problems, I suggest that you use only letters, numbers, dashes, and underscores. If you include spaces, you must have quotes around your labels. For example, you need quotes here

```
label define yesno_v2 1 "1 yes" 0 "0 no"
```

but not here

```
label define yesno_v3 1 1_yes 0 0_no
```

4) Keeping track of where labels are used

The two-step system for labels can cause problems if you do not keep track of which labels are assigned to which variables. Suppose `female` is coded 1 for female and 0 for male and `lfp` is coded 1 for being in the labor force and 0 for not. I could label the values for both variables as yes and no:

```
label define twocat 0 No 1 Yes
label values lfp female twocat
```

When I tabulate the variables, I get the table I want

```
. tabulate female lfp

      R is |  Paid labor force?
   female? |        No        Yes |     Total
-----------+----------------------+----------
        No |       149        196 |       345
       Yes |       176        232 |       408
-----------+----------------------+----------
     Total |       325        428 |       753
```

Later I decide that it would be more convenient to label `female` with `0male` and `1female`. Forgetting that the label `twocat` is also used by `lfp`, I change the label definition:

```
label define twocat 0 0_Male 1 1_Female, modify
```

This works fine for `female` but causes a problem with `lfp`:

```
. tabulate female lfp

      R is |  Paid labor force?
   female? |    0_Male   1_Female |     Total
-----------+----------------------+----------
    0_Male |       149        196 |       345
  1_Female |       176        232 |       408
-----------+----------------------+----------
     Total |       325        428 |       753
```

To keep track of whether a label is used for one variable or many variables, I use these rules:

> *If a value label is assigned to only one variable, the label definition should have the same name as the variable.*

If a value label is assigned to multiple variables, the name of the label definition should begin with L.

For example, I would define `label define female 0 0_Male 1 1_Female` and use it with the variable `female`. I would define `label define Lyesno 1 1_Yes 0 0_No` to remind me that if I change the definition of `Lyesno` I need to verify that the change is appropriate for all the variables using this definition.

5.9.3 Cleaning value labels

There are several commands that make it easier to review and revise value labels. The commands `describe` and `nmlab` list variables along with the name of their value labels. The `codebook, problems` command searches for problems in your dataset, including some related to value labels. I highly recommend using it; see section 6.4.6 for further details. Two other commands provide lists of labels. The `label dir` command lists the names of all value labels that have been defined. For example,

```
. label dir
vignum
serious
female
wc_v3
Lyesno
Ldefnot
Ltenpt
lfp
Lyn
```

This list includes defined labels even if they have not been assigned to a variable with `label values`. The `labelbook` command lists all labels, their characteristics, and the variables to which they are assigned. For example,

```
. labelbook Ltenpt
---------------------------------------------------------------------------
value label Ltenpt
---------------------------------------------------------------------------

      values                                        labels
       range:  [1,10]                     string length:  [6,16]
           N:  5                   unique at full length:  yes
        gaps:  yes                   unique at length 12:  yes
  missing .*:  3                             null string:  no
                                 leading/trailing blanks:  no
                                      numeric -> numeric:  no
  definition
           1   1Not at all Impt
          10   10Vry Impt
          .a   .a_NAP
          .c   .c_Dont know
          .d   .d_No ansr, ref

   variables:  tcfam tc1fam tc2fam tc3fam tc1friend tc2friend tc3friend
               tc1relig tc2relig tc3relig tc1doc tc2doc tc3doc tc1psy tc2psy
               tc3psy tc1mhprof tc2mhprof tc3mhprof
```

5.9.4 Consistent value labels for missing values

Labels for missing values need to be considered carefully. Stata uses the sysmiss `.` and 26 extended missing values `.a`–`.z` (see section 6.2.3 for more information on missing values). Having multiple missing values allows you to code the reason why information is missing. For example,

- The respondent did not know the answer.
- The respondent refused to answer.
- The respondent did not answer the current question because the lead-in question was refused.
- The question was not appropriate for the respondent (e.g., asking children how many cars they own).
- The respondent was not asked the question (e.g., random assignment of who gets asked which questions).

You can prevent confusion by using the same missing-value codes to mean the same things across questions. If you are collecting your own data, you can do this when developing rules for coding the data. If you are using data collected by others, you might find that the same codes are used throughout or you might need to reassign missing values to make them uniform (see section 5.11.4 for an example). In my work, I generally associate these meanings to the missing-values codes in table 5.2:

Table 5.2. Suggested meanings for extended missing-value codes

Letter	Meaning	Example
`.`	Unspecified missing value	Missing data without the reason being made explicit
`.d`	Don't know	Respondent did not know the answer
`.l`	– Do not use this code –	l (lowercase L) is too close to 1 (one) so avoid it
`.n`	Not applicable	Only adults were asked this question
`.p`	Preliminary question refused	Question 5 was not asked because respondent did not answer the lead-in question
`.r`	Refused	Respondent refused to answer question
`.s`	Skipped due to skip pattern	Given answer to question 5, question 6 was not asked
`.t`	Technical problem	Error reading data from questionnaire

5.9.5 Using loops when assigning value labels

The `foreach` command is very effective for adding the same value labels to multiple variables. Suppose that I want to recode the 4-point scales `sdneighb`, `sdsocial`, `sdchild`, `sdfriend`, `sdwork`, and `sdmarry` to binary variables that indicate whether the respondent agrees or disagrees with the question. First, I define a new label (file: `wf5-vallabels.do`):

```
label define Lagree 1 1_Agree 0 0_Disagree
```

Then I use a `foreach` loop to create new variables and add labels:

```
1>  foreach varname in sdneighb sdsocial sdchild sdfriend sdwork sdmarry {
2>      display      _newline "--> Recoding variable `varname'" _newline
3>      clonevar     B`varname' = `varname'
4>      recode       B`varname' 1/2=1 3/4=0
5>      label values B`varname' Lagree
6>      tabulate     B`varname' `varname', miss
7>  }
```

Line 1 creates the local `varname` that holds the name of the variable to recode. The first time through the loop `varname` contains `sdneighb`. Line 2 displays a header indicating which variable is being processed (sample output is given below). The `_newline` directive adds a blank line to improve readability. Line 3 creates the variable `Bsdneighb` as a clone of the source variable `sdneighb`; the variables are identical except for name. Line 4 combines values 1 and 2 into the value 1 and values 3 and 4 into the value 0. Line 5 assigns the value label `Lagree` to `Bsdneighb`. Line 6 tabulates the new `Bsdneighb` with the source `sdneighb`. Line 7 ends the loop. The output for the first pass through the loop is

```
--> Recoding variable sdneighb
(20 missing values generated)
(Bsdneighb: 670 changes made)
```

Q13 Would have X as neighbor	Q13 Would have X as neighbor					
	1Definite	2Probably	3ProbNot	4DefNot	.c_DK	Total
0_Disagree	0	0	133	61	0	194
1_Agree	390	476	0	0	0	866
.c	0	0	0	0	20	20
Total	390	476	133	61	20	1,080

The message 20 `missing values generated` means that when `Bsdneighb` was cloned there were 20 cases with missing values in the source variable. Although `.c` had the label `.c_DK` in the value label used for `sdneighb`, the value labels for the recoded variable do not include a label for `.c`. I could revise the label definition to add this label:

```
label define Lagree 1 1_agree 0 0_disagree .c .c_DK .d .d_NA_ref, modify
```

The message `Bsdneighb: 670 changes made` was generated by `recode` to indicate how many cases were changed when the recodes were made. The program can be improved by adding notes and variable labels:

```
1>  local tag "wf5-vallabels.do jsl 2008-04-03."
2>  foreach varname in sdneighb sdsocial sdchild sdfriend sdwork sdmarry {
3>      display      _newline "--> Recoding variable `varname'" _newline
4>      clonevar     B`varname' = `varname'
5>      recode       B`varname' 1/2=1 3/4=0
6>      label values B`varname' Lagree
7>      notes        B`varname': "Recode of `varname' \ `tag'"
8>      label var    B`varname' "Binary version of `varname'"
9>      tabulate     B`varname' `varname'
10> }
```

Line 1 creates a local used by `notes` in line 7. The variable label in line 8 describes where the variable came from.

5.10 Using multiple languages

The language facility allows you to have multiple sets of labels saved within one dataset. Most obviously you can have labels in more than one language. For example, I have created a dataset with labels in Spanish, English, and French (I discuss how to do this later). If I want labels in English, I select that language and then run the commands as I normally would (file: `wf5-language.do`):

```
. use wf-languages-spoken, clear
(Workflow data with spoken languages \ 2008-04-03)
. label language english
. tabulate male, missing

  Gender of |
 respondent |      Freq.     Percent        Cum.
------------+-----------------------------------
    0_Women |      1,227       53.51       53.51
      1_Men |      1,066       46.49      100.00
------------+-----------------------------------
      Total |      2,293      100.00
```

If I want labels in French, I specify French:

```
. label language french
. tabulate male, missing

   Genre de |
 répondant |      Freq.     Percent        Cum.
------------+-----------------------------------
   0_Femmes |      1,227       53.51       53.51
   1_Hommes |      1,066       46.49      100.00
------------+-----------------------------------
      Total |      2,293      100.00
```

When I first read about `label language`, I thought about it only in terms of languages such as French and German. When documenting and archiving the data collected by Alfred Kinsey, we faced the problem that some of the labels in the original dataset had inconsistencies or small errors. We wanted to fix these, but we also wanted to preserve the original labels. The solution was to use multiple languages. We let `label language original` include the historical labels, whereas `label language revised` incorporated our changes. In the same way, you can create a short and long language for your dataset. The long version could have labels that match the survey instrument. The short version could use labels that are more effective for analysis.

(Continued on next page)

5.10.1 Using label language for different written languages

To create a new language, you indicate the name for the new language and then create labels as you normally would. A simple example shows you how to do this. I start by loading a dataset with only English labels and add French and Spanish labels:

```
. use wf-languages-single, clear
. * french
. label language french, new
. label define male_fr 0 "0_Femmes" 1 "1_Hommes"
. label val male male_fr
. label var male "Genre de répondant"
. * spanish
. label language spanish, new
. label define male_es 0 "0_Mujeres" 1 "1_Hombres"
. label val male male_es
. label var male "Género del respondedor"
```

When you save the dataset, labels are saved for three languages. As far as I know, Stata is the only data format with multiple languages. If you convert a Stata dataset with multiple languages to other formats, you will have to create distinct datasets for each language.

5.10.2 Using label language for short and long labels

Stata's `label language` feature is a great solution to the trade-off between labels that correspond to the data source (e.g., the survey instrument) and labels that are convenient for analysis. For analysis, shorter labels are often more useful, but for documentation, you might want to know exactly how the questions were asked. Here is a simple example of how `label language` can address this dilemma. First, I load the data and set the language to `source` to use the labels based on the source questionnaire (file: `wf5-language.do`):

```
. use wf-languages-analysis, clear
(Workflow data with analysis and source labels \ 2008-04-03)
. label language source
```

Using `describe`, I look at two variables:

```
. describe male warm
              storage  display     value
variable name   type   format      label      variable label
-------------------------------------------------------------------------------
male            byte   %10.0g      Smale      Gender
warm            byte   %17.0g      Swarm      A working mother can establish
                                                just as warm and secure a
                                                relationship with her c
```

The value labels begin with `S` that I used to indicate that these are the source labels. If I tabulate the variables, I get results using source labels:

```
. tabulate male warm, missing

           | A working mother can establish just as warm
           |    and secure a relationship with her c
    Gender |  Strongly      Agree   Disagree  Strongly |     Total
-----------+--------------------------------------------+----------
    Female |       139        323        461        304 |     1,227
      Male |       158        400        395        113 |     1,066
-----------+--------------------------------------------+----------
     Total |       297        723        856        417 |     2,293
```

These labels are too long to be useful. Next I switch to the labels I created for analyzing the data:

```
. label language analysis

. describe male warm

              storage  display     value
variable name   type   format      label      variable label
-------------------------------------------------------------------------------
male            byte   %10.0g      Amale      Gender: 1=male 0=female
warm            byte   %17.0g      Awarm      Mom can have warm relations with
                                                child?
```

The value and variable labels have changed. When I tabulate the variables, the results are much clearer:

```
. tabulate male warm, missing

   Gender: |
    1=male |   Mom can have warm relations with child?
  0=female |      1_SD        2_D        3_A       4_SA |     Total
-----------+--------------------------------------------+----------
   0_Women |       139        323        461        304 |     1,227
     1_Men |       158        400        395        113 |     1,066
-----------+--------------------------------------------+----------
     Total |       297        723        856        417 |     2,293
```

If I need the original labels, I simply change the language with the command `label language source`.

Note on variable and value labels

There is an important difference in how variable and value labels are treated with languages. After changing to the `analysis` language, I simply created new variable labels. For value labels, I had to define labels with different names than they had before. For example, in `wf-languages-analysis.dta`, the label assigned to `warm` was named `Swarm` (where `S` indicates that this is the source label). In the `analysis` language, the label was named `Awarm`. With multiple languages, you must create new value-label definitions for each language.

5.11 A workflow for names and labels

This section provides an extended example of how to revise names and labels using the tools for automation that were introduced in chapter 4. The example is taken from research with Bernice Pescosolido and Jack Martin on a 17-country survey of stigma and mental health. The data we received had nonmnemonic variable names with labels that closely matched the questionnaire. Initial analyses showed that the names were inconsistent and sometimes misleading with labels that were often truncated or unclear in the output. Accordingly, we undertook a major revision of names and labels that took months to complete.[5]

Because we needed to revise 17 datasets with thousands of variables, we spent a great deal of time planning the work and perfecting the methods we used. To speed up the process of entering thousands of `rename`, `label variable`, `label define`, and `label values` commands, we used automation tools to create dummy commands that were the starting point for the commands we needed. To understand the rest of this section, it is essential to understand how dummy commands were used. Suppose that I need the following `rename` commands:

```
rename atdis   atdisease
rename atgenes atgenet
rename ctxfdoc clawdoc
rename ctxfhos clawhosp
rename ctxfmed clawpmed
```

Instead of typing each command from scratch, I create a list of dummy `rename` commands that looks like this:

```
rename atdis   atdis
rename atgenes atgenes
rename ctxfdoc ctxfdoc
rename ctxfhos ctfxhos
rename ctxfmed ctxfmed
```

The dummy commands are edited to create the commands I need. Before getting into the specific details, I want to provide an overview of the five steps and 11 do-files required.

Step 1: Plan the changes

The first step is planning the new names and labels. I start with a list of the current names and labels:

`wf5-sgc1a-list.do`:
List names and labels from the source dataset `wf-sgc-source.dta`.

5. The data from each country consisted of about 150 variables with some variation of content across countries. For the first country, our data manager estimates that it took a month to revise the names and labels and verify the data. Later countries took four to five days. The data used for this example are artificial but have similar names, labels, and content to the real data that have not yet been released.

This information is exported to a spreadsheet used to plan the changes. To decide what changes to make, I check how the current names and labels appear in Stata output:

`wf5-sgc1b-try.do`:
Try the names and labels with `tabulate`.

Step 2: Archive, clone, and rename

Before making any changes, I back up the source dataset. Because I want to keep the original variables in the revised dataset, I create clones:

`wf5-sgc2a-clone.do`:
Add cloned variables and create `wf-sgc01.dta`.

Next I create a file with dummy `rename` commands:

`wf5-sgc2b-rename-dump.do`:
Create a file with `rename` commands.

I edit the file with `rename` commands and use it to rename the variables:

`wf5-sgc2c-rename.do`:
Rename variables and create `wf-sgc02.dta`.

Step 3: Revise variable labels

The original variable labels are used to create dummy commands:

`wf5-sgc3a-varlab-dump.do`:
Use a loop and extended functions to create a file with `label variable` commands.

Before adding new labels, I save the original labels as a second language called `original`. The revised labels are saved in the `default` language:

`wf5-sgc3b-varlab-revise.do`:
Create the `original` language for the original variable labels and save the revised labels in the `default` language to create `wf-sgc03.dta`.

Step 4: Revise value labels

Changing value labels is more complicated than changing variable labels due to the two-step process used to label values. I start by examining the current value labels to determine which variables could share label definitions and how to handle missing values:

`wf5-sgc4a-vallab-check.do`:
List current value labels for review.

To create new value labels, I create dummy `label define` and `label values` commands:

`wf5-sgc4b-vallab-dump.do`:
Create a file with `label define` and `label values` commands.

The edited commands for value labels are used to create a new dataset:

`wf5-sgc4c-vallab-revise.do`:
Add new value labels to the `default` language and save `wf-sgc04.dta`.

Step 5: Verify the changes

Before finishing, I ask everyone on the research team to check the revised names and labels, and then steps 2–4 are repeated as needed.

`wf5-sgc5a-check.do`:
Check the names and labels by trying them with Stata commands.

When everyone agrees on the new names and labels, `wf-sgc04.dta` and the do-files and dataset are posted.

With this overview in mind, we can get into the details of making the changes.

5.11.1 Step 1: Check the source data

Step 1a: List the current names and labels

First, I load the source data and check the data signature (file: `wf5-sgc1a-list.do`):

```
. use wf-sgc-source, clear
(Workflow data for SGC renaming example \ 2008-04-03)
. datasignature confirm
  (data unchanged since 03apr2008 13:25)
. notes _dta
_dta:
  1.  wf-sgc-source.dta \ wf-sgc-support.do jsl 2008-04-03.
```

The `unab` command creates the macro `varlist` with the names of all variables:

```
. unab varlist : _all
. display "`varlist´"
id_iu cntry_iu vignum serious opfam opfriend tospi tonpm oppme opforg atdisease
> atraised atgenes sdlive sdsocial sdchild sdfriend sdwork sdmarry impown imptre
> at stout stfriend stlimits stuncom tcfam tcfriend tcdoc gvjob gvhealth gvhous
> gvdisben ctxfdoc ctxfmed ctxfhos cause puboften pubfright pubsymp trust gender
> age wrkstat marital edudeg
```

Using this list, I loop through each variable and display its name, value label, and variable label. Before generating the list, I `set linesize 120` so that long variable labels are not wrapped. Here is the loop:

```
1> local counter = 1
2> foreach varname in `varlist´ {
3>     local varlabel : variable label `varname´
4>     local vallabel : value label `varname´
5>     display "`counter´." _col(6) "`varname´" _col(19) ///
 >         "`vallabel´" _col(32) "`varlabel´"
6>     local ++counter
7> }
```

Before explaining the loop, it helps to see some of the output:

```
1.   id_iu                     Respondent Number
2.   cntry_iu     cntry_iu     IU Country Number
3.   vignum       vignum       Vignette
4.   serious      serious      Q1 How serious would you consider Xs situation to be?
5.   opfam        Ldummy       Q2_1 What X should do:Talk to family
6.   opfriend     Ldummy       Q2_2 What X should do:Talk to friends
7.   tospi        Ldummy       Q2_7 What X should do:Go to spiritual or traditional healer
8.   tonpm        Ldummy       Q2_8 What X should do:Take nonprescription medication
  (output omitted)
```

Returning to the program, line 1 initiates a counter for numbering the variables. Line 2 begins the loop through the variable names in `varlist` and creates the local `varname` with the name of the current variable. Line 3 is an extended macro function that creates the local `varlabel` with the variable label for the variable in `varname` (see page 159 for further details). Line 4 uses another extended macro function to retrieve the name of the value-label definition. Line 5 displays the results, line 6 adds one to the counter, and line 7 ends the loop.

Although I could use this list to plan my changes, I prefer a spreadsheet where I can sort and annotate the information. To move this information into a spreadsheet, I create a text file, where the columns of data are separated by a delimiter (i.e., a character designated to indicate a new column of data). Although commas are commonly used as delimiters, I use a semicolon because some labels contain commas. The first five lines of the file I created look like this:

```
Number;Name;Value label;Variable labels
1;id_iu;;Respondent Number
2;cntry_iu;cntry_iu;IU Country Number
3;vignum;vignum;Vignette
4;serious;serious;Q1 How serious would you consider Xs situation to be?
```

To create a text file, I need to tell the operating system to open the file named `wf5-sgc1a-list.txt`. The commands that write to this file refer to it by a shorter name, a nickname if you will, called a *file handle*. I chose `myfile` as the file handle. This means that referring to `myfile` is the same as referring to `wf5-sgc1a-list.txt`. Before opening `myfile`, I need to make sure that the file is not already open. I do this with the command `capture file close myfile`, which tells the operating system to close any file named `myfile` that is open. `capture` means that if the file is not open, ignore the error that is generated when you try to close a file that is not open. Next the `file open` command creates the file:

```
capture file close myfile
file open myfile using wf5-sgc1a-list.txt, write replace
```

The options `write` and `replace` mean that I want to write to the file (not just read the file) and if the file exists, replace it. Here is the loop that writes to the file:

```
1>  file write myfile "Number;Name;Value label;Variable labels" _newline
2>  local counter = 1
3>  foreach varname in `varlist´ {
4>      local varlabel : variable label `varname´
5>      local vallabel : value label `varname´
6>      file write myfile "`counter´;`varname´;`vallabel´;`varlabel´" _newline
7>      local ++counter
8>  }
9>  file close myfile
```

Line 1 writes an initial line with labels for each column: Number, Name, Value label, and Variable labels. Lines 2–5 are the same as the commands used in the loop on page 179. Line 6 replaces `display` with `file write`, where `_newline` starts a new line in the file. The string `"`counter´;`varname´;`vallabel´;`varlabel´"` combines three local macros with semicolons in between. Line 7 increments the counter by 1, and line 8 closes the foreach loop. Line 9 closes the file. I import the file to a spreadsheet program, here Excel, where the data look like this (file: `wf5-sgc1a-list.xls`):

	A	B	C	D
1	Number	Name	Value label	Variable labels
2	1	id_iu		Respondent Number
3	2	cntry_iu	cntry_iu	IU Country Number
4	3	vignum	vignum	Vignette
5	4	serious	serious	Q1 How serious would you consider Xs situation to be?
6	5	opfam	Ldummy	Q2_1 What X should do:Talk to family
7	6	opfriend	Ldummy	Q2_2 What X should do:Talk to friends
8	7	tospi	Ldummy	Q2_7 What X should do:Go to spiritual or traditional healer
9	8	tonpm	Ldummy	Q2_8 What X should do:Take non-prescription medication
10	9	oppme	Ldummy	Q2_9 What X should do:Take prescription medication

I use this spreadsheet to plan and document the changes I want to make.

Step 1b: Try the current names and labels

To determine how well the current names and labels work, I start with `codebook, compact` (file: `wf5-sgc1b-try.do`):

```
. codebook, compact
Variable    Obs Unique      Mean      Min      Max  Label
-------------------------------------------------------------------------------
id_iu       200    200   1772875  1100107  2601091  Respondent Number
cntry_iu    200      8    17.495       11       26  IU Country Number
vignum      200     12     6.305        1       12  Vignette
serious     196      4  1.709184        1        4  Q1 How serious would you c...
opfam       199      2  1.693467        1        2  Q2_1 What X should do:Talk...
opfriend    198      2  1.833333        1        2  Q2_2 What X should do:Talk...
  (output omitted)
```

The labels for `opfam` and `opfriend` show that truncation is a problem. Next I use a loop to run `tabulate` with each variable, quickly showing problems with the value labels. I start by dropping the ID variables and `age` because they have too many unique values to tabulate and then create a macro `varlist` with the names of the remaining variables:

```
drop id_iu cntry_iu age
unab varlist : _all
```

The loop is simple:

```
1>  foreach varname in `varlist´ {
2>      display  "`varname´:"
3>      tabulate gender `varname´, miss
4>  }
```

Line 2 prints the name of the variable (because `tabulate` does not tell you the name of a variable if there is a variable label). Line 3 tabulates `gender` against the current variable from the `foreach` loop. I use `gender` as the row variable because it has only two categories, making the tables small. The loop produces tables like this:

```
vignum:

           |                           Vignette
    Gender | Depressiv  Depressiv  Depressiv  Depressiv  Schizophr |     Total
-----------+-------------------------------------------------------+----------
      Male |        15         11          3          4          7 |        90
    Female |         8         12          9          5         13 |       110
-----------+-------------------------------------------------------+----------
     Total |        23         23         12          9         20 |       200
  (output omitted)
```

Clearly, the value labels for `vignum` need to be changed. Here is another example where the truncated category labels are a problem:

(Continued on next page)

sdlive:

Gender	Q13 To have X as a neighbor? Definitel	Probably	Probably	Definitel	.c	Total
Male	39	32	10	4	4	90
Female	45	51	9	5	0	110
Total	84	83	19	9	4	200

Gender	Q13 To have X as a neighbor? .d	Total
Male	1	90
Female	0	110
Total	1	200

Other labels have less serious problems. For example, here I can tell what each category means, but the labels are hard to read:

serious:

Gender	Q1 How serious would you consider Xs situation to be? Very seri	Moderatel	Not very	Not at al	.c	Total
Male	42	37	8	2	1	90
Female	49	38	18	2	3	110
Total	91	75	26	4	4	200

Or, for `trust`, I find the variable label is too long and the value labels are unclear:

trust:

Gender	Q75 Would you say people can be trusted or need to be careful dealing w/people? Most peop	Need to b	.a	.c	.d	Total
Male	14	47	29	0	0	90
Female	13	71	24	1	1	110
Total	27	118	53	1	1	200

As I go through the output, I add notes to the spreadsheet and plan the changes that I want.

5.11.2 Step 2: Create clones and rename variables

When you rename and relabel variables, mistakes can happen. To prevent loss of critical information, I back up the data as described in chapter 8. I also create clones of the original variables that I keep in the dataset to compare them to the variables with revised names and labels. For example, if the source variable is `vignum`, I create the clone `Svignum` (where `S` stands for source variable). I can delete these variables later or keep them in the final dataset. Next I run a pair of programs to rename variables.

Step 2a: Create clones

I start by defining a tag macro to use when adding notes to variables (file: `wf5-sgc2a-clone.do`). The tag includes only that part of the do-file name that is necessary to uniquely identify it:

```
local tag "wf5-sgc2a.do jsl 2008-04-09."
```

Next I load the dataset and check the signature:

```
. use wf-sgc-source, clear
(Workflow data for SGC renaming example \ 2008-04-03)
. datasignature confirm
  (data unchanged since 03apr2008 13:25)
```

To create clones, I use a `foreach` loop that is similar to that used in step 1:

```
1> unab varlist : _all
2> foreach varname in `varlist' {
3>     clonevar S`varname' = `varname'
4>     notes S`varname': Source variable for `varname' \ `tag'
5>     notes `varname': Clone of source variable S`varname' \ `tag'
6> }
```

Line 3 creates a clone whose name begins with `S` and ends with the name of the source variable. Line 4 adds a note to the clone using the local `tag` to add the name of the do-file, the date it was run, and who ran it. Line 5 adds a note to the original variable. (To test your understanding of how `notes` works, think about what would happen if line 5 was placed immediately after line 2.) All that remains is to sign and save the dataset:

```
. note: wf-sgc01.dta \ create clones of source variables \ `tag'
. label data "Workflow data for SGC renaming example \ 2008-04-09"
. datasignature set, reset
  200:90(85238):981823927:1981917236       (data signature reset)
. save wf-sgc01, replace
file wf-sgc01.dta saved
```

Step 2b: Create rename commands

The `rename` command is used to rename variables:

`rename` *old_varname new_varname*

For example, to rename `VAR06` to `var06`, the command is `rename VAR06 var06`. To rename the variables in `wf-sgc01.dta`, I begin by creating a file that contains dummy `rename` commands that I can edit. For example, I create the command `rename atgenes atgenes` that I revise to `rename atgenes atgenet`. I start by loading the dataset and verifying the data signature (file: `wf5-sgc2b-rename-dump.do`):

```
. use wf-sgc01, clear
(Workflow data for SGC renaming example \ 2008-04-09)
. datasignature confirm
  (data unchanged since 09apr2008 14:12)
. notes _dta
_dta:
  1.  wf-sgc-source.dta \ wf-sgc-support.do jsl 2008-04-03.
  2.  wf-sgc01.dta \ create clones of source variables \ wf5-sgc2a.do jsl
      2008-04-09.
```

Next I drop the clones (that I do not want to rename) and alphabetize the remaining variables:

```
drop S*
aorder
```

I use a loop to create the text file `wf5-sgc2b-rename-dummy.doi` with dummy `rename` commands that I edit and include in step 2c:

```
unab varlist : _all
file open myfile using wf5-sgc2b-rename-dummy.doi, write replace
foreach varname in `varlist´ {
    file write myfile "*rename  `varname´" _col(22) "`varname´" _newline
}
file close myfile
```

I use the `file write` command to write commands to the `.doi` file. I preface the commands in the `.doi` file with `*` so that they are commented out. If I want to rename a variable, I remove the `*` and edit the command. The output file looks like this:

```
*rename  age          age
*rename  atdisease    atdisease
*rename  atgenes      atgenes
  (output omitted)
```

I copy `wf5-sgc2b-rename-dummy.doi` to `wf5-sgc2b-rename-revised.doi` and edit the dumped commands.

Step 2c: Rename variables

The do-file to rename variables starts by creating a tag and checking the source data (file: `wf5-sgc2c-rename.do`):

```
local tag "wf5-sgc2c.do jsl 2008-04-09."
use wf-sgc01, clear
datasignature confirm
notes _dta
```

Next I `include` the edited `rename` commands:

```
include wf5-sgc2b-rename-revised.doi
```

For variables that I do not want to rename (e.g., age), I leave the * so that the line is a comment. I could delete these but decide to leave them in case I later change my mind. Here are the names that changed:

Original		Revised
atgenes	⇒	atgenet
ctxfdoc	⇒	clawdoc
ctxfhos	⇒	clawhosp
ctxfmed	⇒	clawpmed
gvdisben	⇒	gvdisab
gvhous	⇒	gvhouse
opforg	⇒	opforget
oppme	⇒	oppremed
pubfright	⇒	pubfrght
sdlive	⇒	sdneighb
stuncom	⇒	stuncmft
tonpm	⇒	opnomed
tospi	⇒	opspirit

Why were these variables renamed? atgenes was changed to atgenet because genet is the abbreviation for genetics used in other names. ctxf refers to "coerced treatment forced", which is awkward compared with claw for "coerced by law". hos was changed to hosp, which is a clearer abbreviation for hospital; med was changed to pmed to indicate psychopharmacological medications. Next the dataset is saved with a new name:

```
. note: wf-sgc02.dta \ rename source variables \ `tag'
. label data "Workflow data for SGC renaming example \ 2008-04-09"
. datasignature set, reset
  200:90(109624):981823927:1981917236        (data signature reset)
. save wf-sgc02, replace
file wf-sgc02.dta saved
```

I check the new names using nmlab, summarize, or codebook, compact.

5.11.3 Step 3: Revise variable labels

Based on my review of variable labels in step 1, I decided to revise some variable labels.

Step 3a: Create variable-label commands

First, I use existing variable labels to create dummy label variables commands (file: wf5-sgc3a-varlab-dump.do). As in step 2b, I load the dataset, drop the cloned variables, sort the remaining variables, and create a local with the names of the variables:

```
use wf-sgc02.dta
datasignature confirm
drop S*
aorder
unab varlist : _all
```

Next I open a text file that will hold the dummy-variable labels. As before, I loop through `varlist` and use an extended macro function to retrieve the variable labels. The `file write` command sends the information to the file:

```
file open myfile using wf5-sgc3a-varlab-dummy.doi, write replace
foreach varname in `varlist´ {
    local varlabel : variable label `varname´
    file write myfile "label var  `varname´ " _col(24) `""`varlabel´""´ _newline
}
file close myfile
```

The only tricky thing is putting double quotes around the variable labels. That is, I want to write `"Current employment status"` not just `Current employment status`. This is done with the code: `` `""`varlabel´""´ ``. At the center, `` "`varlabel´" `` inserts the variable label, such as `Current employment status`, where the double quotes are standard syntax for enclosing strings. To write quote marks as opposed to using them to delimit a string, the characters `` `" `` and `"´` are used. The resulting file looks like this:

```
label var  age          "Age"
label var  atdisease    "Q4 Xs situation is caused by: A brain disease or disorder"
label var  atgenet      "Q7 Xs situation is caused by: A genetic or inherited problem"
label var  atraised     "Q5 Xs situation is caused by: the way X was raised"
label var  cause        "Q62 Is Xs situation caused by depression, asthma, schizophrenia, s
  (output omitted)
```

I copy `wf5-sgc3a-varlab-dummy.doi` to `wf5-sgc3a-varlab-revised.doi` and edit the dummy commands to be used in step 3b.

Step 3b: Revise variable labels

The next do-file adds revised variable labels to the dataset (file: `wf5-sgc3b-varlab-revise.do`). I start by creating a tag, then I load and verify the data:

```
local tag "wf5-sgc3b.do jsl 2008-04-09."
use wf-sgc02, clear
datasignature confirm
notes _dta
```

Although I want to create better labels, I do not want to lose the original labels, so I use Stata's language capability. By default, a dataset uses a language called `default`. I created a second language called `original` (for the original, unrevised variable labels) that is a copy of the `default` language before that language is changed:

```
label language original, new copy
```

With a copy of the original labels saved, I go back to the `default` language where I will change the labels:

```
label language default
```

To document how the languages were created, I add a note:

```
note: language original uses the original, unrevised labels; language ///
    default uses revised labels \ `tag´
```

Next I include the edited file with variable labels:

```
include wf5-sgc3a-varlab-revised.doi
```

The commands in the `include` file look like this:

```
label var age         "Age in years"
label var atdiseas    "Q04 Cause is brain disorder"
label var atgenet     "Q07 Cause is genetic"
label var atraised    "Q05 Cause is way X was raised"
label var cause       "Q62 Xs situation cased by what?"
  (output omitted)
```

With the changes made, I save the data:

```
. note: wf-sgc03.dta \ revised var labels for source & default languages \ `tag´
. label data "Workflow data for SGC renaming example \ 2008-04-09"
. datasignature set, reset
  200:90(109624):981823927:1981917236       (data signature reset)
. save wf-sgc03, replace
file wf-sgc03.dta saved
```

To check the new labels in the `default` language, I use `nmlab`:

```
. nmlab tcfam tcfriend vignum
tcfam     Q43 Family help important?
tcfriend  Q44 Friends help important?
vignum    Vignette number
```

To see the original labels, type

```
. label language original
. nmlab tcfam tcfriend vignum
tcfam     Q43 How Important: Turn to family for help
tcfriend  Q44 How Important: Turn to friends for help
vignum    Vignette
```

If I am not satisfied with the changes, I revise the `include` file and rerun the program.

5.11.4 Step 4: Revise value labels

Revising value labels is more challenging for several reasons: value labels require the two steps of defining and assigning labels; each label definition has labels for multiple values; one value definition can be used by multiple variables; and to create value labels in a new language, you must create new label definitions, not just revise the existing definitions. Accordingly, the programs that follow, especially those of step 4b, are more difficult than those in the earlier steps. I suggest that you start by skimming this section without worrying about the details. Then reread it while working through each do-file, preferably while in Stata where you can experiment with the programs.

Step 4a: List the current labels

I load the dataset and use `labelbook` to list the value labels and determine which variables use which labels definitions (file: `wf5-sgc4a-vallab-check.do`):

```
use wf-sgc03, clear
datasignature confirm
notes _dta
labelbook, length(10)
```

Here is the output for the `Ldist` label definition:

```
--------------------------------------------------------------------------------
value label Ldist
--------------------------------------------------------------------------------

      values                                    labels
       range:  [1,4]                    string length:  [16,20]
           N:  4                unique at full length:  yes
        gaps:  no                 unique at length 10:  no
  missing .*:  0                          null string:  no
                            leading/trailing blanks:  no
                                 numeric -> numeric:  no
  definition
           1   Definitely Willing
           2   Probably Willing
           3   Probably Unwilling
           4   Definitely Unwilling

   in default attached to sdneighb sdsocial sdchild sdfriend sdwork sdmarry
                          Ssdlive Ssdsocial Ssdchild Ssdfriend Ssdwork Ssdmarry

   in original attached to sdneighb sdsocial sdchild sdfriend sdwork sdmarry
                           Ssdlive Ssdsocial Ssdchild Ssdfriend Ssdwork Ssdmarry
```

The first part of the output summarizes the label with information on the number of values defined, whether the values have gaps (e.g., 1, 2, 4), how long the labels are, and more. The most critical information for my purposes is `unique at length 10`, which was requested with the `length(10)` option. This option determines whether the first ten characters of the labels uniquely identify the value associated with the label. For example, the label for `1` is `Definitely Willing` whereas the label for `4` is `Definitely Unwilling`. If I take the first ten letters of these labels, both `1` and `4` are labeled as `Definitely`. Because Stata commands often use only the first ten characters of the value label, this is a big problem that is indicated by the warning `unique at length 10: no`. Next `definition` lists each value with its label. The section `in default attached to` lists variables in the `default` language that use this label, followed by a list of variables that use this label in the `original` language. I review the output and plan my changes.

Step 4b: Create label define commands to edit

To change the value labels, I create a text file with dummy `label define` and `label values` commands. These are edited and included in step 4c. I start by loading the data and dropping the cloned variables (file: `wf5-sgc4b-vallab-dump.do`):

```
use wf-sgc03, clear
datasignature confirm
notes _dta
drop S*
```

Next I create the local `valdeflist` with the names of the label definitions used by all variables except the clones, which have been dropped. Because I only want the list and not the output, I use `quietly`:

```
quietly labelbook
local valdeflist = r(names)
```

The list of label definitions is placed in the local `valdeflist`. Next I create a file with dummy `label define` commands. There are two ways to do this.

Approach 1: Create label define statements with label save

The simplest way to create a file with the `label define` command for the current labels is with `label save`:

```
label save `valdeflist´ using ///
    wf5-sgc4b-vallab-labelsave-dummy.doi, replace
```

This command creates `wf5-sgc4b-vallab-labelsave-dummy.doi` with information that looks like this for the `Ldist` label definition:

```
label define Ldist 1 `"Definitely Willing"´, modify
label define Ldist 2 `"Probably Willing"´, modify
label define Ldist 3 `"Probably Unwilling"´, modify
label define Ldist 4 `"Definitely Unwilling"´, modify
```

I copy `wf5-sgc4b-vallab-labval-dummy.doi` to `wf5-sgc4b-vallab-labval-revised.doi` and make the revisions. I change the name of the definition to `NLdist` because I want to keep the original `Ldist` labels unchanged. The edited definitions look like this:

```
label define NLdist 1 `"1DefWillng"´, modify
label define NLdist 2 `"2ProbWill"´, modify
label define NLdist 3 `"3ProbUnwil"´, modify
label define NLdist 4 `"4DefUnwill"´, modify
```

After revising all definitions, I use the edited file as an `include` file in step 4c.

Approach 2: Create customized label define statements (advanced material)

If you have a small number of label definitions to change, `label save` works fine. Because our project had 17 datasets and hundreds of variables, I wrote a program that creates commands that are easier to edit. Although you can skip this section if you find the commands created by `label save` adequate, you might find the programs useful for learning more about automating your work. First, I run the `uselabel` command:

```
uselabel `valdeflist´, clear
```

This command replaces the data in memory with a dataset consisting of value labels from the definitions listed in `valdeflist` (created above by `labelbook`). Each observation has information about the label for one value from one value-label definition. For example, here are the first four observations with information on the `Ldist` label definition:

```
. list in 1/4, clean

       lname   value                   label   trunc
  1.   Ldist       1      Definitely Willing       0
  2.   Ldist       2        Probably Willing       0
  3.   Ldist       3      Probably Unwilling       0
  4.   Ldist       4    Definitely Unwilling       0
```

Variable `lname` is a string variable containing the name of the value-label definition; `value` is the value being labeled by the current row of the dataset; `label` is the value label; and `trunc` is 1 if the value label has been truncated to fit into the string variable `label`.

Next I open a file to hold dummy `label define` commands that I edit to create an `include` file used in step 4c to create new value labels:

```
file open myfile using wf5-sgc4b-vallab-labdef-dummy.doi, write replace
```

Before examining the loop that creates the commands, it helps to see what the file will look like:

```
//                                 1234567890
label define NLdist       1       "Definitely Willing", modify
label define NLdist       2       "Probably Willing", modify
label define NLdist       3       "Probably Unwilling", modify
label define NLdist       4       "Definitely Unwilling", modify
//                                 1234567890
label define NLdummy      1       "Yes", modify
label define NLdummy      2       "No", modify
  (output omitted)
```

The first line is a comment that includes the numbers `1234567890` that serve as a guide for editing the `label define` commands to create labels that are 10 characters or shorter. These guide numbers are the major advantage of the current approach over approach 1. The next four lines are the `label define` commands needed to create `NLdist`. Another line with the guide is written before the `label define` commands for `NLdummy`, and so on. Here is the loop that produces this output:

```
 1>  local rownum = 0
 2>  local priorlbl ""
 3>  while `rownum´ <= _N {
 4>      local ++rownum
 5>      local lblnm  = lname[`rownum´]
 6>      local lblval = value[`rownum´]
 7>      local lbllbl = label[`rownum´]
 8>      local startletter = substr("`lblval´",1,1)
 9>      if "`priorlbl´"!="`lblnm´" {
10>          file write myfile "//" _col(31) "1234567890" _newline
11>      }
12>      if "`startletter´"!="." {
13>          file write myfile                                      ///
14>              "label define N`lblnm´ " _col(25) "`lblval´"  ///
15>              _col(30) `""`lbllbl´""´ ", modify" _newline
16>      }
17>      local priorlbl "`lblnm´"
18>  }
19>  file close myfile
```

Although the code in this section is complex, I describe what it does in the section below. In addition, I encourage you to try running `wf5-sgc4b-vallab-dump.do` (part of the Workflow package) and experiment with the code.

Lines 1 and 2: define locals. Line 1 creates local `rownum` to count the row of the dataset that is being read. Line 2 defines the local `priorlbl` with the name of the label from the prior row of the dataset. For example, if `rownum` is 9, `priorlbl` contains the name of the label when `rownum` was 8. This is used to compare the label being read in the current row with that in the prior row. If they differ, a new label is started.

Lines 3, 4, 18: loop. Line 3 begins a loop in which local `rownum` increases from 1 through the last row of the dataset (_N indicates the number of rows in the dataset). Line 4 increases the counter `rownum` by 1. The loop ends in line 18.

Lines 5–8: retrieve information on current row of data. These lines retrieve information about the label in the current row of the dataset. Line 5 creates local `lblnm` with the contents of variable `lname` (a string variable with the name of the label for this row) in row `rownum`. For example, in row 1 `lblnm` equals `Ldist`. Lines 6 and 7 do the same thing for the variables `value` and `label`, thus retrieving the value considered in this row and the label assigned to that value. Line 8 creates the local `startletter` with the first letter of the value label (the function `substr` extracts a substring from a string). If `startletter` contains a period, I know that the label is for a missing value.

Lines 9–11: write a header with guide numbers. Line 9 checks if the name of the label in the current row (contained in local `lblnm`) is the same as the name of the label from the prior row (contained in the local `priorlbl`). The first time through the loop, the prior label is a null string which does not match the label for the first row. If the current and prior labels differ, the current row is the first row for the new label. Line 10 adds a comment with guide numbers that help when editing the labels to make them ten characters or less. The `if` condition ends in line 11.

Lines 12–16: write label define command. Line 12 checks if the first letter of the value of the current label is a period. If it is, then the value is a missing value and I do not want to write a label define command. I will handle missing values later in the program. Lines 13–15 write a dummy `label define` command to the file as illustrated in the sample contents of the file listed above. The names of the value labels start with an `N` (standing for new label) followed by the original label name (e.g., label `age` becomes `Nage`). I change the name because I do not want to change the original labels. Line 16 ends the `if` condition.

Line 17: update local priorlbl. Line 17 assigns the current label name to the local `priorlbl`. This information is used in line 9 to determine if the current observation starts a new value label.

Line 19: close the file. Line 19 closes the file `myfile`. Remember that a file is not written to disk until it is closed.

Create label values commands

Next I generate the commands to assign these labels to variables. By now, you should be able to follow what the program is doing:

```
use wf-sgc03, clear
drop S*
aorder
unab varlist : _all
file open myfile using wf5-sgc4b-vallab-labval-dummy.doi, write replace
foreach varname in `varlist´ {
    local lblnm : value label `varname´
    if "`lblnm´"!="" {
        file write myfile "label values `varname´" _col(27) "N`lblnm´" _newline
    }
}
file close myfile
```

The output looks like this:

```
label values  age           Nage
label values  atdisease     NLlikely
label values  atgenet       NLlikely
label values  atraised      NLlikely
label values  cause         Ncause
label values  clawdoc       NLrespons
label values  clawhosp      NLrespons
label values  clawpmed      NLrespons
  (output omitted)
```

The two files created in this step are edited and used to create new labels in step 4c.

Step 4c: Revise labels and add them to dataset

I copy `wf5-sgc4b-vallab-labdef-dummy.doi` to `wf5-sgc4b-vallab-labdef-revised.doi` and revise the label definitions. For example, here are the revised commands for `NLdist`:

```
//                                1234567890
label define NLdist     1        "1Definite", modify
label define NLdist     2        "2Probably", modify
label define NLdist     3        "3ProbNot", modify
label define NLdist     4        "4DefNot", modify
```

The guide numbers verify that the new labels are not too long. Similarly, I copy `wf5-sgc4b-vallab-labval-dummy.doi` to `wf5-sgc4b-vallab-labval-revised.doi` and revise it to look like this:

```
label values  age           Nage
label values  atdiseas      NLlikely
label values  atgenet       NLlikely
label values  atraised      NLlikely
label values  cause         Ncause
label values  clawdoc       NLrespons
  (output omitted)
```

Now I am ready to change the labels. I load the data and confirm that it has the right signature (file: `wf5-sgc4c-vallab-revise.do`):

```
. use wf-sgc03, clear
(Workflow data for SGC renaming example \ 2008-04-09)
. datasignature confirm
  (data unchanged since 09apr2008 17:59)
```

Next I include the files with the changes to the labels:

```
include wf5-sgc4b-vallab-labdef-revised.doi
include wf5-sgc4b-vallab-labval-revised.doi
```

Now I add labels for the missing values. To do this, I need a list of the value labels being used in the noncloned variables. Because I do not want to lose the label definitions I just added, I save a temporary dataset, drop the cloned variables, and use `labelbook` to get a list of value definitions. Then I load the dataset that I had temporarily saved:

```
save x-temp, replace
drop S*
quietly labelbook
local valdeflist = r(names)
use x-temp, clear
```

(*Continued on next page*)

I loop through the value definitions and assign labels for missing values:

```
foreach valdef in `valdeflist' {
    label define `valdef' .a  `".a_NAP"'      , modify
    label define `valdef' .b  `".b_Refuse"'   , modify
    label define `valdef' .c  `".c_DK"'       , modify
    label define `valdef' .d  `".d_NA_ref"'   , modify
    label define `valdef' .e  `".e_DK_var"'   , modify
}
```

Finally, the dataset is documented with a note, a new signature is added, and it is saved:

```
note: wf-sgc04.dta \ revise val labels with source & default languages \ `tag'
label data "Workflow data for SGC renaming example \ 2008-04-09"
datasignature set, reset
save wf-sgc04, replace
```

To check the labels, I look at `tabulate`:

```
. tabulate marital, missing

    Marital |
     status |      Freq.     Percent        Cum.
------------+-----------------------------------
   1Married |        112       56.00       56.00
   2Widowed |         16        8.00       64.00
  3Divorced |         10        5.00       69.00
  4Separatd |          6        3.00       72.00
   5Cohabit |         21       10.50       82.50
    6Single |         34       17.00       99.50
  .d_NA_ref |          1        0.50      100.00
------------+-----------------------------------
      Total |        200      100.00
```

By changing languages, I can compare these labels with the originals:

```
. label language original

. tabulate marital, missing

             Marital status |      Freq.     Percent        Cum.
----------------------------+-----------------------------------
                    Married |        112       56.00       56.00
                    Widowed |         16        8.00       64.00
                   Divorced |         10        5.00       69.00
     Separated, but married |          6        3.00       72.00
Living as a couple/Cohabiting |       21       10.50       82.50
     Single, never married |         34       17.00       99.50
                         .d |          1        0.50      100.00
----------------------------+-----------------------------------
                      Total |        200      100.00
```

5.11.5 Step 5: Check the new names and labels

At this point, I let everything sit for a day or two. After working on a project like this for several hours, any name or label looks better than the prospect of making more changes. After a break, I systematically review what I have done (file: `wf5-sgc5a-check.do`). If necessary, I fine-tune the programs in steps 2–4.

5.12 Conclusions

This is a long chapter that covers both the principles and the practice of naming and labeling your files and data. No matter how simple your dataset, I think it is useful to carefully assess your names and labels before you start the analyses. With better names and labels, you will find that data analysis is simpler and that replication is easier. After working on many projects, I am convinced that the painstaking work described in this chapter is one of the best investments of time that you can make.

6 Cleaning your data

More than half of the work in data analysis involves preparing the data. This is true even with secondary data that has been meticulously prepared, such as the General Social Survey or Add Health. If you fail to clean your data, you risk problems during analysis and even incorrect results. Unfortunately, cleaning data takes a lot of time. The best way to minimize that time is to have a plan, to pay close attention to details, and to work slowly. Although I can spend hours analyzing data, fatigue sets in more quickly when I work on data management. To avoid mistakes, I set out a few well-defined tasks, complete them, and then put the work aside for a few hours or a day. While the work "ages", I document what was done, plan the next steps, and work on other things. When I return to the work, I might find errors, discover unclear comments, or think of a better way to do things. Errors must be fixed, and I almost always add more comments. If I think of a clearer or more reliable approach, I might redo the work.

Preparing data for analysis involves five steps, four of which are considered in this chapter.

1. Importing data: The first step is to import the data into Stata. If the dataset arrives in Stata format, this is simple. If the dataset is in some other format, you must convert it. Because a mistake in conversion affects everything that follows, this step must be approached very carefully.
2. Revision of names and labels: Once the dataset is in Stata format, you should review the names and labels and make revisions as discussed in chapter 5.
3. Verification: Next you should verify that each variable is correct. Verification involves everything from assessing the internal consistency of information (e.g., someone reporting five years of education and working as a doctor) to looking for substantively unreasonable distributions (e.g., 27% report membership in a small religious denomination).
4. Variable creation: A dataset rarely includes all the variables that you need for analysis. Part of data preparation involves adding new variables and verifying that they have been created correctly. For example, you might create a set of indicator variables for educational degree attained based on years of education or add age-squared. Many of the tools from chapter 5 are used in this step.
5. Data extraction: After the source data are clean and new variables added, you might want to extract a subset of variables and/or cases for analysis. You might also want to merge information from several datasets into one dataset for analysis.

When saving datasets, `datasignature` and other commands from chapter 5 are used.[1]

Stata has dozens of commands that are useful for cleaning your data and usually there are multiple ways to accomplish the same thing. The choice among commands is partly a matter of taste and is partly determined by which aspect of a variable you want to focus on. Some users swear by the `assert` command, whereas others (like me) rarely use it. I often use graphs, but you might prefer tables or lists of values. Accordingly, think of the examples in this chapter as illustrations of one among many ways to do something. Although the unequivocal goal of the chapter is to verify your data, there are many ways to attain this goal.

6.1 Importing data

Whether you collect your own data or use data from a secondary source, you often need to convert data from the format in which you received it into Stata format. Although sometimes this is simple, in other cases, especially when missing data are involved, you must be extremely careful that the converted data correspond exactly to the source data. If you mess things up in this step, everything that follows is affected and the only way to correct the problem is to start over. To make things worse, it might take a long time before you realize your mistake. This section begins by reviewing data formats, then discusses ways of importing data into Stata, and ends with an extended example that illustrates useful commands and potential problems.

6.1.1 Data formats

Data are stored in hundreds of formats, and there has been little progress in developing a universal format that can be used by all programs. Among the many formats, there are two basic types: ASCII formats and binary formats.

ASCII data formats

ASCII stands for American Standard Code for Information Interchange and is a standard for data files that originated in 1967. Data are stored as plain text that can be read by any text editor. Although it would be tedious, you could count rows and columns to determine the value of each variable for each observation in the dataset. ASCII data come in two flavors. In a fixed-format ASCII file, each variable is located in fixed (i.e., always the same) columns. This is shown in figure 6.1, where the ID number is in columns 1–7, gender in columns 10–11, age in columns 13–15, and so on. Sometimes the data are accompanied by a dictionary file that describes which variables are in which columns.

1. To conserve space, the examples do not always show the `datasignature` command. Following best practice, the do-files for the chapter check the signature when datasets are loaded.

```
1800001  2  29  1  13  5  8   200  1  1  1  4  4  3
1800002  2  38  1  17  7  1  2500  1  2  3  4  4  2
1800003  2  32  1  14  5  2  3000  4  2  2  4  2  4
1800004  1  20  5  14  6  1  1500  2  3  2  2  2  3
1800005  2  47  2  10  5  1    .a  4  2  3  2  4  1
1800006  1  41  1  11  5  1    .a  1  2  1  4  3  3
1800007  1  47  1  13  5  1    .a  2  1  1  2  2  3
1800008  2  27  1  13  5  1  3000  1  1  2  2  2  3
1800009  2  24  5  15  7  1    .a  2  2  4  4  2  4
1800010  1  24  5  11  5  1  3000  1  1  4  2  4  1
1800011  2  27  1  .b  1  5    .c  1  1  1  1  1  2
1800012  1  79  1  16  7  7  2700  2  2  3  2 .a .a
1800013  1  38  1  11  5  1  6000  1  2  5  2  5  1
1800014  2  20  5  .b  1  6    .c  1  4  1  4  4  1
1800015  1  55  1  15  7  7  1300  4  1  1  1  4  1
1800016  1  70  1  13  5  7  1200  1  1  1  1  2  2
1800017  2  69  2  12  5  7  2000  5  1  1  1  5  5
1800018  1  31  1  10  5  1    .a  2  3  4  4  1  4
1800019  1  41  3  10  5  1    .b  2  3  2  4  4  2
1800020  2  63  2  10  5  7  1200  4  3  3  4  4  4
1800021  2  80  2  11  5  7  2000  2  2  2  4  4  3
```

Figure 6.1. A fixed-format ASCII file

In a free-format ASCII file, values are separated by a delimiter, such as a comma or a tab, with extra spaces removed. Figure 6.2 shows a free-format version of the data in figure 6.1.

```
Vid,Vsex,Vage,Vmarstat,Vedyears,Vedlevel,Vempstat,Vearnings,Vmomwarm
1800001,2,29,1,13,5,8,200,1,1,1,4,4,3
1800002,2,38,1,17,7,1,2500,1,2,3,4,4,2
1800003,2,32,1,14,5,2,3000,4,2,2,4,2,4
1800004,1,20,5,14,6,1,1500,2,3,2,2,2,3
1800005,2,47,2,10,5,1,.a,4,2,3,2,4,1
1800006,1,41,1,11,5,1,.a,1,2,1,4,3,3
1800007,1,47,1,13,5,1,.a,2,1,1,2,2,3
1800008,2,27,1,13,5,1,3000,1,1,2,2,2,3
1800009,2,24,5,15,7,1,.a,2,2,4,4,2,4
1800010,1,24,5,11,5,1,3000,1,1,4,2,4,1
1800011,2,27,1,.b,1,5,.c,1,1,1,1,1,2
1800012,1,79,1,16,7,7,2700,2,2,3,2,.a,.a
1800013,1,38,1,11,5,1,6000,1,2,5,2,5,1
1800014,2,20,5,.b,1,6,.c,1,4,1,4,4,1
1800015,1,55,1,15,7,7,1300,4,1,1,1,4,1
1800016,1,70,1,13,5,7,1200,1,1,1,1,2,2
1800017,2,69,2,12,5,7,2000,5,1,1,1,5,5
1800018,1,31,1,10,5,1,.a,2,3,4,4,1,4
1800019,1,41,3,10,5,1,.b,2,3,2,4,4,2
1800020,2,63,2,10,5,7,1200,4,3,3,4,4,4
1800021,2,80,2,11,5,7,2000,2,2,2,4,4,3
```

Figure 6.2. A free-format ASCII file with variable names

In the fixed-format file, **Vearnings** uses seven columns even if someone's earnings needs only one column to report them. In the free-format file, the unneeded spaces are removed. For example, for **Vid** 1800001, **Vearnings** is 200, which requires only three columns. Because extra space is removed, free-format files are generally smaller than their fixed-format counterparts.

The disadvantage of ASCII files is that they do not contain variable labels, value labels, or other metadata.[2] You must add this information, which is time consuming and error prone. Binary formats allow you to include this information within the file.

Binary-data formats

Binary datasets store information in a more complex format that requires special software to decode. Within the file each group of eight bits (i.e., a string of eight 0s or 1s) is translated into a single character. Sometimes this character appears on the screen as a recognizable character, and in other cases the bits represent a character that cannot be viewed and appear as a □. Figure 6.3 shows a portion of the Stata file corresponding to the ASCII files in figures 6.1 and 6.2.

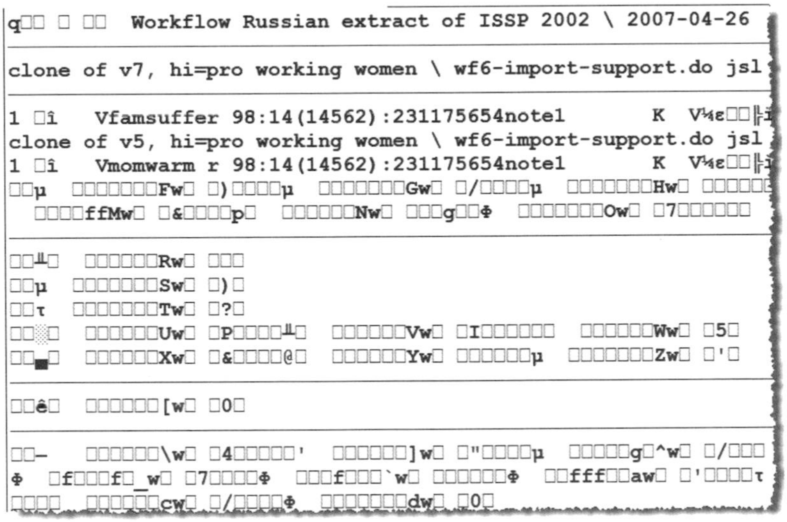

```
q□□ □ □□  Workflow Russian extract of ISSP 2002 \ 2007-04-26
clone of v7, hi=pro working women \ wf6-import-support.do jsl
1 □î    Vfamsuffer 98:14(14562):231175654note1       K  V¼ε□□╠î
clone of v5, hi=pro working women \ wf6-import-support.do jsl
1 □î    Vmomwarm r 98:14(14562):231175654note1       K  V¼ε□□╠î
□□µ  □□□□□□□Fw□ □)□□□□µ  □□□□□□□Gw□ □/□□□□µ  □□□□□□□Hw□ □□□□□□
  □□□□ffMw□ □&□□□□p□  □□□□□□Nw□ □□□g□□Φ  □□□□□□□Ow□ □7□□□□□□
□□╨□  □□□□□□Rw□ □□□
□□µ  □□□□□□□Sw□ □)□
□□τ  □□□□□□□Tw□ □?□
□□░□  □□□□□□Uw□ □P□□□□╨□  □□□□□□Vw□ □I□□□□□□  □□□□□□Ww□ □5□
□□■□  □□□□□□Xw□ □&□□□□@□  □□□□□□Yw□ □□□□□□µ  □□□□□□□Zw□ □'□
□□ê□  □□□□□□[w□ □0□
□□-   □□□□□□\w□ □4□□□□□'  □□□□□□]w□ □"□□□□µ  □□□□□g□^w□ □/□□□
Φ  □f□□□f□_w□ □7□□□□Φ  □□□f□□□`w□ □□□□□□Φ  □□fff□□aw□ □'□□□□τ
□□□□  □□□□□□cw□ □/□□□□Φ  □□□□□□□dw□ □0□
```

Figure 6.3. A binary file in Stata format

Binary files contain metadata such as variable labels, value labels, and notes. In most binary formats, the data are compressed to make the file physically smaller. Unfortunately, almost every statistical package uses its own format and often a new release of a program comes with a new variation of the format. If you save data in a binary

2. Metadata means "data about data". It includes anything that describes characteristics of a dataset.

format that becomes obsolete, you risk having the physical file but not being able to make sense of what it contains. You have preserved the bits without preserving the information. This problem is considered in greater detail in chapter 8.

Converting data from one binary format to another discards information that is not compatible with the destination format. For example, Stata 10 allows variable names that are 32 characters in length, whereas some programs restrict names to eight characters. This means that the carefully chosen name `mom_warm_5point_scale` in Stata is changed to the obscure `mom_w~01`. When you convert Stata to other formats, you lose the metadata from `notes` and languages.

6.1.2 Ways to import data

Stata can read ASCII files, SAS transport format, Stata format, and a few less common formats. If the data you have are in Stata format, you can simply `use` the data (e.g., `use binlfp2, clear`) and skip the rest of this section. With other formats, you need to import the data. There are three general approaches.

1. If the data are in a format that Stata can read (e.g., SAS transport, ASCII), you can use Stata to read the data.
2. You can load the data with some other program that can read the dataset and use that program to export the data to a format that Stata can read.
3. You can use a program like Stat/Transfer (http://www.stattransfer.com) or DBMS/Copy (http://www.dataflux.com) that converts data among many formats.

The rest of this section considers these options, along with examples that show how to import and convert data.

Stata commands to import data

Several Stata commands read data that are in non-Stata formats. Because these commands are fully documented in the Stata manuals and with `help`, I discuss them only briefly.

Stata commands for ASCII formats

Stata has three "`in`" commands to read ASCII files.

1. `insheet` reads tab-delimited or comma-delimited files that are often created by spreadsheet programs.
2. `infile` reads space-, tab-, or comma-delimited free-format files and also reads fixed-format files where a dictionary specifies the location of variables.
3. `infix` reads fixed-format files that do not have a dictionary.

Unless you have a dictionary file, importing ASCII data is tedious and error prone because you must specify the location of each variable within the file. Plus, you need to add variable and value labels.

To illustrate how these commands work, I use `insheet` to read the free-format ASCII file shown in figure 6.2 (file: `wf6-import.do`):

```
. insheet using wf6-import-free.txt, clear
(14 vars, 1798 obs)
```

Because the first row of `wf6-import-free.txt` contains the names of the variables, `insheet` requires only the name of the input file. Listing the first seven cases, you can see that the data match those in figures 6.1 and 6.2:

```
. list vid-vempstat in 1/7, clean

           vid   vsex   vage   vmarstat   vedyears   vedlevel   vempstat
  1.   1800001      2     29          1         13          5          8
  2.   1800002      2     38          1         17          7          1
  3.   1800003      2     32          1         14          5          2
  4.   1800004      1     20          5         14          6          1
  5.   1800005      2     47          2         10          5          1
  6.   1800006      1     41          1         11          5          1
  7.   1800007      1     47          1         13          5          1
```

For further details on the "`in`" commands, use `help`. For information on using a dictionary file, see http://www.stata.com/support/faqs/data/dict.html.

Importing SAS XPORT files

If your data are in SAS XPORT transport format, you can import the data with the `fdause` command. For example (file: `wf6-import.do`),

```
fdause wf6-import-fdause.xpt, clear
```

The command is called `fdause` because this format is used by the U.S. Food and Drug Administration (FDA) for new drug and device applications. `fdause` can also read value labels from a `formats.xpf` XPORT file.

Importing ODBC files

Open DataBase Connectivity (ODBC) is a standard format for exchanging data between programs. For example, Microsoft Access can save data in ODBC format. The `odbc` command reads ODBC data.

Importing XML data

The `xmluse` command reads an emerging standard know as Extensible Markup Language (XML), which is designed to be highly portable. `xmlsave` saves data in XML format, which is sometimes used when archiving data.

Using other statistical packages to export data

Many statistical packages allow you to save data in alternative formats. For example, SPSS can write data in Stata format. SAS can write data in SAS XPORT format. This can be an effective way to convert data between formats.

Using a data conversion program

Often the simplest approach to getting your dataset into Stata format is to use a data conversion program such as Stat/Transfer or DBMS/Copy. These programs convert data among dozens of data formats. Although these are excellent programs, you should not assume that they convert the data accurately. Almost every statistical package uses its own format and often the format changes with each version of the program. Although Stat/Transfer and DBMS/Copy do an excellent job of keeping up with the latest changes, problems can occur, especially with more recent formats.

6.1.3 Verifying data conversion

Things can go wrong when you convert data from one format to another. If you are importing an ASCII file in fixed format, it is easy to make a mistake when specifying columns. Fortunately, this type of error generally leads to obvious problems in the imported data. When using other methods to convert data, the problems can be more subtle. Accordingly, I recommend the following steps to verify that your dataset has been converted correctly.

Step 1: Compare descriptive statistics from the source and destination

Use the source statistical package and the unconverted source data to compute univariate, descriptive statistics for all variables. For variables with few categories, I recommend a frequency distribution. For variables with many categories, you can check the mean, standard deviation, minimum, and maximum. Next compute the same information using Stata with the converted data. For example, if the source data are in SAS format, compute descriptive statistics in SAS. Next use the converted data in Stata to compute the same statistics. Verify that everything matches. If the first few variables convert correctly, do not assume that later variables are also correct (as illustrated on page 207).

Step 2: Examine the distribution of missing values

Compare the distribution of missing values for each variable using both the source program with the source data and Stata with the converted data. Because descriptive statistics exclude missing values, you should tabulate values with `tab1` *variable-list*`, missing`. The most frequent problem I have seen in conversion is when multiple missing-value codes in the source data are merged into one value when converted.

Step 3: Convert the data two ways

For example, use SPSS to export the data into Stata format. Then use Stat/Transfer to convert the same source data into Stata format. In Stata, use the `cf` command to compare the two files. If the files match exactly, you can be more confident in the conversions. If they do not, verify where the differences occur. For example, if one dataset uses uppercase names and the other uses lowercase names, `cf` will show that the files differ even though the numeric values in the data are identical.

Converting the ISSP 2002 data from Russia

To illustrate the problems that can occur when you import data, I use the Russian sample from the International Social Survey Program Family and Changing Gender Roles III Study (International Social Survey Program 2004).

Examining the data in SPSS

The source file `04106-0001-Data.por` is an SPSS Portable file that includes variable labels, value labels, and information on missing values.[3] I opened the dataset with SPSS version 14. Variable `v3` is the country where the data were collected. I selected cases with `v3` equal to 18, the code for Russia, and saved an SPSS binary file as `wf6-isspru-spss01.sav`.

While still in SPSS, I computed the descriptive statistics (see figure 6.4).

3. I thank Tait Medina for suggesting this dataset. The source file is 24 megabytes and is not included in the Workflow package but is available as Study No. 4106 from ICPSR (http://www.icpsr.umich.edu).

Descriptive Statistics

	N	Minimum	Maximum	Mean	Std. Deviation
ZA Study Number	1798	3880	3880	3880.00	.000
Respondent Number	1798	1800001	1801798	1800900	519.182
Country	1798	18	18	18.00	.000
Workg mom: warm relation child ok	1765	1	5	2.32	1.148
Workg mom: pre school child suffers	1755	1	5	2.37	1.050
Workg woman: family life suffers	1759	1	5	2.40	1.098
What women really want is home & kids	1717	1	5	2.64	1.112
Household satisfies as much as paid job	1680	1	5	2.86	1.102
Work is best for womens independence	1710	1	5	2.07	.952
Both should contribute to hh income	1773	1	5	2.00	.939
Mens job is work,womens job household	1772	1	5	2.40	1.078

Figure 6.4. Descriptive statistics from SPSS

Because these results do not show missing values, I also computed frequency distributions. For example, variable v4 asks if a working mother can have as warm of a relationship with her children as mothers who do not work (see figure 6.5).

Workg mom: warm relation child ok

		Frequency	Percent	Valid Percent	Cumulative Percent
Valid	Strongly agree	464	25.8	26.3	26.3
	Agree	712	39.6	40.3	66.6
	Neither agree nor disagree	197	11.0	11.2	77.8
	Disagree	336	18.7	19.0	96.8
	Strongly disagree	56	3.1	3.2	100.0
	Total	1765	98.2	100.0	
Missing	Cant choose	26	1.4		
	Na, refused	7	.4		
	Total	33	1.8		
Total		1798	100.0		

Figure 6.5. Frequency distribution from SPSS

Converting data to Stata format

Next, using Stat/Transfer version 8, I click on the Transfer tab and enter options to create a Stata dataset named `wf6-isspru-sttr01.dta` (see figure 6.6).

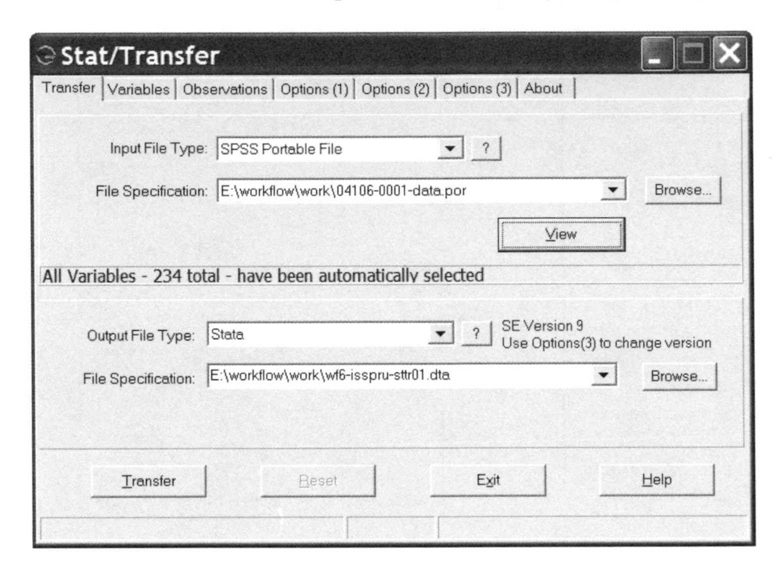

Figure 6.6. Transfer tab from the Stat/Transfer dialog box

The *File Specification* is the source file `04106-0001-Data.por`, not the file `wf6-isspru-spss01.sav` that was converted to SPSS format. Always stay as close to the original data as possible rather than converting an already converted dataset. Switching to the Observations tab, I select cases from Russia (see figure 6.7).

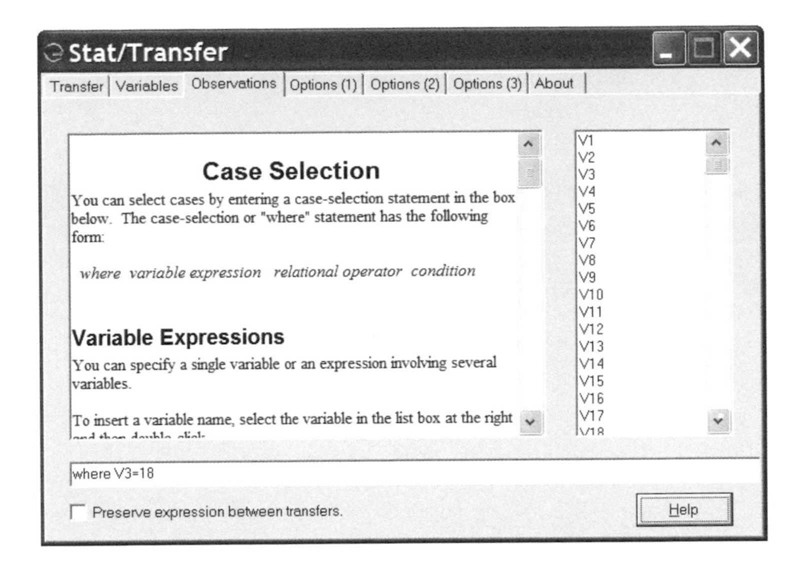

Figure 6.7. Observations tab from the Stat/Transfer dialog box

Returning to the Transfer tab, I click on the **Transfer** button to create `wf6-isspru-sttr01.dta` in Stata format. If asked if you want single or double precision, choose double to make sure that there is no rounding error in the conversion. The only disadvantage is that the Stata file will initially be larger; however, by running `compress` (see page 264), you can make the file smaller without risking data loss.

Verifying the transfer

To check the transferred data, I use `codebook` (file: `wf6-isspru-sttr01.do`):

```
. codebook, compact
Variable  Obs Unique      Mean      Min      Max  Label
------------------------------------------------------------------------
V1       1798      1      3880     3880     3880  ZA Study Number
V2       1798   1798   1800900  1800001  1801798  Respondent Number
V3       1798      1        18       18       18  Country
V4       1765      5  2.324646        1        5  Workg mom: warm relation child ok
V5       1755      5  2.373789        1        5  Workg mom: pre school child suffers
V6       1759      5  2.401933        1        5  Workg woman: family life suffers
V7       1717      5  2.639487        1        5  What women really want is home & kids
V8       1680      5  2.855357        1        5  Household satisfies as much as paid job
V9       1710      5  2.071345        1        5  Work is best for womens independence
V10      1773      5  2.003948        1        5  Both should contribute to hh income
V11      1772      5  2.401806        1        5  Mens job is work,womens job household
  (output omitted)
```

The results match those from SPSS except that the conversion changed the variable names to uppercase (e.g., `V4` not `v4`).

Because descriptive statistics like the mean exclude missing values, I use `tab1` to verify that missing values have been converted correctly. The results for `V4` match those from SPSS that were shown earlier:

```
-> tabulation of V4
```

Workg mom: warm relation child ok	Freq.	Percent	Cum.
Strongly agree	464	25.81	25.81
Agree	712	39.60	65.41
Neither agree nor disagree	197	10.96	76.36
Disagree	336	18.69	95.05
Strongly disagree	56	3.11	98.16
.c	26	1.45	99.61
.n	7	0.39	100.00
Total	1,798	100.00	

Unfortunately, there are 233 additional variables that need to be checked. You should never assume that because some of the variables were converted correctly that all the variables were converted correctly. Indeed, I found a problem with one of the last variables I checked:

```
-> tabulation of V239
```

R: Current employment status	Freq.	Percent	Cum.
Employed-full time	855	47.55	47.55
Employed-part time	87	4.84	52.39
Empl-< part-time	35	1.95	54.34
Unemployed	69	3.84	58.18
Studt,school,vocat.trng	59	3.28	61.46
Retired	531	29.53	90.99
Housewife,home duties	72	4.00	94.99
Permanently disabled	34	1.89	96.89
Oth,not i labour force	23	1.28	98.16
.a	33	1.84	100.00
Total	1,798	100.00	

The 33 cases coded `.a` in Stata correspond to 3 cases of `Dont know` plus 30 cases of `Na` in SPSS (see figure 6.8).

R: Current employment status

		Frequency	Percent	Valid Percent	Cumulative Percent
Valid	Employed-full time	855	47.6	48.4	48.4
	Employed-part time	87	4.8	4.9	53.4
	Empl-< part-time	35	1.9	2.0	55.4
	Unemployed	69	3.8	3.9	59.3
	Studt,school,vocat.trng	59	3.3	3.3	62.6
	Retired	531	29.5	30.1	92.7
	Housewife,home duties	72	4.0	4.1	96.8
	Permanently disabled	34	1.9	1.9	98.7
	Oth,not i labour force	23	1.3	1.3	100.0
	Total	1765	98.2	100.0	
Missing	Dont know	3	.2		
	Na	30	1.7		
	Total	33	1.8		
Total		1798	100.0		

Figure 6.8. Combined missing values in frequencies from SPSS

What happened? Stat/Transfer 8 has several options for treating missing values, with the default option *Use all, Map to extended (a-z) missing values*. I tried other options and found that the option *Use None, Map to extended (a-z) missing values* did not combine missing values. Using this option, I created `wf6-isspru-sttr02.dta`. The frequencies for `V239` looked like this (file: `wf6-isspru-sttr02.do`):

```
-> tabulation of V239

  R: Current employment |
                 status |      Freq.     Percent        Cum.
------------------------+-----------------------------------
     Employed-full time |        855       47.55       47.55
     Employed-part time |         87        4.84       52.39
       Empl-< part-time |         35        1.95       54.34
             Unemployed |         69        3.84       58.18
Studt,school,vocat.trng |         59        3.28       61.46
                Retired |        531       29.53       90.99
  Housewife,home duties |         72        4.00       94.99
   Permanently disabled |         34        1.89       96.89
 Oth,not i labour force |         23        1.28       98.16
              Dont know |          3        0.17       98.33
                     Na |         30        1.67      100.00
------------------------+-----------------------------------
                  Total |      1,798      100.00
```

The 33 cases were correctly split into two categories, but when I examined the converted values without the labels by adding the `nolabel` option to `tab1`, I found that missing values were coded as 98 and 99 instead of extended missing values:

```
-> tabulation of V239

   R: Current
  employment
      status |      Freq.     Percent        Cum.
-------------+-----------------------------------
           1 |        855       47.55       47.55
           2 |         87        4.84       52.39
           3 |         35        1.95       54.34
           5 |         69        3.84       58.18
           6 |         59        3.28       61.46
           7 |        531       29.53       90.99
           8 |         72        4.00       94.99
           9 |         34        1.89       96.89
          10 |         23        1.28       98.16
          98 |          3        0.17       98.33
          99 |         30        1.67      100.00
-------------+-----------------------------------
       Total |      1,798      100.00
```

Because I wanted missing-value codes rather than numbers, I tried DBMS/Copy and encountered the same problem. Next I used SPSS to save the data directly to Stata format. Although SPSS did not combine categories, it also did not use extended missing-value codes. I tried various other approaches before concluding that the only solution was to convert the missing data from SPSS into numeric codes and then to change these into extended missing values using Stata (e.g., `recode V239 98=.a 99=.b`).

6.2 Verifying variables

Before statistical analysis begins, you should verify each variable that you plan to use. Although some aspects of verification were discussed in section 6.1 on importing data and in chapter 5 on names and labels, here I provide a more comprehensive review that covers four, sometimes overlapping, aspects of data verification (see figure 6.9).

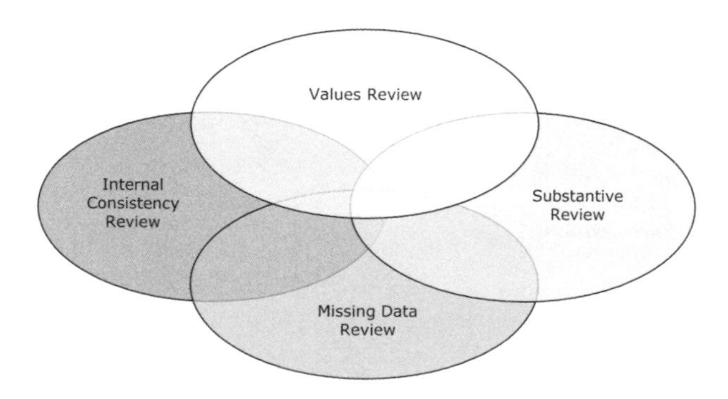

Figure 6.9. Four overlapping dimensions of data verification

Values review

A values review examines whether the values held by each variable are appropriate and whether all values that should be represented are found in your data. For example, if you expect that some individuals will have a PhD, but the variable for education has no cases in that category, something is probably wrong.

Substantive review

Although all the values that you find might be appropriate, is the distribution of values substantively reasonable? Are there too many or too few cases for some values? Does the range correspond to what you expect? Is the variable appropriately related to other variables? Although this step is similar to a values review, it requires a substantive understanding of the data.

Missing data review

Do the data have the types of missing data that you expect and are the number of cases for each type reasonable? Are the skip patterns from the survey reflected in the missing-value codes? For example, if a respondent indicates that she did not attend college, the question on the type of college attended should be skipped. This can lead to complex patterns of missing data that you should verify.

Internal consistency review

Information on one variable can have logical implications for the values of other variables. For example, someone who did not attend college should have missing values for a question about how she paid for college. In other situations, there might be strong evidence from past research on how two variables will be associated. Checking that responses are consistent across variables can uncover subtle problems with the data.

Each type of review is discussed along with examples using Stata to uncover problems.

6.2.1 Values review

The simplest thing to check is that all variables have values in the proper range. For this, I prefer `codebook, compact`, which shows the number of nonmissing observations, the number of unique values, the mean, the minimum, the maximum, and the variable label. You could use `summarize`, which includes the standard deviation, but it does not list the variable label. If everything looks fine with `codebook`, I check the specific values for each variable. The `inspect` command creates a small histogram and indicates the number of negative, zero, and positive values; the number of integers and nonintegers; the number of unique values; and the number of missing values. If I want to see the distribution of all values, I use tab1. The advantage of tab1 over tabulate is that tab1 lets you specify a list of variables (e.g., tab1 V4-V22, missing) whereas tabulate requires a separate command for each variable. If a variable has many values, I might use dotplot or stem to create a histogram. The advantage of stem is that it produces a text plot that is part of the log file while dotplot creates a graphic file.

Values review of data about the scientific career

This example uses data on the job placement of biochemists who obtained their degrees in the late 1950s and early 1960s. First, I look at the range of values (file: wf6-review-biochem.do):

```
. use wf-acjob, clear
(Workflow data on academic biochemists \ 2008-04-02)
. codebook, compact
Variable    Obs Unique       Mean  Min        Max  Label
-------------------------------------------------------------------------------
job         408     80   2.233431    1        4.8  Prestige of first job
fem         408      2   .3897059    0          1  Gender: 1=female 0=male
phd         408     89   3.200564    1        4.8  PhD prestige
ment        408    123   45.47058    0   531.9999  Citations received by mentor
fel         408      2   .6176471    0          1  Fellow: 1=yes 0=no
art         408     14   2.276961    0         18  # of articles published
cit         408     87   21.71569    0        203  # of citations received
-------------------------------------------------------------------------------
```

The maximum of 18 for the number of papers published at the time of the PhD seems high, but I know from other work that collaboration is common among these scientists. The ranges for job and phd match, which I expected because scientists are recruited from the same group of departments that produce new PhDs. For the binary variables fem or fel, the minimum, maximum, and mean show that these variables are reasonable, but I still need to verify that they contain only 0s and 1s.

Next I examine the distribution of values. Because art has only 18 distinct values, I can use tab1 without getting too much output:

```
. tab1 art, missing

-> tabulation of art

       # of |
   articles |
  published |      Freq.     Percent        Cum.
------------+-----------------------------------
          0 |         85       20.83       20.83
          1 |        102       25.00       45.83
          2 |         72       17.65       63.48
          3 |         49       12.01       75.49
          4 |         45       11.03       86.52
          5 |         25        6.13       92.65
          6 |         13        3.19       95.83
          7 |          9        2.21       98.04
          8 |          2        0.49       98.53
          9 |          1        0.25       98.77
         10 |          2        0.49       99.26
         12 |          1        0.25       99.51
         15 |          1        0.25       99.75
         18 |          1        0.25      100.00
------------+-----------------------------------
      Total |        408      100.00
```

This looks fine. Although I could tabulate `job`, `phd`, `ment`, and `cit`, a histogram is easier because these variables have many unique values. The `stem` command creates a type of histogram known as a stem-and-leaf plot:

```
. stem cit

Stem-and-leaf plot for cit (# of citations received)

  0* | 0000000000000000000000000000000000000000000000000000000000000 ... (212)
  1* | 0000001111222222233333333344444455555556778888888899999
  2* | 000001111122223333334444445566666777789
  3* | 0000122222333444556667789
  4* | 122335566777788888889
  5* | 1144556677789
  6* | 0034555556667
  7* | 01145778
  8* | 012368
  9* |
 10* | 0057
 11* | 3
 12* | 03
 13* |
 14* | 069
 15* | 4
 16* | 39
 17* |
 18* |
 19* |
 20* | 113
```

The "stem" is the number on the left axis. For example, the stem `5*` indicates that this row is for values between 50 and 59 (i.e., `5*`). The "leaves" are the digits to the right of the stem. For the `5*` stem, the leaves are `1144556677789`. Each digit or leaf corresponds to one observation. For example, the first `1` represents a case with the value

51; ditto for the second `1`. The two `4`s are cases with values of 54, and so on. If you prefer a graph, I like the `dotplot` command. For example, `dotplot cit` produces

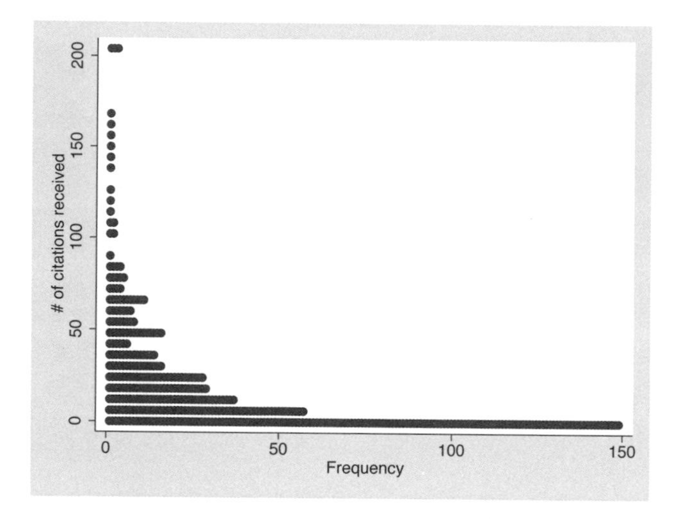

To create plots for all variables that are not binary, I use a loop:

```
foreach var in art cit phd job ment {
    dotplot `var´
    graph export wf6-review-biochem-`var´.png, replace
}
```

Quick and dirty graphs

If Stata takes a long time to create your graphs, which can be a problem when running Stata on a network, you can use version 7 graphics. For example, the command `version 7: dotplot cit` almost instantly produces the following plot that is fine for data cleaning:

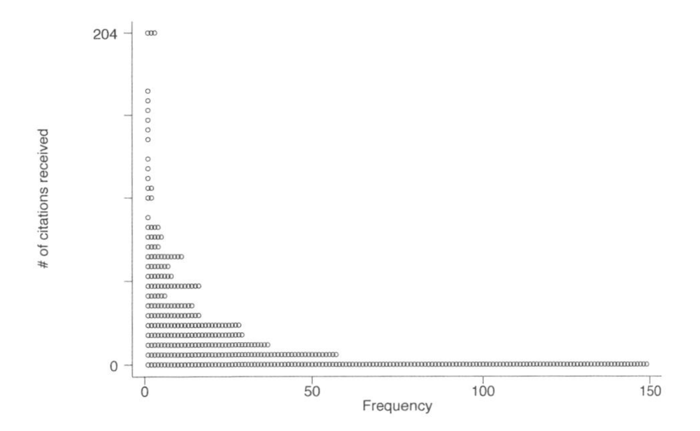

Values review of data on family values

The 2002 General Social Survey (GSS) included a module on family values. Repeating a question from the 1977 and 1989 surveys, the 2002 GSS asked: "To what extent do you agree or disagree that a working mother can establish just as warm and secure a relationship with her children as a mother who does not work?" (Davis, Smith, and Marsden 2007). Because I used this question from the 1977 and 1989 surveys to illustrate the ordinal logit model (Long and Freese 2006), I wondered if things had changed by 2002. First, I looked at the frequency distribution (file: `wf6-review-gss.do`):

```
. use wf-gsswarm, clear
(Workflow data from 2002 GSS on women and work \ 2008-04-02)
. tabulate v4, miss
```

Workg mom: warm relation child ok	Freq.	Percent	Cum.
Strongly agree	468	39.97	39.97
Agree	383	32.71	72.67
Neither agree nor disagree	124	10.59	83.26
Strongly disagree	184	15.71	98.98
Cant choose	11	0.94	99.91
Na, refused	1	0.09	100.00
Total	1,171	100.00	

At first glance, everything looked fine. Then I realized that the category for disagree was missing. I checked the values associated with the value labels (this is an example of why I prefer value labels that include the value as discussed in chapter 5, page 167):

(*Continued on next page*)

```
. tabulate v4, miss nolabel
 Workg mom: |
       warm |
   relation |
   child ok |      Freq.     Percent        Cum.
------------+-----------------------------------
          1 |        468       39.97       39.97
          2 |        383       32.71       72.67
          3 |        124       10.59       83.26
          5 |        184       15.71       98.98
          8 |         11        0.94       99.91
          9 |          1        0.09      100.00
------------+-----------------------------------
      Total |      1,171      100.00
```

Indeed, category 4 is missing. Because the source data were distributed by ICPSR as an SPSS portable file, I used SPSS to confirm that the problem was not caused by data conversion. Next I went to the codebook and found: "USA: In the GSS2002 original questionnaire, the fourth response category for these items (V4 to V13) 'disagree' is omitted. Therefore, 'disagree' should be considered as being collapsed into the fifth category of 'strongly disagree' assuming that answers on 'strongly disagree' may reflect both 'disagree' and 'strongly disagree'." This means that the 2002 survey is not comparable to the 1977 and 1989 surveys.

6.2.2 Substantive review

You should also consider the substantive meaning of variables, not just whether the values are valid. That is, does the distribution of cases among values make substantive sense? This may seem obvious, but I know from experience that when reviewing thousands of lines of output it is easy to forget what the data are really about. Here are several examples that explore the substantive meaning of variables using the `tab1`, `stem`, `dotplot`, and `graph scatter` commands.

What does time to degree measure?

An example of the importance of evaluating each variable substantively occurred when I was writing a report on career outcomes in science for the National Academy of Sciences (Long 2002). Although the data used in the report are not publicly available, I can show you what happened using similar data. A standard predictor of scientific productivity is the number of years a scientist was enrolled in graduate school, where enrolled time is negatively associated with later productivity. The data analyst for the project sent me output showing a surprising, positive effect of enrolled time (file: `wf6-review-timetophd.do`):

```
. use wf-acpub, replace
(Workflow data on scientific productivity \ 2008-04-04)
. nbreg pub enrol phd female, nolog irr
Negative binomial regression                      Number of obs   =        278
                                                  LR chi2(3)      =      23.54
Dispersion     = mean                             Prob > chi2     =     0.0000
Log likelihood = -606.28466                       Pseudo R2       =     0.0190

------------------------------------------------------------------------------
         pub |        IRR   Std. Err.      z    P>|z|     [95% Conf. Interval]
-------------+----------------------------------------------------------------
       enrol |   1.056071   .0156467     3.68   0.000     1.025845    1.087188
         phd |   1.103679   .0654233     1.66   0.096     .9826206    1.239652
      female |      .7533   .0968775    -2.20   0.028     .5854637    .9692504
-------------+----------------------------------------------------------------
    /lnalpha |  -.4592692   .1471825                     -.7477416   -.1707969
-------------+----------------------------------------------------------------
       alpha |   .6317451   .0929818                      .4734346    .8429928
------------------------------------------------------------------------------
Likelihood-ratio test of alpha=0:  chibar2(01) =  172.26 Prob>=chibar2 = 0.000
```

The coefficient for `enrol` indicates that for each additional year in graduate school a scientist's rate of productivity is expected to increase by 5.6% (i.e., a factor of 1.056), holding other variables constant. After many discussions with the analyst and requests for alternative specifications to understand this peculiar result, I asked for frequency distributions for all the variables. For `enrol`, I found a handful of values that were too large to be enrolled time. Further investigation determined that `enrol` was the elapsed time between the bachelor's degree and the PhD, with large values for those scientists who had careers after the bachelor's and later decided to complete their doctorate. These outliers turned the relationship from negative to positive. Using the correct variable for enrolled time, things turned around as expected:

```
. nbreg pub enrol_fixed phd female, nolog irr
Negative binomial regression                      Number of obs   =        278
                                                  LR chi2(3)      =      26.97
Dispersion     = mean                             Prob > chi2     =     0.0000
Log likelihood = -604.5674                        Pseudo R2       =     0.0218

------------------------------------------------------------------------------
         pub |        IRR   Std. Err.      z    P>|z|     [95% Conf. Interval]
-------------+----------------------------------------------------------------
 enrol_fixed |     .82013    .037127    -4.38   0.000     .7504973    .8962233
         phd |   1.112075   .0666021     1.77   0.076     .9889072    1.250582
      female |   .7450266   .0964034    -2.27   0.023     .5781357     .960094
-------------+----------------------------------------------------------------
    /lnalpha |  -.4493616   .1428516                     -.7293456   -.1693777
-------------+----------------------------------------------------------------
       alpha |   .6380353   .0911444                      .4822244      .84419
------------------------------------------------------------------------------
Likelihood-ratio test of alpha=0:  chibar2(01) =  211.39 Prob>=chibar2 = 0.000
```

If I had begun with a review of the data, a simple histogram would have pointed out the problem and saved hours:

```
. label variable enrol_fixed "enroll_fixed: enrolled time"
. label variable enrol_fixed "enroll_fixed: enrolled time"
. label variable enroll "enrol: elapsed time"
. dotplot enrol_fixed enrol, ytitle("Years", size(medium))
> xlabel(, labsize(medium))
```

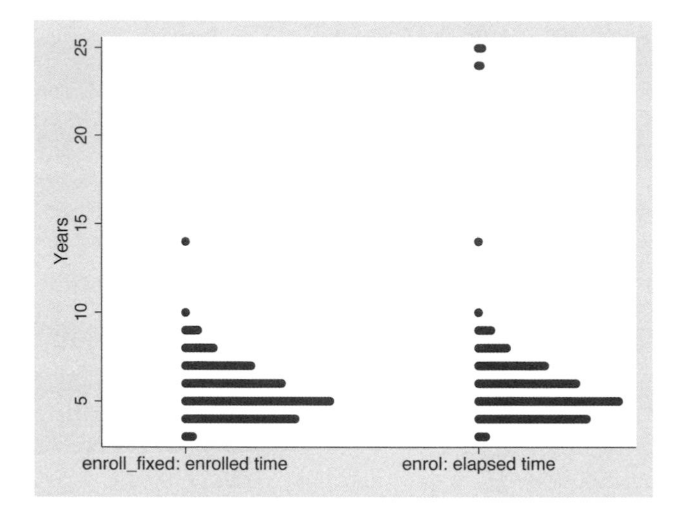

Examining high-frequency values

A second example from my work on the scientific career involves data on the prestige of the doctoral department for a sample of biochemists. The descriptive statistics look fine (file: `wf6-review-phdspike.do`):

```
. use wf-acjob, clear
(Workflow data on academic biochemists \ 2008-04-02)
. summarize phd
```

Variable	Obs	Mean	Std. Dev.	Min	Max
phd	408	3.200564	.9537509	1	4.8

The histogram shows a number of spikes in the data:

```
. dotplot phd
```

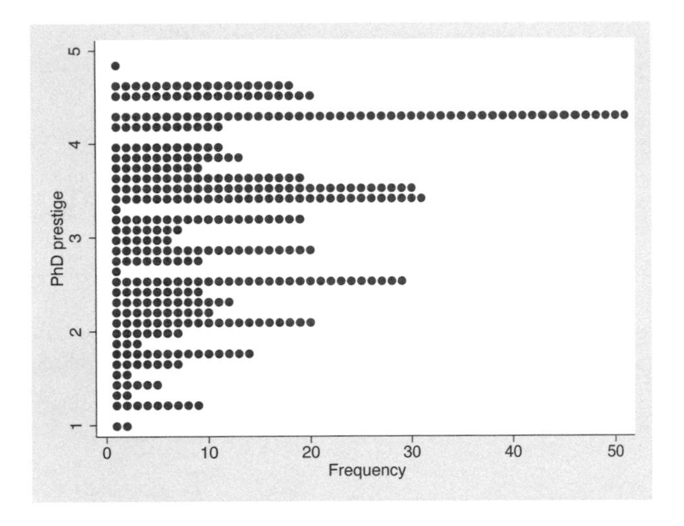

Although I expected the data to be lumpy because multiple PhDs can graduate from the same department and hence have the same PhD prestige, the spike between 4 and 5 looked very large. When I tabulated **phd** for cases between 4 and 4.5, I found the following:

```
. tabulate phd if phd>4 & phd<4.5
```

PhD prestige	Freq.	Percent	Cum.
4.14	3	4.35	4.35
4.16	8	11.59	15.94
4.25	1	1.45	17.39
4.29	37	53.62	71.01
4.32	9	13.04	84.06
4.34	4	5.80	89.86
4.48	7	10.14	100.00
Total	69	100.00	

Thirty-seven cases, nearly 10% of the sample, came from a school with a prestige of 4.29. Further investigation determined that this was the University of Wisconsin at Madison, which had several large departments that produced PhDs in biochemistry during the 1950s and 1960s. Later substantive analyses looked at this group of PhDs more carefully.

(Continued on next page)

Links among variables

Looking at pairs of variables together is an important part of cleaning your data. Here I examine the prestige of a scientist's doctoral department and the prestige of the first academic job. I start by looking at the descriptive statistics (file: `wf6-review-jobphd.do`):

```
. use wf-acjob, clear
(Workflow data on academic biochemists \ 2008-04-02)
. codebook phd job, compact
Variable    Obs Unique      Mean  Min  Max  Label
-------------------------------------------------------------------------------
phd         408     89  3.200564    1  4.8  PhD prestige
job         408     80  2.233431    1  4.8  Prestige of first job
-------------------------------------------------------------------------------
```

These statistics look reasonable. I expect the ranges to be the same with the mean for `job` to be smaller because there are fewer faculty employed than PhDs produced in more prestigious institutions. To compare the distributions, I use `dotplot`, where I change the variable labels to include the variable names and increase their font size in the graph:

```
. label var phd "phd: PhD prestige"
. label var job "job: Prestige of first job"
. dotplot phd job, xlabel(,labsize(medium))
```

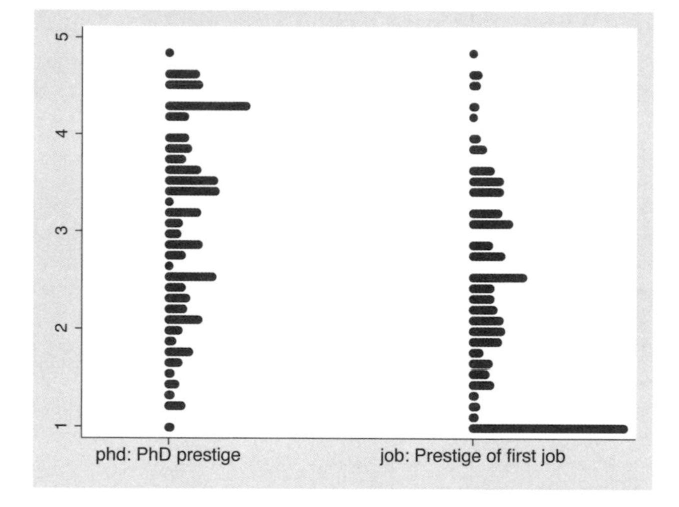

The distribution for `job` shows a spike at 1 because unrated departments were given the minimum prestige code. Clearly, this spike needs to be taken into account in substantive analyses. I also want to see how these variables are related. First, I create a simple scatterplot using the default options:

```
. scatter job phd
```

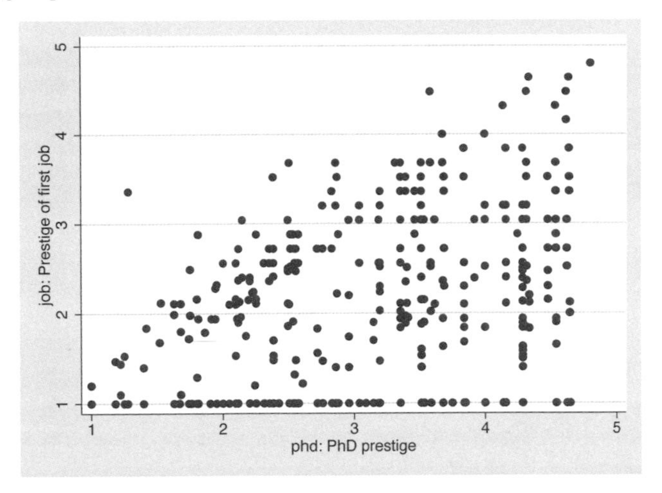

The biggest problem with this graph is that I cannot tell if there are multiple cases that correspond to a given point. For example, I know that 37 PhDs came from Madison with a prestige score of 4.29, but I cannot see 37 distinct observations with this PhD prestige. The solution is to jitter the data by adding a small amount of noise to each point with the `jitter()` option. To avoid black blobs when multiple points are close together, I use hollow circles for the plot symbol by adding the `msymbol(circle_hollow)` option. Finally, because the prestige scales for `phd` and `job` are the same, it makes sense to make the graph square by specifying an aspect ratio of 1. Using these options, we create

```
. scatter job phd, msymbol(circle_hollow) jitter(8) ylabel(, grid)
> xlabel(, grid) aspectratio(1)
```

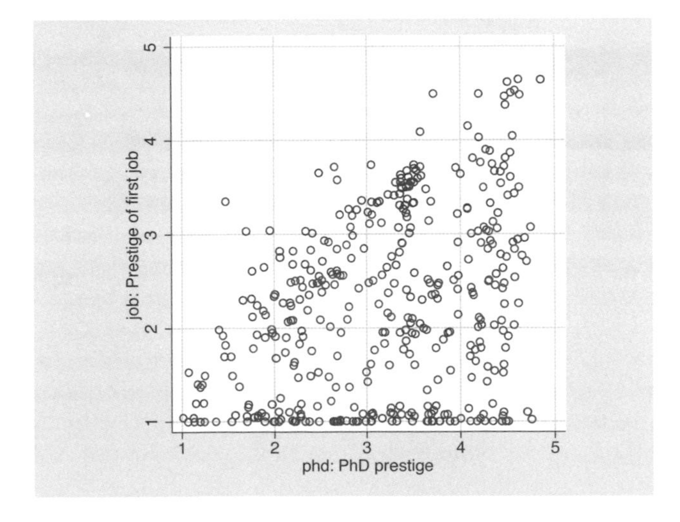

The distinctly triangular grouping of points demonstrates that mobility in science is largely downward after the PhD.

Scatterplots for all pairs of variables

I also want to look at other pairs of variables. One way to do this is with the graph matrix command:

```
. use wf-acjob, clear
(Workflow data on academic biochemists \ 2008-04-02)
. graph matrix job phd ment art cit fem fel, jitter(3) half
> msymbol(circle_hollow)
```

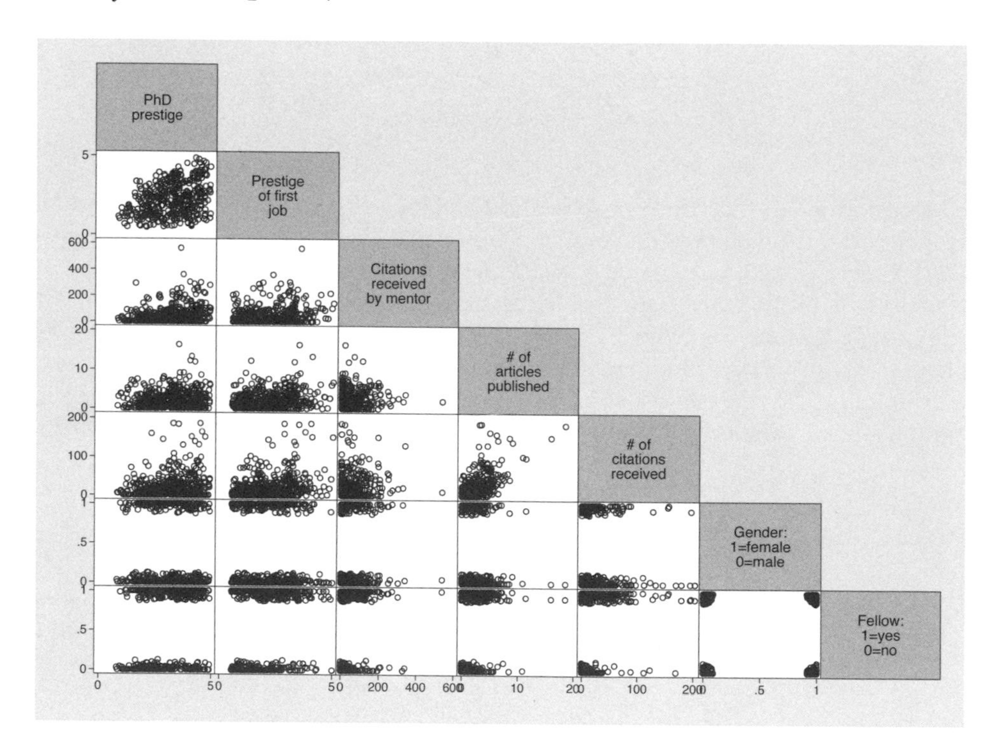

Each cell corresponds to a bivariate scatterplot. For example, compare the upper-left cell of the scatterplot matrix with the scatterplot for phd and job that was created above.

If there are many observations or a lot of variables in a scatterplot matrix, the graph can be too dense to be useful. When this happens, I generate individual plots for each pair of variables. This can be done with a pair of foreach loops:

```
 1> local varlist "phd job ment art cit fem fel"
 2> local nvars : word count `varlist´
 3> forvalues y_varnum = 1/`nvars´ {
 4>     local y_var : word `y_varnum´ of `varlist´
 5>     local y_lbl : variable label `y_var´
 6>     label var `y_var´ "`y_var´: `y_lbl´"
 7>     local x_start = `y_varnum´ + 1
 8>     forvalues x_varnum = `x_start´/`nvars´ {
 9>         local x_var : word `x_varnum´ of `varlist´
10>         local x_lbl : variable label `x_var´
11>         label var `x_var´ "`x_var´: `x_lbl´"
12>         scatter `y_var´ `x_var´, msymbol(circle_hollow) jitter(8) ///
  >             ylabel(, grid) xlabel(, grid) aspectratio(1)
13>         graph export wf6-review-jobphd-`y_var´-`x_var´.png, replace
14>         label var `x_var´ "`x_lbl´"
15>     }
16> }
```

This example is more complex than most. Although the following paragraphs explain what each line does, to fully understand the program I encourage you to experiment with `wf6-review-jobphd.do`. For example, to see how the content of a local macro changes, add a `display` command to the loop (e.g., `display "x_var is:  `x_var´"`).

Line 1 creates a list of variables, whereas line 2 counts the variables in the list. Line 3 begins a loop for the variable on the y axis, where `y_varnum` is the number of the variable to be plotted. Line 4 uses `y_varnum` to select the variable name from the list of variables. Using commands discussed in chapter 5, line 5 retrieves the variable label and line 6 changes the label by adding the variable's name to the front. For example, `phd` starts with the variable label `PhD prestige`. The new label is `phd: PhD prestige`. This allows me to label the axes with both the name and variable label. To see what we are doing with the labels, take a look at the `job` by `phd` graph:

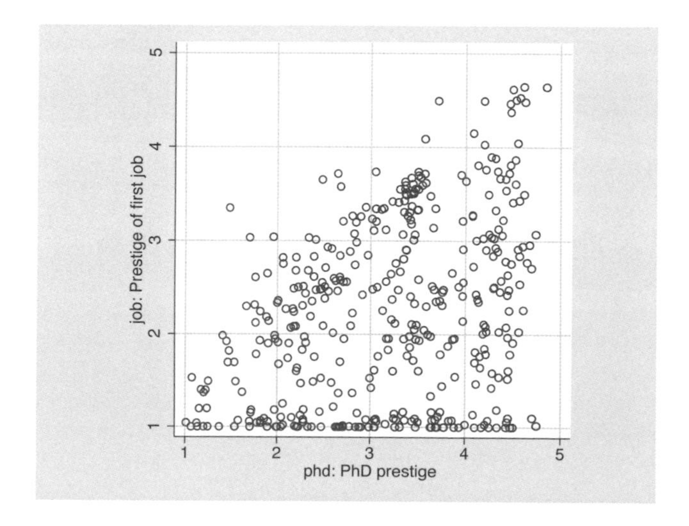

Line 7 determines the number of the variable for the x axis. For example, if the y axis contains the first variable, I do not want to plot that variable against itself. Instead, I want to plot it against the second variable in the list. Then line 8 loops through the variables beginning with the start variable through the rest of the list. When `y_varnum` is 1, I plot this variable against the x variables numbered 2–7. When `y_varnum` is 2, I plot the x variables numbered 3–7. And so on. Lines 9–11 adjust the label for the x variable. Lines 12 and 13 create and export the graph. Line 14 resets the label of the x variable to what it was before line 11. (As an exercise, figure out why this line is needed. If you are not sure, run the do-file after commenting out this line.) Lines 15 and 16 end the loops. When the loop finishes, 21 graphs have been created. To review these, I use a file viewer such as Irfanview (http://www.irfanview.com) or the viewer in the operating system to look at thumbnails for two-way graphs (see figure 6.10).

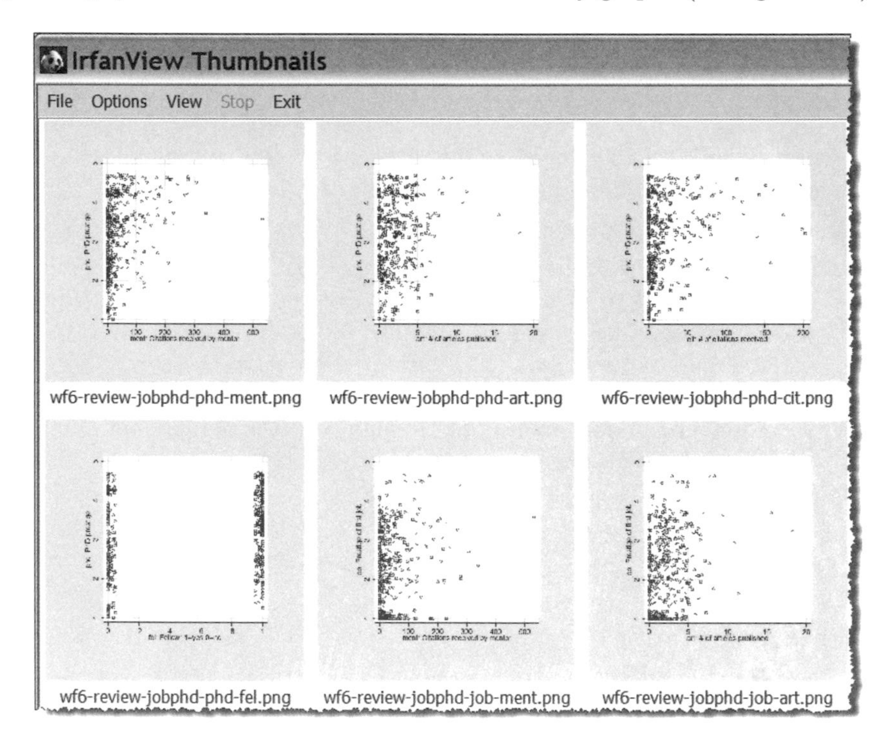

Figure 6.10. Thumbnails of two-way graphs

If I want to look more closely at one of the graphs, I click on it and a full-size version is displayed.

Changes in survey questions

The next example illustrates the importance of looking at univariate distributions to uncover changes and possible problems with panel data.[4] For a study of the effects of religion on sexual activity, Chavez (2007) used data from the National Longitudinal Study of Adolescent Health (Add Health). In wave 1, she found that 10.7% reported their religion was “Christian Church/Disciples of Christ”. In wave 3, the percentage dropped to 0.5%. Although other sources of data confirmed that there was a decline in membership for the Disciples of Christ, the numbers from the Add Health survey seemed extreme. In the codebook, she discovered a change in wording between waves. In wave 1, this category was labeled “Christian Church (Disciples of Christ)”; in wave 3, the label was simply “Disciples of Christ”. Chavez inferred that in wave 1 respondents who were Christian, but not necessarily Disciples of Christ, selected this category. This type of problem is easy to overlook if you focus on the output from complex models rather than basic descriptive statistics.

6.2.3 Missing-data review

Because problems with missing data occur in many subtle ways, the distribution of missing values should be checked carefully. I begin by discussing how Stata handles missing values. The system missing value, sometimes called the *sysmiss*, is indicated by `.`. For example,

```
generate var01 = .
```

assigns all observations for `var01` to be missing. Stata also has 26 extended missing values indicated as `.a`, `.b`, ..., `.z`. When missing values are included in numerical expressions, the result of the expression is a missing value. For example, if you add `.a` to another number, the result is a missing value. Because missing values are not valid numbers, commands such as `summarize` or `regress` cannot include a missing value in numerical computations. This is in contrast to software, especially older programs, where any number (e.g., 9 or 99) could be used as a missing value and the user had to indicate which numbers were missing. If you forgot to specify which values were missing, missing values were included in the computations with sometimes disastrous results. If you import data from a program in which valid numbers can be used to indicate missing values, be careful that these values are recoded as missing values in Stata.

Comparisons and missing values

Missing values can be used with the comparison operators `>`, `>=`, `==`, `>`, `>=`, and `!=`. In comparisons, missing values are treated as numbers that are larger than any nonmissing values with `.z` > `.y` > ··· > `.b` > `.a` > `.`. For example, if you check whether a missing value is greater than 1,000 (e.g., `. > 1000`), the answer is yes. This can easily lead to mistakes. For example, I begin by looking at the distribution of articles, which includes 19 cases with missing data (file: `wf6-review-missing.do`):

4. I thank Christine Chavez for bringing this example to my attention.

```
. use wf-missing, clear
(Workflow data to illustrate missing values review \ 2008-04-02)
. tabulate art, missing

       # of
   articles
  published        Freq.     Percent        Cum.

          0          102       41.98       41.98
          1           72       29.63       71.60
          2           25       10.29       81.89
          3           13        5.35       87.24
          4            9        3.70       90.95
          9            1        0.41       91.36
         12            1        0.41       91.77
         15            1        0.41       92.18
          .           19        7.82      100.00

      Total          243      100.00
```

Next I create a variable where values 9, 12, and 15 are truncated to 5:

```
. generate art_tr5 = art
(19 missing values generated)
. replace art_tr5 = 5 if art>5
(22 real changes made)
. label var art_tr5 "trunc at 5 # of articles published"
```

The `generate` command reports 19 missing values, which is the correct number. The `replace` command reports 22 changes, but only three values should have changed. To see what happened, I tabulate `art` and `art_tr5` using the `missing` option:

```
. tabulate art art_tr5, missing

       # of
   articles                                art_tr5
  published          0          1          2          3          4          5        Total

          0        102          0          0          0          0          0          102
          1          0         72          0          0          0          0           72
          2          0          0         25          0          0          0           25
          3          0          0          0         13          0          0           13
          4          0          0          0          0          9          0            9
          9          0          0          0          0          0          1            1
         12          0          0          0          0          0          1            1
         15          0          0          0          0          0          1            1
          .          0          0          0          0          0         19           19

      Total        102         72         25         13          9         22          243
```

The values 9, 12, and 15 of `art` were correctly changed to 5 in `art_tr5`, but the 19 missing values in `art` were also changed to 5. This happened because `replace art_tr5 = 5 if art>5` says that any value of `art` greater than 5 should be changed to 5. Because missing values are treated as very large numbers in comparisons, they are larger than 5.

The solution is to request changes only if values are not missing. That is, I want to change values of **art** to 5 only if two conditions are true. First, **art** must be greater than 5 (i.e., **art>5**). This is true for values such as 19 but is also true for missing values. Second, **art** must not be missing. There are two ways to select cases that are not missing. First, I can select values less than the sysmiss. For example, **art<.** selects cases that are valid numbers because in comparisons valid numbers are less than the sysmiss, which is less than all the extended missing values. Using this **if** condition, I get the correct result:

```
. generate art_tr5V2 = art
(19 missing values generated)

. replace art_tr5V2 = 5 if art>5 & art<.
(3 real changes made)

. label var art_tr5V2 "trunc at 5 # of articles published"

. notes art_tr5V2: created using art<.

. tabulate art_tr5V2, missing
```

trunc at 5 # of articles published	Freq.	Percent	Cum.
0	102	41.98	41.98
1	72	29.63	71.60
2	25	10.29	81.89
3	13	5.35	87.24
4	9	3.70	90.95
5	3	1.23	92.18
.	19	7.82	100.00
Total	243	100.00	

I selected cases using **art<.** rather than **art!=..** The later condition selects cases that are not equal to the sysmiss, so extended values would also be selected. Getting in the habit of using **<.** instead of **!=.** can prevent problems when your variables have extended missing values.

The second way to select valid cases is with the **missing()** function. For example, the function **missing(art)** returns a 1 if **art** is missing (either sysmiss or an extended missing), else 0. To select cases that are not missing, I use **!missing(art)**, where **!** means not. To create a truncated version of **art**, I can use these commands

```
. generate art_tr5V3 = art
(19 missing values generated)

. replace art_tr5V3 = 5 if art>5 & !missing(art)
(3 real changes made)

. label var art_tr5V3 "trunc at 5 # of articles published"

. notes art_tr5V3: created using !missing(art)
```

I prefer **!missing(art)** over **art<.** because nonprogrammers are more likely to understand what your program is doing. Plus, **missing(art)** avoids the mistake of using **art>.** when you mean **art>=..**

Creating indicators of whether cases are missing

Sometimes I need an indicator variable equal to 1 when a variable is missing, else 0. This indicator can be used to examine whether being missing on a given variable is associated with other variables. To create such an indicator for `art` (file: `wf6-review-missing.do`), use the commands

```
generate art_ismiss = missing(art)
label var art_ismiss "art is missing?"
label def Lismiss 0 0_valid 1 1_missing
label val art_ismiss Lismiss
```

To see how the new variable is related to the source variable, type

```
. tabulate art art_ismiss, missing
     # of  |
  articles |     art is missing?
 published |   0_valid  1_missing |     Total
-----------+----------------------+----------
         0 |       102          0 |       102 
         1 |        72          0 |        72 
         2 |        25          0 |        25 
         3 |        13          0 |        13 
         4 |         9          0 |         9 
         9 |         1          0 |         1 
        12 |         1          0 |         1 
        15 |         1          0 |         1 
         . |         0         19 |        19 
-----------+----------------------+----------
     Total |       224         19 |       243
```

In this example `art` only has the sysmiss, but if it had observations that were extended missing values, these would have been coded as 1 in `art_ismiss`.

Using extended missing values

Data can be missing for many reasons and extended missing values can be used to differentiate among the reasons for being missing. For example, a response could be missing because

- the respondent refused to answer the question.
- the respondent did not know the answer.
- the question was not appropriate for the respondent (e.g., the question "Where did you go to college?" is not applicable if a person did not go to college).
- the respondent was not asked the question (e.g., a split-ballot design).

Three commands are useful for examining the distribution of cases among types of missing values. The `tabulate` and `tab1` commands with the `missing` option tabulate both numeric and missing values. If I want the distribution of only missing values (e.g., for a variable that has thousands of valid values), I can add an `if` condition (file: `wf6-review-missing.do`):

```
tab1 phd if missing(phd), miss
```

By conditioning on missing(phd), tab1 tabulates only missing values. Unfortunately, this approach does not work to compute the distribution of missing values for multiple variables using a single tab1 command. For example, the command

```
tab1 phd art cit if missing(phd), missing
```

creates univariate tabulations for phd, art, and cit and includes only those observations in which phd is missing. The solution is to use a loop

```
foreach varname in phd art cit {
    tab1 `varname´ if missing(`varname´), missing
}
```

which leads to

```
-> tabulation of phd if missing(phd)
        PhD |
   prestige |      Freq.     Percent        Cum.
------------+-----------------------------------
    a_NonUS |          7       36.84       36.84
 b_Unranked |         12       63.16      100.00
------------+-----------------------------------
      Total |         19      100.00
-> tabulation of art if missing(art)
       # of |
   articles |
  published |      Freq.     Percent        Cum.
------------+-----------------------------------
          . |         19      100.00      100.00
------------+-----------------------------------
      Total |         19      100.00
  (output omitted)
```

Verifying and expanding missing-data codes

In some datasets, you will have variables with missing values where the reason the data are missing is not indicated. For example, the variable car_type might have missing observations but you do not know if the values are missing because a person refused to answer the question or because the person did not have a car. When this happens, you might be able to determine why the information is missing by examining how the variable is related to other variables. To illustrate how this is done, I use an example from a study of human sexuality at The Kinsey Institute (http://www.kinseyinstitute.org/research/kins.html). The data were collected both with a standard survey and with a web-based survey. Because a primary goal of the study was to compare the two modalities of data collection, it was essential to understand why responses were missing.[5]

5. The data in this example were simulated to reflect the properties of the original data, which are not publicly available.

In the original data, variables were coded `.a` if a question was refused and `.b` if a question was not applicable. No explanation was given for why a response was refused or a question was not applicable. For example, sometimes an answer was not applicable because a person was male and only women were asked the question. Sometimes a question was not applicable because of the respondent's answer to a previous question. Refusals also occurred for a variety of reasons. A person might simply refuse to answer the current question, but sometimes a person refused a previous question so the current question was not asked. In all, there were nine reasons why answers could be missing:

Missing Code	Value Label	Meaning
`.c`	`catskip`	Categorical response was not needed[a]
`.d`	`nodebrief`	Refused to answer debriefing questions
`.f`	`femskip`	Women were not asked this question
`.m`	`maleskip`	Men were not asked this question
`.p`	`priorref`	Question not asked because previous question refused
`.r`	`refused`	Current question was refused
`.s`	`single`	Not asked because respondent was single
`.x`	`nosxrel`	Not asked because respondent was not in sexual relationship
`.z`	`prior_0`	Not asked because answer to previous question was zero

[a] Some questions asked how many times an event occurred. If a person refused to give a specific number, she was asked to categorize the frequency of the event (e.g., never, a few times).

These values were defined with the command

```
label def missdat .c "c_catskip"   .d "d_nodebrief"  .f "f_femskip"  ///
                  .m "m_maleskip"  .p "p_priorref"   .r "r_refused"  ///
                  .s "s_single"    .x "x_nosxrel"    .z "z_prior_0"
```

To recode the original `.a`'s and `.b`'s into the refined missing-value types required several steps.

I began by reviewing the survey to determine which types of missing values were possible for each question. This information was kept in a spreadsheet (file: `wf6-review-misstype-registry.xls`); see figure 6.11.

		Possible reasons for missing data								
		.c	.d	.f	.m	.p	.r	.s	.x	.z
Question	*Variable*	*catskip*	*nodebrief*	*femskip*	*maleskip*	*priorrefuse*	*refused*	*single*	*nosexrel*	*priorzero*
19	reldevtn						r			
20	politics						r			
21	married						r			
22	maryear					p	r	s		
22	marmth					p	r	s		
23	sxrelin						r			
24	sxrelyear					p	r	s		
24	sxrelmth					p	r	s		
25	des4w						r			
26	des1y4w						r			
27	sxrel4w					p	r		x	
28	sxrelwry					p	r		x	
29	attract1st						r			
30	ownsx4w						r			
31	ownwry4w						r			
32	sxactptn						r			
33	intcrs4w					p	r			z
34	orlgive					p	r			z
35	orlrecv					p	r			z
36	arousal					p	r			z
37	ejack			f		p	r			z
38	erect			f		p	r			z
39	ejackqk			f		p	r			z
40	org4w				m	p	r			z
41	lube				m	p	r			z
42	pain				m	p	r			z
43	sxptn1y						r			

Figure 6.11. Spreadsheet of possible reasons for missing data

Some variables, such as `politics`, can only be missing if a person refuses to answer the question. Other variables allowed multiple reasons. For example, `maryear` can be missing if the respondent refused to indicate years married, because she refused the previous question on whether she was married, or because she was single and the question was not asked. I sorted the registry so that variables that required similar data processing appeared together (see figure 6.12).

(*Continued on next page*)

		Possible reasons for missing data								
		.c	.d	.f	.m	.p	.r	.s	.x	.z
Question	Variable	catskip	nodebrief	femskip	maleskip	priorrefuse	refused	single	nosexrel	priorzero
43a	sxptn1y_cat	c					r			
44a	femptn18_cat	c					r			
45a	maleptn18_cat	c					r			
46a	sxrellast_cat	c					r			
69	survey		d			p	r			
70	rembrpast		d			p	r			
71	difansr1y		d			p	r			
72	difansr4w		d			p	r			
73	difansrlt		d			p	r			
74	difansr1sx		d			p	r			
75	uncmfqstn		d			p	r			
76	face2face		d			p	r			
37	ejack			f		p	r			z
38	erect			f		p	r			z
39	ejackqk			f		p	r			z
60	erectprbtm			f		p	r			z
59	erectprb1y			f			r			
40	org4w				m	p	r			z

Figure 6.12. Spreadsheet with variables that require similar data processing grouped together

I processed all variables within a group using essentially the same commands, as shown next.

Simple refusals

The easiest variables to clean were those where a simple refusal, coded `.r`, is the only reason for missing data. I start by cloning the original variable. For example, `acttv` contains responses to a questions about the activity of watching television (file: `wf6-review-misstype.do`):

```
. clonevar acttvV2 = acttv
(1 missing value generated)
```

Looking at the frequencies, I find one missing value that is labeled `ref` for refused:

```
. tab1 acttvV2, missing

-> tabulation of acttvV2

  Q1, impt: wtch |
    TV or movies |      Freq.     Percent        Cum.
-----------------+-----------------------------------
Not at all impt1 |         12        5.50        5.50
               2 |         12        5.50       11.01
  (output omitted)
               9 |          5        2.29       91.28
     vry impt10 |         18        8.26       99.54
             ref |          1        0.46      100.00
-----------------+-----------------------------------
           Total |        218      100.00
```

To check the value associated with the label `ref`, I use the `nolabel` option:

```
. tab1 acttvV2, missing nolabel
-> tabulation of acttvV2
   Q1, impt:
  wtch TV or
      movies |      Freq.     Percent        Cum.
-------------+-----------------------------------
           1 |         12        5.50        5.50
           2 |         12        5.50       11.01
  (output omitted)
           9 |          5        2.29       91.28
          10 |         18        8.26       99.54
          .a |          1        0.46      100.00
-------------+-----------------------------------
       Total |        218      100.00
```

To change `.a` to `.r`, I use `recode`:

```
recode acttvV2 .a = .r
```

Equivalently, I could use `replace`:

```
replace acttvV2 = .r if acttvV2==.a
```

To verify that the recoding is correct, I tabulate the variables:

```
tabulate acttvV2 acttv, missing
```

Because everything is fine, I do not show the output.

Multiple missing values

Other variables had multiple reasons for being missing. Consider the question on years married. Here I want to code simple refusals as `.r`, single respondents as `.s`, and married respondents who refused a lead-in question as `.p`. I use the commands

```
1>  clonevar maryearV2 = maryear
2>  replace  maryearV2 = .r if maryear==.a
3>  replace  maryearV2 = .s if married==2
4>  replace  maryearV2 = .p if married==.a
```

Lines 1 and 2 are similar to those used for `acttv`. Line 3 changes `maryearV2` to `.s` if the person is single (i.e., `married==2`). Line 4 assigns the missing value `.p` if a respondent refused to answer the previous question on marital status and consequently was not asked how long she or he was married. A similar approach was used for other questions where multiple variables were needed to determine the specific reason a response was missing.

Missing values that are not missing

You should also look for values that are coded as missing but that should not be missing. Here is an example. The time a person was married was determined by two questions: 1) How many years have you been married? 2) How many months have you been married? The total months of marriage was obtained by combining these questions:

```
. generate martotal = (maryearV2*12) + marmthV2
(109 missing values generated)
. label var martotal "Total months married"
```

Missing codes were added for those who are single or who refused the previous questions:

```
. replace martotal = .s if married==2
(86 real changes made, 86 to missing)
. replace martotal = .p if married==.a
(1 real change made, 1 to missing)
```

Next `martotal` was coded `.r` for refused if either `maryearV2` or `marmthV2` were refused:[6]

```
. replace martotal = .r if marmthV2==.r | maryearV2==.r
(22 real changes made, 22 to missing)
```

After adding labels, I look at the distribution of missing values and discovered that substantially more respondents refused these questions than any other questions in the survey:

```
. label val martotal missdat
. tab1 martotal if missing(martotal), missing
-> tabulation of martotal if missing(martotal)
```

Total months married	Freq.	Percent	Cum.
p_priorref	1	0.92	0.92
r_refused	22	20.18	21.10
s_single	86	78.90	100.00
Total	109	100.00	

6. The symbol | means "or", so the `if` statement is checking whether any of the comparisons are true.

To figure out why, I listed those cases coded as `.r`:

```
. list martotal maryearV2 marmthV2 if martotal==.r, clean
         martotal   maryearV2    marmthV2
 12.    r_refused          53   r_refused
 34.    r_refused          31   r_refused
 37.    r_refused   r_refused          11
 38.    r_refused          33   r_refused
 40.    r_refused   r_refused           8
 45.    r_refused          54   r_refused
  (output omitted)
173.    r_refused          46   r_refused
190.    r_refused   r_refused           4
198.    r_refused          26   r_refused
206.    r_refused          24   r_refused
210.    r_refused          13   r_refused
214.    r_refused          28   r_refused
```

Nobody refused both questions. The most reasonable explanation, given that respondent's had answered more intimate questions about their marriage, is that some people rounded to the nearest year and did not report months. Those who were married less than a year skipped the year question rather than answering zero. Using these assumptions, I created a new version of `martotal`:

```
. generate martotalV2 = .
(218 missing values generated)
. label var martotalV2 "Total months married"
```

If both year and month are given, I used this formula:

```
. replace martotalV2 = (12*maryearV2) + marmthV2 ///
>     if !missing(maryearV2) & !missing(marmthV2)
(109 real changes made)
```

If only month is missing, I use the year information:

```
. replace martotalV2 = 12*maryearV2 if !missing(maryearV2) & marmthV2==.r
(19 real changes made)
```

These 19 cases had been coded as missing in the original version of the variable. Similarly, if year is refused, I only use the month:

```
. replace martotalV2 = marmthV2 if maryearV2==.r & !missing(marmthV2)
(3 real changes made)
```

Three cases that were originally missing now have valid values. Finally, I added missing codes for those who are single and or who refused the previous question:

```
. replace martotalV2 = .s if married==2
(86 real changes made, 86 to missing)
. replace martotalV2 = .p if married==.a
(1 real change made, 1 to missing)
. label val martotalV2 missdat
```

The revised distribution of missing values looks reasonable:

```
. tab1 martotalV2 if missing(martotalV2), miss
-> tabulation of martotalV2 if missing(martotalV2)
```

Total months married	Freq.	Percent	Cum.
p_priorref	1	1.15	1.15
s_single	86	98.85	100.00
Total	87	100.00	

Because constructing this variable involved several steps, I checked my work by sorting on `martotalV2` and listing relevant variables:

```
. sort martotalV2
. list martotalV2 maryearV2 marmthV2, clean

        martotalV2    maryearV2      marmthV2
  1.             2            0             2
  2.             4    r_refused             4
  3.             8    r_refused             8
  4.            11    r_refused            11
  5.            16            1             4
  6.            16            1             4
  7.            16            1             4
  8.            24            2     r_refused
  9.            25            2             1
  (output omitted)
127.           648           54     r_refused
128.           651           54             3
129.           715           59             7
130.           721           60             1
131.           735           61             3
132.    p_priorref   p_priorref    p_priorref
133.      s_single     s_single      s_single
134.      s_single     s_single      s_single
  (output omitted)
```

Everything looks fine and I add a note to document what was done:

```
notes martotalV2: marmthV2+(12*maryearV2) if both parts answered; ///
    marmthV2 if years is missing; maryearV2 if month is missing \ `tag´
```

where the local `tag` includes information on the name of the do-file, who ran it, and when.

Using include files

Recoding missing values was a big job for this dataset. To make the task easier and less error prone, I grouped variables according to the reasons data could be missing as shown in the spreadsheet above. I then used `include` files to simplify the programming (see chapter 4, page 107 for details on include files). For example, to code simple refusals, such as acttv above, I included the file wf6-review-misstype-refused.doi. This file assumes that the variable being processed is named by the local macro varnm (e.g., local varnm acttv). The include file has two commands:

```
clonevar `varnm´V2 = `varnm´
recode `varnm´V2 .a = .r
```

To use this file to recode actalk, use the commands

```
local varnm acttalk
include wf6-review-misstype-refused.doi
```

This is equivalent to the commands

```
clonevar acttalkV2 = acttalk
recode acttalkV2 .a = .r
```

To use the include file for a series of variables that have the same pattern of missing data, my commands are

```
local varnm acttalk
include wf6-review-misstype-refused.doi
local varnm actexer
include wf6-review-misstype-refused.doi
local varnm acthby
include wf6-review-misstype-refused.doi
```

For variables that had multiple reasons for missing data, I used multiple include files. For example, the variable sxrel4w was asked only if a respondent was in an active sexual relationship. The include file for this was wf6-review-misstype-ifsxrel.doi:

```
replace `varnm´V2 = .x if sxrelinV2==2
replace `varnm´V2 = .p if sxrelinV2==.r
```

To create the missing-value codes for sxrel4w, I used the include file for refusals as well as the include file for being in an active relationship:

```
local varnm sxrel4w
include wf6-review-misstype-refused.doi
include wf6-review-misstype-ifsxrel.doi
```

The advantage of include files is that if I change my mind about how to assign codes, I need only to change the include files and the change is applied to all variables that are being recoded. It also has the advantage that I need to enter the name of the variable only once. With hundreds of variables to check, it is easy to make an error when entering the names of variables. For example,

```
clonevar var01V2 = var01V2
replace var01V2 = .r if var01==.a
replace var01V2 = .x if sxrelinV2==2
replace var01V2 = .p if sxrelinV2==.r
clonevar var02V2 = var02V2
replace var02V2 = .r if var01==.a
replace var01V2 = .x if sxrelinV2==2
replace var02V2 = .p if sxrelinV2==.r
```

It is not hard to find the error here, but with hundreds of variables, it can be difficult.

6.2.4 Internal consistency review

When there are logical links among variables, you should verify that your data are consistent. For example, if I assume that formal education does not begin until age five, years of education should be at least five years less than a person's age. If not, there is probably an error in the data. If someone is not in the labor force, they should not report a wage. In other situations, I expect a particular relationship between variables although the relationship is not logically required. For example, I expect responses to questions about attitudes on working mothers to be consistent in the sense that people who are positive on one question will be positive on related questions. I do not expect a scientist to be hired in a department that is substantially more prestigious than her doctoral origin. Here I illustrate some ways to look for such problems.

Consistency in data on the scientific career

In data about careers in science, there are logical relationships among some variables and I expect certain patterns of relationship among other variables. For example, if a person does not publish, he logically cannot have citations. To check if this is true, I use the `assert` command (file: `wf6-review-consistent.do`):

```
. use wf-acjob, clear
(Workflow data on academic biochemists \ 2008-04-02)
. assert cit==0 if art==0
```

I am asserting that `cit` will equal 0 if `art` is 0. Because the assertion is true, `assert` does not produce any output. Here is an example of an assertion that does not hold (and logically need not be true):

```
. assert art==0 if cit==0
21 contradictions in 106 observations
assertion is false
r(9);
end of do-file
r(9);
```

The first `r(9)` is the return code from `assert` indicating that the assertion is false. Because the return code is not 0, the do-file terminates as indicated by the message `end of do-file` and the second return code of `r(9)`. If I am cleaning multiple variables and expect some assertions to be false, I want a record of the false assertion but do not want the do-file to terminate. To do this, I add the option `rc0`, which forces `assert` to return a code of 0 even if the assertion is false. For example,

```
. assert art==0 if cit==0, rc0
21 contradictions in 106 observations
assertion is false
```

I get the same message that the assertion is false, but because the return code is zero the do-file does not terminate. Another way to check for consistencies is with `tabulate` using an `if` condition:

```
. tabulate cit if art==0, miss

       # of |
  citations |
   received |      Freq.     Percent        Cum.
------------+-----------------------------------
          0 |         85      100.00      100.00
------------+-----------------------------------
      Total |         85      100.00
```

This is what I expected, so everything is fine.

I expect that a scientist's job prestige will usually be lower than her doctoral prestige although this is not logically required. I can check this with the `compare` command:

```
. compare job phd

                                        ---------- difference ----------
                            count       minimum      average     maximum
------------------------------------------------------------------------
job<phd                       288         -3.64    -1.462847        -.02
job=phd                        48
job>phd                        72           .01     .3709723        2.08
                       ----------
jointly defined               408         -3.64    -.9671323        2.08
                       ----------
total                         408
```

This shows that most of the time job prestige is less than PhD prestige (i.e., `job < phd`). If I want more detailed information, I can create a variable equal to the difference between the two variables:

```
. generate job_phd = job - phd
. label var job_phd "job-phd: >0 if better job"
```

(Continued on next page)

Next I **inspect** the new variable:

```
. inspect job_phd
job_phd:  job-phd: >0 if better job              Number of Observations
--------------------------------------          ------------------------------
                                                 Total   Integers  Nonintegers
|            #           Negative                 288        3         285
|        #   #           Zero                      48       48           -
|    #   #   #           Positive                  72        -          72
|    #   #   #                                  -----    -----       -----
|#   #   #   #           Total                    408       51         357
|#   #   #   #   .       Missing                    -
+----------------------                          -----
-3.64            2.08                              408
(More than 99 unique values)
```

To get even more detail, I can create a histogram with **stem**, or I can list those cases where the job is substantially more prestigious than the PhD. I start by sorting the data so that the list is ordered by the magnitude of the difference:

```
. sort job_phd
```

Next I list the large, positive differences:

```
. list job_phd art ment fem cit fel job phd if job_phd>.6, clean
        job_phd   art        ment        fem   cit           fel    job    phd
398.   .6000001     0           9   1_Female     0   0_NotFellow   2.72   2.12
399.   .6200001     3           6   1_Female    15      1_Fellow   2.88   2.26
400.   .6500001     1          36   1_Female    18   0_NotFellow   3.52   2.87
401.        .74     2           6     0_Male    19      1_Fellow   2.49   1.75
402.   .8200002     0          20   1_Female     0   0_NotFellow   3.68   2.86
403.   .8899999     0           9     0_Male     0   0_NotFellow   3.04   2.15
404.   .8900001     0          20     0_Male     0      1_Fellow   4.48   3.59
405.       1.07     4         233     0_Male    22      1_Fellow   2.88   1.81
406.       1.13     0           0   1_Female     0   0_NotFellow   3.52   2.39
407.       1.17     4    69.99999     0_Male    41      1_Fellow   3.68   2.51
408.       2.08     1    3.999999   1_Female    32      1_Fellow   3.36   1.28
```

The only large difference is for someone who had one article with 32 citations. Perhaps important work from the dissertation lead to a prestigious job. Next I look at those whose jobs are much less prestigious than their PhD:

```
. list job_phd art ment fem cit fel job phd if job_phd<-2, clean
       job_phd   art        ment        fem   cit           fel   job    phd
  1.     -3.64     0           2   1_Female     0      1_Fellow     1   4.64
  2.     -3.64     3          16     0_Male    24      1_Fellow     1   4.64
  3.     -3.62     0    87.99999     0_Male     0      1_Fellow     1   4.62
  4.     -3.54     1          23   1_Female     5   0_NotFellow     1   4.54
  5.     -3.54     5    47.00001     0_Male    27   0_NotFellow     1   4.54
  6.     -3.29     2         204   1_Female     9   0_NotFellow     1   4.29
  (output omitted)
```

Before you start substantive analysis, I encourage you to explore your data. Not only can this help find errors, but it also provides information that is useful in your substantive analysis.

6.2.5 Principles for fixing data inconsistencies

It would be nice if your data arrived with everything coded the way you wanted and with all inconsistencies resolved. In my experience, this never happens. When you examine your data, you will find problems that might have been caused by the survey instrument, by mistakes in data processing before the data were released, by inconsistencies introduced by the respondents when answering the questions, or by your own data processing. If you are lucky, the solution will be obvious. For example, the codebook might say that `gender` is coded 1 and 2, but in the dataset, the values 0 and 1 are used. An email to the data provider should resolve the issue. Unfortunately, things are rarely that simple. Often you will not get a definitive answer that explains an anomaly, and you must decide what to do based on the best available, albeit imperfect, information. In the example of years and months of marriage (page 234), I made a judgment call that some missing values were zeros. This decision could be wrong, but I decided that the potential distortion introduced by treating these values as missing was greater than that from coding these values as zero. This type of problem occurs frequently when cleaning data. In a study of organizations, the question "Does your organization have any revenues?" might not be answered while the question "How much revenue does your organization receive?" is answered with a positive value. Or a survey of organizations might include several yes/no questions about the availability of resources such as computers or Internet access. Suppose that five resources are checked as yes, but four others are left blank. Is this a nonresponse or simply a shortcut taken by the person filling out the survey?[7] When cleaning data you often need to make decisions to resolve issues where the data are ambiguous or inconsistent. Coding such cases as missing can introduce more distortion than if you make a reasonable judgment based on a careful assessment of all available information. Although uncomfortable, I think that such informal imputation is often appropriate. Of course, you should state in the methods section of any paper that uses the data how you treated such problems.

6.3 Creating variables for analysis

Datasets rarely arrive with all the variables that you need. For example, a dataset might include years of education, but you also need binary indicators of the highest degree attained. Alternatively, a dataset includes age, but you also need age-squared and age-cubed in some models. I start by discussing principles for creating variables, then review commands for creating variables. Next I discuss methods for verifying that new variables are created correctly, and I end with several examples.

7. I thank Curtis Child for raising this issue and providing me with several examples.

6.3.1 Principles for creating new variables

There are four simple principles for creating new variables.

1. If a variable is new, give it a new name.
2. Verify that new variables are constructed correctly.
3. Document new variables with notes and labels.
4. Keep the source variables used to create new variables.

Each principle is now considered (file: `wf6-create.do`).

New variables get new names

The most basic principle is that you should create a new variable rather than changing an existing variable. For example, if you need the log of `x`, create a new variable, say `xlog`, rather than replacing the current values of `x` with the log of `x`; that is, do not `replace x = log(x)`. If you change an existing variable, you risk doing something like the following. I start by fitting a model that includes the log of wages `lwg`:

```
use wf-lfp, clear
logit lfp k5 k618 age wc hc lwg inc
estimates store model_1
```

To explore the effect of using wages without taking the log, I replace the variable `lwg`:

```
replace lwg = exp(lwg)
```

Then I fit a model that includes the revised variable and drops `inc`:

```
use wf-lfp, clear
logit lfp k5 k618 age wc hc lwg
estimates store model_2
```

To compare the models, I list the estimates:

```
. estimates table _all, stats(N bic) eform b(%9.3f) t(%6.2f)
```

Variable	model_1	model_2
k5	0.232	0.237
	-7.43	-7.44
k618	0.937	0.916
	-0.95	-1.31
age	0.939	0.934
	-4.92	-5.49
wc	2.242	1.999
	3.51	3.10
hc	1.118	0.867
	0.54	-0.73
lwg	1.831	1.752
	4.01	3.77
inc	0.966	
	-4.20	
_cons	24.098	18.889
	4.94	4.67
N	753	753
bic	958.258	971.139

legend: b/t

Looking at the table, I could easily conclude that the change in the magnitude and significance of `lwg` is due to excluding `inc` from the model. In fact, the changes in `model_2` are caused by transforming `lwg`. In a simple example like this, it is easy to see what is going on. But with complex models and pages of output from weeks of analysis, it is easy to get confused. The best and simplest solution is to never change an existing variable, but instead create a new variable with a new name.

Verify that new variables are correct

When you create a variable, you should immediately verify that the variable was created correctly, paying particular attention to missing values. For example, suppose that I create a variable by taking the log of income,

```
. generate inclog = log(inc)
(1 missing value generated)
. label var inclog "log(inc)"
```

The output indicates that the new variable has one observation that is missing. I might incorrectly conclude that the missing value is because of a missing value in the source variable `inc`. In fact, the missing value is caused by `inc` having one observation that is less than zero, which results in a missing value because the log is undefined:

```
. list inc inclog if inc<0, clean

             inc   inclog
373.   -.0290001        .
```

When doing exploratory data analysis, it is easy to get caught up in your work and assume that you will check what you did later. In my experience, it is faster and more reliable to check a new variable immediately after creating it.

Document new variables

As soon as you create a variable, you should add a variable label, include a note to document the variable's origin, and attach value labels if appropriate. For example,

```
generate inc_log5 = ln(inc+.5) if !missing(inc)
label var inc_log5 "Log(inc+.5)"
notes inc_log5: log(inc+.5) \ wf6-create.do jsl 2008-04-04.
```

Adding notes makes it much easier to later verify how a variable was created. If the variable is categorical, be sure to add value labels as well.

Keep the source variables

Keeping source variables in your dataset allows you to later verify your work, fix errors in your new variables, and create additional variables. There are two exceptions to the rule. First, if the size of your dataset is a major concern, you might need to delete variables. You need a very large dataset for this to be an issue. Second, if the source variables had errors that were corrected in newer variables, you might delete the problem variables so that there is no chance of using them by mistake. If you decide to delete a variable, I suggest adding a note, such as

```
notes: inc was deleted from binlfp3.dta due to a coding error; use incV2 ///
    instead. If you need inc, see binlfp2.dta \ myjob01.do jsl 2007-04-19.
```

Alternatively, you could keep the problem variables but add a note and change the variable label. For example,

```
notes inc: inc should not be used; incorrect coding of high incomes ///
    Variable incV2 fixes the error \ myjob01.do jsl 2007-04-19.
label var inc: "DO NOT USE; see incV2 \ myjob01.do jsl 2007-04-19."
```

6.3.2 Core commands for creating variables

`generate`, `clonevar`, and `replace` are the basic commands for creating variables. Indeed, many other commands are built using these commands. I review these commands with examples that emphasize issues related to missing data (file: `wf6-create.do`).

The generate command

The **generate** command creates a new variable using the syntax

```
generate newvar = exp [if] [in]
```

where *exp* is a Stata expression, for example, `generate agesqrt = sqrt(age)`. Optionally, `if` and `in` conditions can be used to determine which observations to use when creating the variable, with excluded observations coded as missing. For example, `generate agesqrt = sqrt(age) if age>5` takes the square root of values of `age` greater than five with missing values assigned for other values of age. The `generate` command is used so often that it can be abbreviated briefly as `g agesqrt = sqrt(age)`, although I most often use the longer abbreviation `gen`.

The clonevar command

The `clonevar` command creates a duplicate of an existing variable and has the syntax:

```
clonevar newvar = sourcevar [if] [in]
```

The cloned variable has the same values, variable label, and value labels (among other things) as *sourcevar*. To illustrate the difference between `generate` and `clonevar`, I created copies of `lfp` using both commands:

```
. use wf-lfp, clear
(Workflow data on labor force participation \ 2008-04-02)
. generate lfp_gen = lfp
. clonevar lfp_clone = lfp
```

I am not adding variable labels because I want to show how `clonevar` copies the labels from the source variable. The values of the three variables are identical:

```
. summarize lfp*
    Variable |       Obs        Mean    Std. Dev.       Min        Max
-------------+--------------------------------------------------------
         lfp |       753    .5683931    .4956295          0          1
     lfp_gen |       753    .5683931    .4956295          0          1
   lfp_clone |       753    .5683931    .4956295          0          1
```

However, if I describe the variables, I see that `lfp_gen` does not have the same storage type or labels as the original `lfp`:

```
. describe lfp*
              storage  display     value
variable name   type   format      label      variable label
---------------------------------------------------------------------
lfp             byte   %9.0g       lfp        In paid labor force? 1=yes 0=no
lfp_gen         float  %9.0g
lfp_clone       byte   %9.0g       lfp        In paid labor force? 1=yes 0=no
```

The replace command

The replace command changes the values of an existing variable. The syntax is

replace *newvar* = *exp* [*if*] [*in*]

For example, if edyears is years of education, I can use replace to create a variable with categories of education based on years of education. I start by creating educcat, which is equal to edyears (I do not use replace on edyears because it is the source variable). I use generate instead of clonevar because I do not want to retain the original labels:

```
. use wf-russia01, clear
(Workflow data to illustrate creating variables \ 2008-04-02)
. generate educcat = edyears
(159 missing values generated)
. label var educcat "Categorized years of education"
```

I change educcat to 1 if edyears is within the range 0–8:

```
. replace educcat = 1 if edyears>=0  & edyears<=8    // no HS
(278 real changes made)
```

Similarly, I create the other values for educcat:

```
. replace educcat = 2 if edyears>=9  & edyears<=11  // some HS
(501 real changes made)
. replace educcat = 3 if edyears==12                // HS
(205 real changes made)
. replace educcat = 4 if edyears>=13 & edyears<=15  // some college
(517 real changes made)
. replace educcat = 5 if edyears>=16 & edyears<=24  // college plus
(135 real changes made)
```

Next I label the categories:

```
. label def educcat 1 1_NoHS 2 2_someHS 3 3_HS 4 4_someCol 5 5_ColPlus
> .b b_Refused .c c_DontKnow .d d_AtSchool .e e_AtCollege
> .f f_NoFrmlSchl
. label val educcat educcat
```

The new variable has the following values:

```
. tab1 educcat, missing
-> tabulation of educcat

Categorized |
   years of |
  education |      Freq.     Percent        Cum.
------------+-----------------------------------
     1_NoHS |        281       15.63       15.63
   2_someHS |        501       27.86       43.49
       3_HS |        205       11.40       54.89
  4_someCol |        517       28.75       83.65
  5_ColPlus |        135        7.51       91.16
  b_Refused |          3        0.17       91.32
 c_DontKnow |         61        3.39       94.72
 d_AtSchool |          7        0.39       95.11
e_AtCollege |         73        4.06       99.17
f_NoFrmlSchl |        15        0.83      100.00
------------+-----------------------------------
      Total |      1,798      100.00
```

6.3.3 Creating variables with missing values

The most common problems I see when creating new variables involve missing values. It is easy to create missing values that should not be missing or to incorrectly turn missing values into seemingly valid values. The following example illustrates how easy it is to make this type of mistake (file: `wf6-create.do`).

The variable `marstat` classifies marital status into five categories and has 19 observations with missing values. I want to create an indicator variable that is 1 if a person is married and 0 if a person is widowed, divorced, separated, or single. First, I look at `marstat` using the `missing` option:

```
. use wf-russia01, clear
(Workflow data to illustrate creating variables \ 2008-04-02)

. tab1 marstat, miss

-> tabulation of marstat

    Marital |
     status |      Freq.     Percent        Cum.
------------+-----------------------------------
  1_married |        931       51.78       51.78
  2_widowed |        321       17.85       69.63
 3_divorced |        215       11.96       81.59
4_separated |         33        1.84       83.43
   5_single |        279       15.52       98.94
         .b |         19        1.06      100.00
------------+-----------------------------------
      Total |      1,798      100.00
```

Because 1 indicates a person is married, I use an equality comparison to create an indicator of being married:

```
. generate ismar_wrong = (marstat==1)
. label var ismar_wrong "Is married created incorrectly"
. label def Lyesno 0 0_no 1 1_yes
. label val ismar_wrong Lyesno
```

To check the new variable, I tabulate it with the source variable:

```
. tabulate marstat ismar_wrong, miss

             |  Is married created
     Marital |      incorrectly
      status |      0_no      1_yes |     Total
-------------+----------------------+----------
   1_married |         0        931 |       931
   2_widowed |       321          0 |       321
  3_divorced |       215          0 |       215
 4_separated |        33          0 |        33
    5_single |       279          0 |       279
          .b |        19          0 |        19
-------------+----------------------+----------
       Total |       867        931 |     1,798
```

The problem is that missing values in **marstat** were assigned 0 in **ismar_wrong**. This is because a missing value is not equal to 1 and anything that is not equal to 1 is set equal to 0. Because I want missing values to stay missing, I need to exclude missing values from the comparison by adding the condition **if !missing(marstat)**:

```
. generate ismar_right = (marstat==1) if !missing(marstat)
(19 missing values generated)
. label var ismar_right "Is married?"
. label val ismar_right Lyesno
```

This changes the extended missing **.b** to the sysmiss:

```
. tabulate marstat ismar_right, miss

     Marital |            Is married?
      status |      0_no      1_yes          . |     Total
-------------+---------------------------------+----------
   1_married |         0        931          0 |       931
   2_widowed |       321          0          0 |       321
  3_divorced |       215          0          0 |       215
 4_separated |        33          0          0 |        33
    5_single |       279          0          0 |       279
          .b |         0          0         19 |        19
-------------+---------------------------------+----------
       Total |       848        931         19 |     1,798
```

To retain the extended missing value **.b** rather than converting it to sysmiss, I can use a **replace** command:

```
. replace ismar_right = .b if marstat==.b
(19 real changes made, 19 to missing)

. tabulate marstat ismar_right, miss

   Marital |           Is married?
    status |      0_no      1_yes         .b |     Total
-----------+---------------------------------+----------
 1_married |         0        931          0 |       931
 2_widowed |       321          0          0 |       321
3_divorced |       215          0          0 |       215
4_separated|        33          0          0 |        33
  5_single |       279          0          0 |       279
        .b |         0          0         19 |        19
-----------+---------------------------------+----------
     Total |       848        931         19 |     1,798
```

If the source variable included multiple extended missing values (e.g., `.a`, `.b`, `.c`), I would need a `replace` command for each type of missing value. (As an exercise, write a `foreach` loop that would do this for you.)

6.3.4 Additional commands for creating variables

There are many other commands that create variables. Here I discuss three commands that I find particularly useful (file: `wf6-create.do`).

The recode command

The `recode` command is a quick way to recode the values of a variable and/or combine values into new categories. The easiest way to understand this is by extending our last example in which I created an indicator for being married. I want the value of the variable I am creating to equal 1 when the source variable is 1. With `recode` I indicate this with the simple expression `1=1`. I want values 2 through 5 in the source variable to be recoded as 0, which is specified as `2/5=0`. My command is

```
recode marstat 1=1 2/5=0, generate(ismar2_right)
```

The `generate()` option indicates the name of the variable being created. Here are the results I obtain

```
. recode marstat 1=1 2/5=0, gen(ismar2_right)
(848 differences between marstat and ismar2_right)

. label var ismar2_right "Is married?"

. tabulate marstat ismar2_right, miss

   Marital |           Is married?
    status |         0          1         .b |     Total
-----------+---------------------------------+----------
 1_married |         0        931          0 |       931
 2_widowed |       321          0          0 |       321
3_divorced |       215          0          0 |       215
4_separated|        33          0          0 |        33
  5_single |       279          0          0 |       279
        .b |         0          0         19 |        19
-----------+---------------------------------+----------
     Total |       848        931         19 |     1,798
```

Importantly, with recode, any values that are not explicitly recoded retain their original values. Thus you can easily recode valid values while keeping the missing values unchanged.

The recode command has many options that you can learn about with help recode or in [D] **recode**. I provide only a few examples of key features. Earlier, I created educcat with a series of replace commands. I could have done exactly the same thing more easily with a single recode command:

```
. recode edyears 0/8=1 9/11=2 12=3 13/15=4 16/24=5, generate(educcat2)
(1636 differences between edyears and educcat2)
```

I am specifying that source values from 0 through 8 be recoded to 1; values 9 through 11 be recoded to 2; and so on. The generate() option indicates that I want to create a variable named educcat2.

To recode 1 to 0 and change all other values, including missing, to 1 I use * to indicate "all other values":

```
. recode edyears 1=0 *=1, generate(edtest1)
(1798 differences between edyears and edtest1)
```

If I do not want missing values to be recoded, I exclude those cases with an if condition:

```
. recode edyears 1=0 *=1 if !missing(edyears), generate(edtest2)
(1639 differences between edyears and edtest2)
```

To leave 1–5 unchanged and recode 6–24 to 6

```
. recode edyears 6/24=6 if !missing(edyears), generate(edtest3)
(1546 differences between edyears and edtest3)
```

To change the values 1, 3, 5, 7, and 9 to −1 while leaving other values unchanged

```
. recode edyears 1 3 5 7 9=-1, generate(edtest4)
(182 differences between edyears and edtest4)
```

To recode values from 6 to the maximum to 6 while leaving other values unchanged

```
. recode edyears 6/max=6, generate(edtest5)
(1546 differences between edyears and edtest5)
```

max stands for the maximum nonmissing value; min stands for the minimum value.

The egen command

The command egen stands for extended generate command. There are dozens of egen commands that allow you to make complex transformations easily. To whet your appetite for how easy and powerful egen commands can be, here is a simple example. Suppose that I want to standardize the variable age by subtracting the mean and dividing by the standard deviation. I could do this by using returned results from summarize:

```
. use wf-lfp, clear
(Workflow data on labor force participation \ 2008-04-02)

. summarize age

    Variable |       Obs        Mean    Std. Dev.       Min        Max
-------------+--------------------------------------------------------
         age |       753    42.53785    8.072574         30         60

. generate agestd = (age - r(mean)) / r(sd)

. label var agestd "Age standardized using generate"
```

With `egen`, I could do this without needing to run `summarize`:

```
. egen agestdV2 = std(age)

. label var agestdV2 "Age standardized using egen"
```

The results are identical:

```
. summarize agestd agestdV2

    Variable |       Obs        Mean    Std. Dev.       Min        Max
-------------+--------------------------------------------------------
      agestd |       753   -7.05e-09           1  -1.553141   2.163145
    agestdV2 |       753   -7.05e-09           1  -1.553141   2.163145
```

Although `egen` commands are extremely powerful, it takes some time to fully understand how each one works. Accordingly, I do not use `egen` commands unless it is something I do frequently or the transformation is difficult to make any other way. For example, suppose that I want to create the variable `count0` that counts how many of the variables `lfp`, `k5`, `k618`, `age`, `wc`, `hc`, `lwg`, and `inc` are equal to 0. I can do this with the `egen anycount` command:

```
. egen count0 = anycount(lfp k5 k618 age wc hc lwg inc), values(0)

. label var count0 "# of 0's in lfp k5 k618 age wc hc lwg inc"

. tabulate count0, miss

# of 0's in |
lfp k5 k618 |
  age wc hc |
    lwg inc |      Freq.     Percent        Cum.
------------+-----------------------------------
          0 |         11        1.46        1.46
          1 |         94       12.48       13.94
          2 |        157       20.85       34.79
          3 |        251       33.33       68.13
          4 |        169       22.44       90.57
          5 |         71        9.43      100.00
------------+-----------------------------------
      Total |        753      100.00
```

This tells me that for 11 observations none of the variables are equal to 0; for 94 observations exactly one of the variables is equal to 0; and so on. To make sure I understood what the `anycount` command was doing, I created another variable that counted 0s using a `foreach` loop:

```
generate count0v2 = 0
label var count0v2 "v2:lfp k5 k618 age wc hc lwg inc == 0"
foreach var in lfp k5 k618 age wc hc lwg inc {
    replace count0v2 = count0v2 + 1 if `var´==0
}
```

I use `compare count0 count0v2` to confirm that the variables are identical.

If you do much data management or encounter a difficult problem in constructing a variable, it is well worth the time it takes to read [D] **egen**.

The tabulate, generate() command

It is often necessary to create indicator variables for the categories of a variable. The easy way to do this is with the `tabulate` command using the `generate()` option:

`tabulate` *varname* [*if*] [*in*], `generate(`*newstub*`)` [`missing`]

The `generate(`*newstub*`)` option requests that indicator variables be created for each category of *varname*. Each indicator starts with the prefix *newstub* and ends with the value of the source variable. If the `missing` option is included, indicators are created for the missing-value categories as well. For example, `marstat` has five categories of marital status. For a regression model, I want dummy variables for each marital status (except for the excluded or reference category). I could do this with a series of `generate()` commands, but it is easier to use `tabulate`:

```
. use wf-russia01, clear
(Workflow data to illustrate creating variables \ 2008-04-02)
. tabulate marstat, gen(ms_is)

    Marital
     status |      Freq.     Percent        Cum.
------------+-----------------------------------
  1_married |        931       52.33       52.33
  2_widowed |        321       18.04       70.38
 3_divorced |        215       12.09       82.46
4_separated |         33        1.85       84.32
   5_single |        279       15.68      100.00
------------+-----------------------------------
      Total |      1,779      100.00
```

Five variables are created:

```
. codebook ms_is*, compact
Variable    Obs Unique      Mean  Min  Max  Label
---------------------------------------------------------------------
ms_is1     1779      2  .5233277    0    1  marstat==1_married
ms_is2     1779      2  .1804384    0    1  marstat==2_widowed
ms_is3     1779      2  .1208544    0    1  marstat==3_divorced
ms_is4     1779      2  .0185497    0    1  marstat==4_separated
ms_is5     1779      2  .1568297    0    1  marstat==5_single
---------------------------------------------------------------------
```

The means correspond to the percentages from the tabulation of `marstat` shown above. The variable labels explain how each variable is defined. To verify the new variables, I can tabulate them against the source variable. For example,

```
. tabulate marstat ms_is1, miss
   Marital |        marstat==1_married
    status |         0          1          . |     Total
-----------+---------------------------------+----------
 1_married |         0        931          0 |       931
 2_widowed |       321          0          0 |       321
3_divorced |       215          0          0 |       215
4_separated|        33          0          0 |        33
  5_single |       279          0          0 |       279
        .b |         0          0         19 |        19
-----------+---------------------------------+----------
     Total |       848        931         19 |     1,798
```

Although `tabulate, generate()` creates the indicators very quickly, I generally want to improve the names and labels that are used. For example, I can rename `ms_is1` and add some other details

```
label def Lyesno 0 0_no 1 1_yes
rename ms_is1 ms_married
note ms_married: Source var is marstat \ `tag'
label var ms_married "Married?"
label val ms_married Lyesno
```

and so on for other variables.

6.3.5 Labeling variables created by Stata

When commands such as `predict` create variables, the default labels are usually brief and generic. To document such variables, it is good practice to add your own labels and notes. This is particularly important when the variables are based on complex procedures such as `ice` (Royston 2004) for imputing missing data or `predict` when making postestimation predictions. Here is an example that illustrates how the default labels can be confusing. I start by fitting two models and computing predicted probabilities (file: `wf6-create.do`):

```
. use wf-lfp, clear
(Workflow data on labor force participation \ 2008-04-02)
. * model 1
. logit lfp k5 k618 age wc hc lwg inc
  (output omitted)
. predict prm1
(option pr assumed; Pr(lfp))
. * model 2
. logit lfp age wc hc lwg inc
  (output omitted)
. predict prm2
(option pr assumed; Pr(lfp))
```

The `predict` command created `prm1` and `prm2`:

```
. codebook prm*, compact
Variable     Obs Unique        Mean         Min         Max  Label
-------------------------------------------------------------------------------
prm1         753    753   .5683931    .0139875    .9621198  Pr(lfp)
prm2         753    753   .5683931    .1012935    .8985487  Pr(lfp)
-------------------------------------------------------------------------------
```

Although these labels are accurate, they are not useful for distinguishing between the two variables. To avoid confusion, I now add my own labels along with a note about how the predictions were computed:

```
. use wf-lfp, clear
(Workflow data on labor force participation \ 2008-04-02)
. logit lfp k5 k618 age wc hc lwg inc
  (output omitted)
. predict prm1
(option pr assumed; Pr(lfp))
. label var prm1 "Pr(lfp|m1=k5 k618 age wc hc lwg inc)"
. notes prm1: m1=logit lfp k5 k618 age wc hc lwg inc \ wf6-create.do jsl 2008-04-05.
. logit lfp age wc hc lwg inc
  (output omitted)
. predict prm2
(option pr assumed; Pr(lfp))
. label var prm2 "Pr(lfp|m2=age wc hc lwg inc)"
. notes prm2: m2=logit lfp age wc hc lwg inc \ wf6-create.do jsl 2008-04-05.
```

Looking at the labels and notes, I can easily tell what each variable is.

```
. codebook prm*, compact
Variable     Obs Unique        Mean         Min         Max  Label
-------------------------------------------------------------------------------
prm1         753    753   .5683931    .0139875    .9621198  Pr(lfp|m1=k5 k618 age wc...
prm2         753    753   .5683931    .1012935    .8985487  Pr(lfp|m2=age wc hc lwg ...
-------------------------------------------------------------------------------

. notes prm*
prm1:
  1.  m1=logit lfp k5 k618 age wc hc lwg inc \ wf6-create.do jsl 2008-04-05.
prm2:
  1.  m2=logit age wc hc lwg inc \ wf6-create.do jsl 2008-04-05.
```

6.3.6 Verifying that variables are correct

When you create new variables, there are lots of ways that things can go wrong and you are bound to make mistakes along the way. This is not a problem, unless you do not find those mistakes. Accordingly, a critical part of the workflow for creating variables is verifying that things have been done correctly. Earlier examples illustrated several ways to do this. Here I review those approaches to verification and provide additional suggestions (file: `wf6-verify.do`).

Checking the code

Sometimes the best way to find an error is to check the commands in your do-file. My preferred workflow for complicated data management is to write the programs without being vigilant about including comments, notes, and labels. I return to the program later to check the code and add additional notes, comments, and labels. Often I find errors.

Listing variables

Sometimes the easiest way to verify a variable is to list the source and created variables. This is particularly useful when you know something is wrong, but you are not sure why. Using information in the list, you can check the calculations by hand or move the data into a spreadsheet for review. Suppose that `fincome` categorizes income into ranges such as from $70,000 to $90,000. I want to create `finc_mid` where the categories are replaced with the midpoint of the range:

```
recode finc_mid ///
     1=1.5    2=4     3=6     4=8     5=9.5    6=10.5  7=11.5  8=12.5    ///
     9=13.5  10=14.5 11=16   12=18.5 13=21    14=23.5 15=23.5 16=32.5    ///
    17=37.5  18=42.5 19=47.5 20=55   21=67.5  22=82.5 23=97.5 24=131.25
```

To verify the transformation, I list the values of `fincome` and `finc_mid` for some observations. Although it is easy to list the first 100 observations (e.g., `list fincome finc_mid in 1/100`), observations are often systematically arranged in a dataset. For example, the first 100 cases might all have `fincome` equal to 1. Accordingly, I take a random sampling of cases. I start by creating a random variable with values that range from 1 to the sample size:

```
set seed 1951
generate xselect = int( (runiform()*_N)+ 1 )
label var xselect "Random numbers from 1 to _N"
```

I set the seed so that I can get the same selection of variables if I run the program again. In the `generate` command, the `runiform()` function creates random numbers from 0 to 1. To convert this to the range for the observations, I multiply by `_N` (the sample size), add 1, and use the `int()` function to keep only the integer part of the number. I sort the data by `fincome` and list 20 random observations:

```
. sort fincome
. list fincome finc_mid if xselect<20, clean
          fincome   finc_mid
  92.      2_3-5K          4
 211.      3_5-7K          6
 242.      4_7-9K          8
 333.     5_9-10K        9.5
 479.    8_12-13K       12.5
 727.   12_17-20K       18.5
 819.   13_20-22K         21
 876.   14_22-25K       23.5
 930.   14_22-25K       23.5
1105.   15_25-30K       23.5
1118.   15_25-30K       23.5
1174.   16_30-35K       32.5
1236.   16_30-35K       32.5
1338.   17_35-40K       37.5
  (output omitted)
```

I use this list to verify the values of the new variable.

Plotting continuous variables

For continuous variables, a plot is a useful way to verify a transformation. For example,

```
generate inc_sqrt = sqrt(inc) if !missing(inc)
label var inc_sqrt "Square root of inc"
scatter inc_sqrt inc, msymbol(circle_hollow)
```

produces the graph

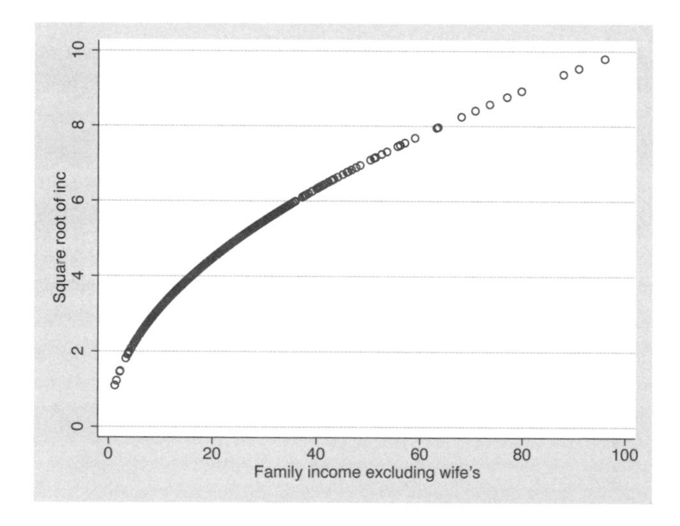

Graphs can also be useful when recoding an ordinal variable. For example, `fincome` measures family income grouped into ranges and assigned values from 1 to 24:

```
. tabulate fincome, miss

     Income |      Freq.     Percent        Cum.
------------+-----------------------------------
      1_<3K |         67        2.69        2.69
     2_3-5K |         70        2.81        5.51
     3_5-7K |         84        3.38        8.89
  (output omitted)
  22_75-90K |         85        3.42       86.97
 23_90-105K |         42        1.69       88.66
   24_>105K |         78        3.14       91.80
          . |        202        8.12       99.92
         .a |          2        0.08      100.00
------------+-----------------------------------
      Total |      2,487      100.00
```

I create `finc_mid` by recoding values 1 through 24 to the midpoint of the range:

```
. generate finc_mid = fincome
(204 missing values generated)
. label var finc_mid "Income coded at the midpoint"
. notes finc_mid: midpoints for fincome; upper range is 1.25X ///
> truncation point \ wf6-verify.do jsl 2008-10-18.
. recode finc_mid ///
>      1=1.5    2=4     3=6     4=8     5=9.5   6=10.5  7=11.5  8=12.5     ///
>      9=13.5 10=14.5 11=16   12=18.5 13=21   14=23.5 15=23.5 16=32.5     ///
>     17=37.5 18=42.5 19=47.5 20=55   21=67.5 22=82.5 23=97.5 24=131.25
(finc_mid: 2283 changes made)
```

To check the recoding, I can plot `finc_mid` against `fincome`

```
scatter finc_mid fincome, msymbol(circle_hollow)
```

resulting in

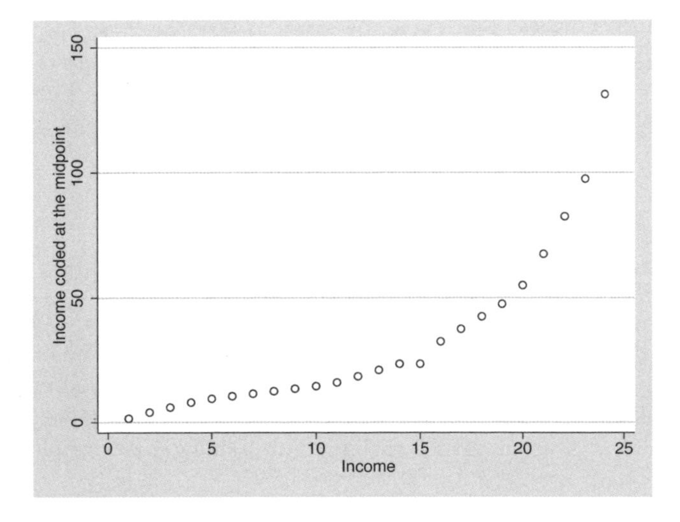

This looks pretty good, although there might be a problem at `fincome` equal to 15. When using graphs to check variables, I find that it is often useful to run both an x–y plot and a y–x plot. If I reverse the axis, using `scatter fincome finc_mid, msymbol(circle_hollow)`, the problem is much easier to see:

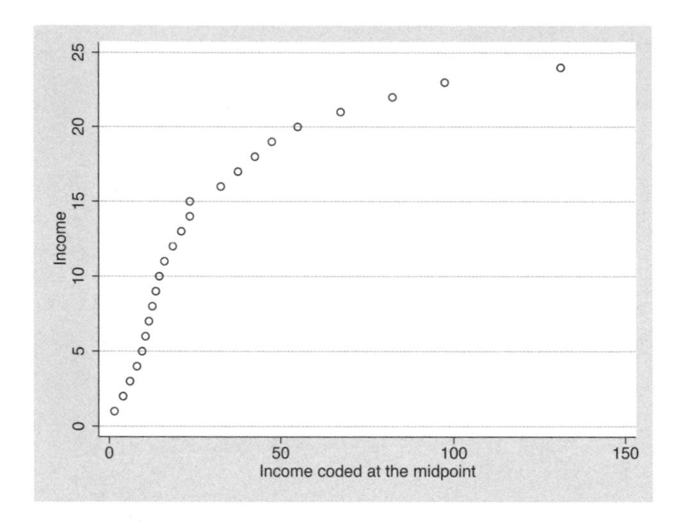

I suspect that few readers noticed the problem in the `recode` command on page 257, which is why you must verify all transformations. Indeed, I did not notice the problem until I plotted the values as an example.

Tabulating variables

The `tabulate` command is a quick way to verify a new variable if there are relatively few categories. When using `tabulate`, be sure to use the `missing` option (which I abbreviate as `miss`) so you can verify that missing values are created correctly. When the variables have many values, `tabulate` is not as useful because the table is large and hard to check. For example, suppose that I am transforming income to the square root of income:

```
. generate inc_sqrt = sqrt(inc)
(1764 missing values generated)
. label var inc_sqrt "Square root family income excluding wife´s"
```

This is such a simple transformation that I am tempted not to check it. Even with simple transformations, there can be problems with missing values. For example, negative values could be valid for `inc`. Taking the square root of a negative number leads to a missing value, which is not the answer I want. To check missing values, I can use an `if` statement to select cases where either variable is missing. For example,

```
. tabulate inc inc_sqrt if missing(inc) | missing(inc_sqrt), miss

                     Square
                      root
                     family
     Family          income
     income        excluding
  excluding          wife´s
     wife´s               .        Total

  -.0290001               1            1
          .           1,742        1,742
         .a               5            5
         .b              16           16

      Total           1,764        1,764
```

The negative value of inc is changed to a missing value in inc_sqrt. Because this is not what I want, I must find another way to construct inc_sqrt.

Constructing variables multiple ways

If the steps required to create a variable are complicated or I am using commands that I am not familiar with, I often create the same variable two ways as illustrated above when using the egen anycount command. Then I use compare to verify that the two variables are identical. For example, suppose that I have not worked much with recode and want to verify that I am using it correctly. I start by creating a variable that converts years of education into categories using recode:

```
. recode edyears 0/8=1 9/11=2 12=3 13/15=4 16/24=5, gen(educcat)
(1636 differences between edyears and educcat)
```

Next I create what should be the same variable but this time I use replace:

```
. generate educcatV2 = edyears
(848 missing values generated)

. replace educcatV2 = 1 if edyears>=0  & edyears<=8   // no HS
(278 real changes made)

. replace educcatV2 = 2 if edyears>=9  & edyears<=11  // some HS
(501 real changes made)

. replace educcatV2 = 3 if edyears==12                // HS
(205 real changes made)

. replace educcatV2 = 4 if edyears>=13 & edyears<=15  // some college
(517 real changes made)

. replace educcatV2 = 5 if edyears>=16 & edyears<=24  // college plus
(135 real changes made)

. label var educcat2 "categorize educ using replace"
```

I compare the two variables:

```
. compare educcat educcatV2

                                          ---------- difference ----------
                          count       minimum      average     maximum
-------------------------------------------------------------------------
educcat=educcatV2          1639
                        ----------
jointly defined            1639             0            0           0
jointly missing             848
                        ----------
total                      2487
```

The line `educcat=educcatV2` indicates that these variables are equal in 1,639 observations. `jointly defined` reports that 1,639 values are defined and the three 0s indicate that the values are the same. `jointly missing` indicates that 848 cases have missing values for both variables. In short, everything matches.

6.4 Saving datasets

When you initially receive a dataset, you need to examine it carefully and will often need to make changes before you begin your substantive analysis, including renaming variables, correcting errors in coding, revising labels, creating new variables, adding notes, and selecting cases. As discussed on page 127, this work is part of your workflow for data management. When you complete this work, you should create a new dataset that contains these changes and use the new dataset for statistical analyses. You do not want to rerun the same data-management commands each time you run statistical analyses. I often consult with people who have multiple do-files that each start with a long sequence of data-management commands followed by the commands for statistical analyses. Although the data-management commands are supposed to be the same in all do-files, it is difficult to make sure that any changes you make when debugging your programs are consistently applied to all do-files. It is simpler and less error prone to create a new dataset that incorporates your data-management revisions and use this dataset for your statistical analysis. This section discusses issues involved in saving these datasets.

Steps in saving a dataset

Although saving a dataset is as simple as `save` *filename*`, replace`, there are several things you might want to do before saving your data.

1. Drop variables and/or observations.
2. Create new variables.
3. Rearrange variables within the dataset.
4. Add internal documentation that indicates when and how the dataset was created.
5. Compress the data to minimize the file size.

6. Run diagnostics to look for data problems.
7. Add a data signature to guard against inadvertent changes to the data.

I consider each of these issues in turn, with the exception of creating new variables which was discussed in the last section.

6.4.1 Selecting observations

You can select observations with the `keep if` and `drop if` commands. The syntax is

`keep if` *exp*

`drop if` *exp*

where *exp* is an expression that selects which cases to keep or drop. For example, `keep if female==1` keeps all women in the sample (or, equivalently, `keep if female` because `if female` implies `if female==1`). The command `keep if age>40 & !missing(age)` keeps cases with `age` greater than 40 where `age` is not missing. If you use `keep if age>40`, observations where `age` is missing would be kept because missing values are greater than 40. Dropping cases works the same way. For example, to drop those who are older than 40 or who have missing values for `age`, I can run `drop if age>40`. Cases with `age` missing are dropped because those values are greater than 40.

You can also use an `in` condition to select observations based on their start and end row in the dataset, although I do not find I need this very often:

`keep in` *start/end* [`if` *exp*]

`drop in` *start/end* [`if` *exp*]

Deleting cases versus creating selection variables

Selecting observations raises the question of whether you should create different datasets for different groups that you plan to analyze. For example, should you create one dataset with only men and a second dataset with only women? Is it better to keep all observations in the dataset?

If there are observations that you will never need, I suggest that you drop them. For example, if I am studying attitudes toward working women in Russia, I do not need respondents from other countries in my dataset. By dropping non-Russian respondents, I have a physically smaller dataset and eliminate the chance of mistakenly including non-Russians in my analyses.

If there are subsets of observations that I plan to analyze separately (e.g., men and women), I do not create separate datasets. Rather, I use sample-selection variables that define the samples. For example, if I plan to analyze men and women separately, I create two variables:

```
generate sampfem = (female==1)
label var sampfem "Sample: females only"
generate sampmale = (female==0)
label var sampmale "Sample: males only"
label def Lsamp 0 0_Drop 1 1_InSample
label val sampfem Lsamp
label val sampmale Lsamp
```

Then I can easily load the full dataset and select the cases I want. For example,

```
use mydata, clear
keep if sampfem  // women only
summarize age inc
use mydata, clear
keep if sampmale // men only
summarize age inc
```

Alternatively, I can use `if` conditions with other commands:

```
use mydata, clear
summarize age inc if sampfem  // women only
summarize age inc if sampmale // men only
```

Creating selection variables is particularly useful when you have complex criteria for selecting observations. For example, if I often analyze women (`female==1`) between ages 20 and 40 (`age>19 & age<41`) who lived in large cities (`urban==1`), I could add this variable to the dataset:

```
generate sampf2040urb = (female==1) & (age>19 & age<41) & (urban==1)
label var sampf2040urb "Is female, 20-40, and urban?"
label val sampf2040urb Lsamp
```

Then I can select the sample easily:

```
logit lfp income age wage kid5 if sampf2040urb
```

After loading the dataset, I can keep only those observations:

```
keep if sampf2040urb
```

6.4.2 Dropping variables

Before saving a dataset, you might want to drop some variables. While I do not recommend dropping source variables used to create analysis variables, you might have variables that you will never use. These can be dropped with the command

`drop` *varlist*

Or you can specify the variables you want to keep:

`keep` *varlist*

You can also use a combination of `keep` and `drop` commands.

Selecting variables for the ISSP 2002 Russian data

The full dataset for the ISSP 2002 has 234 variables, many of which are meaningless for the Russian sample. An easy way to find variables that do not apply to the Russian sample is with the `codebook, problems` command, which lists variables that have no variation (file: `wf6-save.do`):

```
. use wf-isspru01, clear
(Workflow data from Russian ISSP 2002 \ 2008-04-02)
. codebook, problems
  Potential problems in dataset   wf-isspru01.dta
              potential problem   variables
-------------------------------------------------------------------------------
  constant (or all missing) vars   v1 v3 v206 v207 v208 v209 v210 v211 v212
                                   v213 v214 v215 v216 v217 v218 v219 v220 v221
                                   v222 v223 v224 v225 v226 v227 v228 v229 v230
                                   v231 v233 v234 v235 v236 v237 v238 v248 v280
                                   v287 v290 v291 v337 v358 v359 v360 v362
        incompletely labeled vars   v36 v37 v69 v71 v201 v204 v240 v243 v249
                                   v250 v361
-------------------------------------------------------------------------------
```

To drop the variables that are constant, I first retrieve the names of those variables using a returned value provided by `codebook`:

```
. local dropvars = r(cons)
```

Using this local, I drop the variables that are constants:

```
. drop `dropvars´
```

Now I save the dataset including only those variables that are appropriate for the Russian sample.

6.4.3 Ordering variables

The order in which variables are arranged in a dataset is shown in the Variables window and in the Data Editor. It also affects the output from some commands. Before saving your dataset, it is worth thinking about the order in which you want the variables to appear. For example, you might want the most commonly used variables at the front so that they are easy to select from the Variables window. You might also want to move variables that you are unlikely to use to the end of the dataset, or you might simply prefer to have the variables arranged alphabetically. The order in which variables appear can be specified with the `order` and `aorder` commands, which are discussed on page 155.

6.4.4 Internal documentation

Before saving a dataset, you should add metadata that documents the dataset. To do this, you can use the `label data` and `notes` commands that were discussed on page 138. To review briefly, to add a label to a dataset, use the command (file: `wf6-save.do`)

```
label data "label"
```

For example,

```
label data "Workflow data from Russian ISSP 2002 \ 2008-04-02"
```

To add additional internal documentation, you can use the `notes` command

```
notes: text
```

For example,

```
notes: wf-isspru02.dta \ workflow ch 6 \ wf6-save.do jsl 2008-04-05.
```

The label is echoed when you load the dataset and notes are displayed by `notes _dta`.

6.4.5 Compressing variables

Before I save a file, I routinely compress the data with the command

```
compress
```

This command can make the file smaller, sometimes substantially smaller, without losing any information. To understand what `compress` does, you need to know something about how data are stored. By default, a new variable is stored as a floating-point number. This type of variable can have values from −1.70141173319 ∗ 10^38 to 1.70141173319 ∗ 10^38. Many variables do not require such a large range and can be stored in a way that uses less space than floating-point numbers. Stata has five storage types that are illustrated here where each square represents a byte of storage:

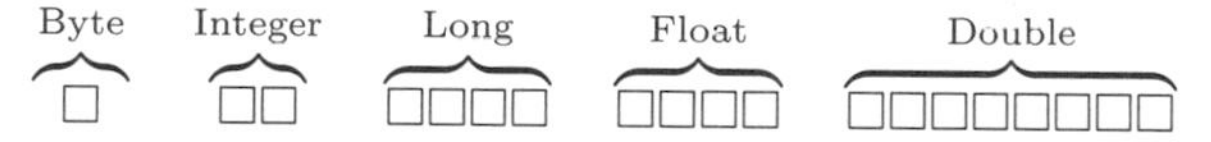

A byte variable can range from −127 to 100 and uses 1 byte of storage; an integer variable can range from −32,767 to 32,740 and uses two bytes. A double-precision variable can include very large and small numbers with many digits of precision, but it takes 8 times more space than a byte variable. For further details, type `help data types`. The `compress` command determines the most compact storage type that can hold all a variable's information and then changes the variable into that type.

Here is an example from the full ISSP data for all countries that was converted from SPSS to Stata without compression, leaving all variables in double precision. The file is 83.4 megabytes:

```
. use 04106-0001-data, clear

. dir 04106-0001-data.dta
 83.4M 3/11/06 9:42 04106-0001-data.dta
```

When I run `compress`, each variable is evaluated and the output shows if a variable's storage type was changed:

```
. compress
v1 was double now int
v2 was double now long
  (output omitted)
v361 was double now long
v362 was double now byte
```

When I converted the data, each variable was stored in double precision. By compressing, the variables are stored in formats that use less space and as a result the file is only 11.3 megabytes:

```
. save x-temp, replace
file x-temp.dta saved

. dir x-temp.dta
11.3M 4/14/07 11:02 x-temp.dta
```

Because I do not need to know which variables are compressed, I often use `quietly compress` to suppress the output.

6.4.6 Running diagnostics

Stata has several commands that look for problems in your data. I often run these commands the first time I use a dataset (e.g., immediately after converting it from SPSS) or when I create a revised version of the data. The `codebook, problems` command can check for a variety of potential problems, whereas the `isid` and `duplicates` commands check if the ID variable is unique.

The codebook, problems command

The `codebook, problems` command looks for three characteristics that might indicate a problem.

1. Variables that have no variation are listed. This is not necessarily a problem. For example, you might have a variable `female` in a dataset that contains only women.
2. Variables with nonexisting value labels occur if a variable has a value label assigned but the label has not been defined. If you want value labels for such variables, you need to add a value label with a `label define` command.
3. Incompletely labeled variables are those where there is a value label assigned to a variable, but one or more values for that variable have not been labeled. To fix this problem, you need to change the value labels with a `label define` command.

Here is an example of how the command works (file: `wf6-save.do`):

```
. use wf-diagnostics, clear
(Workflow data to illustrate data diagnostics \ 2008-04-05)
. codebook, problems
  Potential problems in dataset   wf-diagnostics.dta
              potential problem   variables
--------------------------------------------------------------------------
  constant (or all missing) vars  v3 v256 v265 v274 v283 v294 v303 v312
     vars with nonexisting label  v7
       incompletely labeled vars  v36 v37
--------------------------------------------------------------------------
```

Eight variables are listed as being constant. For some variables, this is not a problem. For example, because our sample includes only those from Russia, we do not want any variation in the country variable:

```
. tab1 v3, miss
-> tabulation of v3
    Country |      Freq.     Percent        Cum.
------------+-----------------------------------
        RUS |        100      100.00      100.00
------------+-----------------------------------
      Total |        100      100.00
```

On the other hand, `v256` should be deleted from the dataset because the variable is useful only for the Bulgarian sample:

```
. tab1 v256, miss
-> tabulation of v256
R: Party affiliation: |
             Bulgaria |      Freq.     Percent        Cum.
----------------------+-----------------------------------
                  NAV |        100      100.00      100.00
----------------------+-----------------------------------
                Total |        100      100.00
```

The second error message indicates that `v7` has been assigned a label definition that does not exist. I use `describe` to check the variable:

```
. describe v7
              storage  display     value
variable name   type   format      label      variable label
---------------------------------------------------------------------------
v7              byte   %10.0g      labv7      What women really want is home &
                                               kids
```

I realize that I assigned `labv7` to variable `v7` but meant to assign the value label `v7`. The third error message points to gaps in a value label. This problem is illustrated with `v37`:

```
. tab1 v37, miss

-> tabulation of v37

How many hrs spouse,partner |
               works on hh |      Freq.     Percent        Cum.
----------------------------+-----------------------------------
             NAP,no partner |         50       50.00       50.00
   1 hour or less than 1 hr |          1        1.00       51.00
                      2 hrs |          1        1.00       52.00
                            |          2        2.00       54.00
                          7 |          6        6.00       60.00
                          8 |          1        1.00       61.00
                          9 |          1        1.00       62.00
                            |          4        4.00       66.00
                         14 |          3        3.00       69.00
  (output omitted)
```

This occurs because value labels were defined for only some of the values found in `v37`. To fix the problem, I need to revise `label define` for this variable.

Checking for unique ID variables

If you have an ID variable, you will normally want the IDs to be unique. You can check this with the `isid` command:

`isid` *varlist*

For example, to check whether `id` uniquely identifies observations, (file: `wf6-save.do`) type

```
. use wf-diagnostics, clear
(Workflow data to illustrate data diagnostics \ 2008-04-05)
. isid id
variable id does not uniquely identify the observations
r(459);
end of do-file
r(459);
```

The output from `isid` tells you that multiple observations have the same value for `id`. This is not necessarily a problem. For example, with panel data, each person might have multiple records, one for each panel. If `id` is supposed to be unique within each panel, you could check with the command

```
isid id panel
```

You would get an error if any values of `id` occurred more than once within the same panel. If `isid` finds that observations are uniquely defined, the command has no output.

The `duplicates` command is more general than `isid` and makes it easy to find duplicate observations. For example,

```
. duplicates report id
Duplicates in terms of id

--------------------------------------
   copies | observations       surplus
----------+---------------------------
        1 |           98             0
        2 |            2             1
--------------------------------------
```

This shows that two observations have the same value for `id` (i.e., there is one surplus value). To find out more, type

```
. duplicates examples id, clean
Duplicates in terms of id

    #   e.g. obs        id
    2          1   1800007
```

This tells me that the value 1800007 is duplicated.

Aside

The `duplicates` command can also find observations that are legitimately the same on one or more variables. For example, here I load a dataset on occupational attainment where the independent variables are race, years of education, and years of experience. The `duplicates` command tells me that quite a few observations are identical:

```
. use nomocc2, clear
(1982 General Social Survey)

. duplicates report

Duplicates in terms of all variables

--------------------------------------
   copies | observations       surplus
----------+---------------------------
        1 |          241             0
        2 |           72            36
        3 |           12             8
        4 |           12             9
--------------------------------------
```

I can use `duplicates` to examine the duplicate observations:

```
. duplicates examples, clean

Duplicates in terms of all variables

   group:   #    e.g. obs        occ   white   ed   exper
       1    2          10     Menial       1   12       3
       2    2          65    BlueCol       1    6      46
       3    2          43    BlueCol       1   10       6
       4    2          54    BlueCol       1   11       4
       5    4          36    BlueCol       1   12       4
       6    2          60    BlueCol       1   12       5
  (output omitted)
```

Here it is not surprising that some combinations of these variables are repeated in the dataset.

6.4.7 Adding a data signature

The `datasignature` command adds metadata to a saved dataset that helps determine if you have the correct, unchanged data. I recommend that you add a data signature to every dataset you create. For your convenience, quickly review the materials on page 139. To create a signature and save your dataset, use the commands

```
. datasignature set
  753:8(54146):1899015902:1680634677        (data signature set)

. save wf-datasig02, replace
```

To verify the signature when loading the dataset, use the commands

```
. use wf-datasig02, clear
(Workflow dataset for illustrating datasignature \ 2008-03-09)
. datasignature confirm
  (data unchanged since 09mar2008 12:40)
```

If you load a dataset with a signature, make changes to the data, and create a new signature with `datasignature set`, you will get the following error:

```
. datasignature set
  data signature already set -- specify option -reset-
r(110);
```

You need to add the `reset` option:

```
. datasignature set, reset
  753:9(85320):1280133433:4173826113       (data signature reset)
```

6.4.8 Saving the file

Now you are ready to save the file, which is the simplest step:

`save` *filename* [`, replace`]

The *filename* needs to be in quotes if it has spaces in the name. The `replace` option specifies that you can overwrite the dataset if it already exists. If you are sharing a dataset with someone using an older version of Stata, you can use

`saveold` *filename* [`, replace`]

For example, although Stata 9 cannot read a dataset saved in Stata 10 by using `save mydata, replace`, it can read a dataset saved by using `saveold mydata, replace`. If you try to read a dataset from a later version of Stata than you are using, you receive an error like this:

```
. use mydata, clear
file mydata.dta not Stata format
r(610);
```

The only remaining issue is what to name the dataset. As discussed in chapter 5, if you change a dataset, give it a new name. I prefer to end filenames with two-digit numbers, adding one to the number each time I revise the dataset. For example, if `wf-russia01.dta` is changed, I would create a new dataset `wf-russia02.dta`. When a dataset is temporary, I begin the filename with `x-` (e.g., `x-job01a.dta`). My rule is that any dataset beginning with `x-` can be deleted any time I want to recover the disk space.

6.4.9 After a file is saved

After a file is saved is a good time to check your documentation (see chapter 2), to decide if you are ready to post the file (see page 125), and to back up your files (see chapter 8).

6.5 Extended example of preparing data for analysis

This example combines the tools described above to illustrate how variables are created in a more realistic application. Using data from the Russian ISSP for 2002, I create the analysis variables in three steps: 1) control variables; 2) binary measures of attitudes toward working women; and 3) ordinal measures of these attitudes.

Creating control variables

I begin by loading the data and checking the data signature (file: `wf6-create01-controls.do`):

```
. use wf-russia01, replace
(Workflow data to illustrate creating variables \ 2008-04-02)
. datasignature confirm
  (data unchanged since 02apr2008 13:29)
```

Because I plan to add a note to each variable that I create, I use a local macro with information about the program name, author, and date:

```
local tag "wf6-create01.do jsl 2008-04-05."
```

The existing variable `gender` equals 1 for men and 2 for women. For analysis, I want a pair of binary variables for gender. I create `female` as follows:

```
gen female = gender - 1
label var female "Female?"
label def female 0 0_male 1 1_female
label val female female
notes female: based on gender \ `tag´
```

I use `tabulate` to verify the new variable:

```
. tabulate gender female, miss
   Gender:  |
   1=male,  |        Female?
  2=female  |   0_male   1_female |     Total
------------+---------------------+----------
   1. Male  |      695          0 |       695
 2. Female  |        0      1,103 |     1,103
------------+---------------------+----------
     Total  |      695      1,103 |     1,798
```

Next I create `male` by typing

```
gen male = 1 - female
label var male "Male?"
label def male 1 1_male 0 0_female
label val male male
notes male: based on gender \ `tag´
tabulate gender male, miss
```

To create an indicator of being married from a variable for marital status, I use `recode` as discussed earlier in the chapter:

```
. recode marstat (1 2 3 4=1) (5=0), gen(married)
(848 differences between marstat and married)

. label def married 1 1_married 0 0_never

. label val married married

. label var married "Ever married?"

. notes married: recoding of marstat \ married includes married, ///
> widowed, divorced, separated \ `tag´

. tabulate marstat married, miss

    Marital |           Ever married?
     status |   0_never  1_married         .b |     Total
------------+---------------------------------+----------
  1_married |         0        931          0 |       931
  2_widowed |         0        321          0 |       321
 3_divorced |         0        215          0 |       215
4_separated |         0         33          0 |        33
   5_single |       279          0          0 |       279
         .b |         0          0         19 |        19
------------+---------------------------------+----------
      Total |       279      1,500         19 |     1,798
```

Similarly, I create an indicator of having a college or higher degree and being employed full time:

```
recode edlevel (1 2 3 4 5=0)(6 7=1)(99=.n), gen(hidegree)
label var hidegree "Any higher education?"
label def hidegree 0 0_not 1 1_high_ed
label val hidegree hidegree
notes hidegree: recode of edlevel \ `tag´
tabulate edlevel hidegree, miss

recode empstat (1 7=1)(2 3 5 6 8 9 10=0)(98=.d)(99=.n), gen(fulltime)
label def fulltime 1 1_fulltime 0 0_not
label val fulltime fulltime
label var fulltime "Ever worked full time?"
notes fulltime: recoding of empstat; includes fulltime & retired \ `tag´
tabulate empstat fulltime, miss
```

Before saving the file, I check the new variables:

```
. codebook female-fulltime, compact
Variable     Obs Unique      Mean  Min  Max  Label
--------------------------------------------------------------------------
female      1798      2  .6134594    0    1  Female?
male        1798      2  .3865406    0    1  Male?
married     1779      2  .8431703    0    1  Ever married?
hidegree    1795      2  .2250696    0    1  Any higher education?
fulltime    1765      2  .7852691    0    1  Ever worked full time?
--------------------------------------------------------------------------
```

I do a bit of housecleaning and save the file:

```
. sort id
. quietly compress
. label data "Workflow example of adding analysis variables \ 2008-04-05"
. note: wf-russia02.dta \ `tag´
. datasignature set, reset
  1798:19(15177):591800297:459057199       (data signature reset)
. save wf-russia02, replace
file wf-russia02.dta saved
```

After saving a file, I like to reload it and make sure things are working properly:

```
. use wf-russia02, clear
(Workflow example of adding analysis variables \ 2008-04-05)
. datasignature confirm
  (data unchanged since 05apr2008 19:50)
. notes
_dta:
  1.  wf-russia01.dta \ wf-isspru01.dta \ wf-russia01-support.do jsl 2008-04-02.
  2.  wf-russia02.dta \ wf6-create01.do jsl 2008-04-05.
id:
  1.  clone of v2 \ wf-russia01-support.do jsl 2008-04-02.
  (output omitted)
married:
  1.  recoding of marstat \ married includes married, widowed, divorced,
> separated \ wf6-create01.do jsl 2008-04-05.
hidegree:
  1.  recode of edlevel \ wf6-create01.do jsl 2008-04-05.
fulltime:
  1.  recoding of empstat; includes fulltime & retired \ wf6-create01.do jsl
      2008-04-05.
```

(Continued on next page)

When sequentially building a dataset, I often use the `cf` command to check what has changed in the new dataset. `cf` compares the dataset in memory to a saved file and indicates which variables differ in the two datasets. Here I compare the file I just created (which is still in memory) with the file I started with by typing

```
. cf _all using wf-russia01
          female:  does not exist in using
            male:  does not exist in using
         married:  does not exist in using
        hidegree:  does not exist in using
        fulltime:  does not exist in using
r(9);
```

The output shows me that the new file differs by including those variables that I intended to add. Because `cf` returned an error code because the two files are not identical, the do-file ends without closing the log file. Accordingly, I need to type `log close` in the command line or include it in my next do-file.

Creating binary indicators of positive attitudes

The next step is to create binary indicators of attitudes toward working mothers. I load and check the dataset (file: `wf6-create02-binary.do`):

```
. use wf-russia02, replace
(Workflow example of adding analysis variables \ 2008-04-05)
. datasignature confirm
  (data unchanged since 05apr2008 19:50)
```

Next I check the six questions about working women:

```
. codebook momwarm kidsuffer famsuffer wanthome housesat workbest, compact
Variable     Obs Unique       Mean  Min  Max  Label
-----------------------------------------------------------------------------------
momwarm     1765      5   2.324646    1    5  Working mom can have warm relatio...
kidsuffer   1755      5   2.373789    1    5  Pre-school child suffers?
famsuffer   1759      5   2.401933    1    5  Family life suffers?
wanthome    1717      5   2.639487    1    5  Women really want is home & kids?
housesat    1680      5   2.855357    1    5  Housework satisfies like paid job?
workbest    1710      5   2.071345    1    5  Work best for women´s independence?
-----------------------------------------------------------------------------------
```

Each variable is a five-point scale. For example,

```
. tab1 momwarm, miss
-> tabulation of momwarm
```

Working mom can have warm relations w kids?	Freq.	Percent	Cum.
1StAgree	464	25.81	25.81
2Agree	712	39.60	65.41
3Neither	197	10.96	76.36
4Disagree	336	18.69	95.05
5StDisagree	56	3.11	98.16
a_Can´t choose	26	1.45	99.61
b_Refused	7	0.39	100.00
Total	1,798	100.00	

The variable **momwarm** and **workbest** are coded so that agreeing with a question indicates a positive attitude toward working mothers; **kidsuffer**, **famsuffer**, **wanthome**, and **housesat** are coded so that agreeing indicates a negative attitude toward working mothers. When variables are coded in different directions, it is easy to get things confused. One way to verify that you understand which way each variable is coded is to run a correlation among the variables. Although Pearson correlations are not appropriate for ordinal variables, they work fine as a rough indicator of whether variables are positively or negatively associated. I use the **obs** options, which reports the number of nonmissing values shared by each pair of variables:

```
. pwcorr momwarm kidsuffer famsuffer wanthome housesat workbest, obs
```

	momwarm	kidsuf~r	famsuf~r	wanthome	housesat	workbest
momwarm	1.0000					
	1765					
kidsuffer	-0.2494	1.0000				
	1736	1755				
famsuffer	-0.2517	0.5767	1.0000			
	1737	1738	1759			
wanthome	-0.1069	0.2357	0.2977	1.0000		
	1698	1688	1698	1717		
housesat	-0.0148	0.1465	0.1921	0.4133	1.0000	
	1664	1657	1662	1649	1680	
workbest	0.0624	-0.0220	-0.0717	-0.1369	-0.2019	1.0000
	1691	1684	1690	1659	1636	1710

This confirms what I thought. Notice that **pwcorr** truncates column names at eight characters.

I want to recode each variable so that 1 indicates a positive attitude and 0 a negative attitude. I begin by creating the value label for the variables I plan to create. My first attempt is

```
label def Lagree 1 1_agree 0 0_not .a a_Unsure .b b_Refused .n n_Neutral
```

On reflection, I decide that this is confusing because someone can agree with a positive statement and can also agree with a negative statement. To make it explicit that a 1 indicates a positive attitude toward working women, I decided on these labels:

```
label def Lprowork 1 1_yesPos 0_noNeg .a a_Unsure .b b_Refused .n n_Neutral
```

Next I dichotomize momwarm and add labels and notes. I am treating neutral attitudes as missing (i.e., 3=.) and add a note to remind myself of this decision:

```
. * momwarm: 1=SA working mom can have warm relationship
. * Bwarm:   1=agree (not reversed)
. recode momwarm (1/2=1) (4/5=0) (3=.n), gen(Bwarm)
(1301 differences between momwarm and Bwarm)
. label var Bwarm "Working mom can have warm relations?"
. label val Bwarm Lprowork
. notes Bwarm: 3=neutral in source was coded .n \ `tag´
```

To verify the recode, type

```
. tabulate Bwarm momwarm, miss
   Working
   mom can
 have warm        Working mom can have warm relations w kids?
relations?   1StAgree     2Agree   3Neither  4Disagree  5StDisagr      Total

   0_noNeg          0          0          0        336         56        392
  1_yesPos        464        712          0          0          0      1,176
  a_Unsure          0          0          0          0          0         26
 b_Refused          0          0          0          0          0          7
 n_Neutral          0          0        197          0          0        197

     Total        464        712        197        336         56      1,798

   Working   Working mom can have
   mom can     warm relations w
 have warm          kids?
relations?   a_Can´t c  b_Refused      Total

   0_noNeg          0          0        392
  1_yesPos          0          0      1,176
  a_Unsure         26          0         26
 b_Refused          0          7          7
 n_Neutral          0          0        197

     Total         26          7      1,798
```

Because each variable is transformed in the same way, I can use a local to automate the work. For example,

```
local vin  momwarm
local vout Bwarm
recode `vin´ (1/2=1)(4/5=0)(3=.n), gen(`vout´)
label var `vout´ "Working mom can have warm relations?"
label val `vout´ Lprowork
notes `vout´: 3=neutral in source was coded .n \ `tag´
tabulate `vout´ `vin´, miss
```

Then, to recode `workbest`, type

```
local vin  workbest
local vout Bindep
recode `vin´ (1/2=1)(4/5=0)(3=.n), gen(`vout´)
label var `vout´ "Agree work creates independence?"
label val `vout´ Lprowork
notes `vout´: 3=neutral in source was coded .n \ `tag´
tabulate `vout´ `vin´, miss
```

and so on for each variable. After finishing the recodes, I check that the binary variables are coded in the same direction:

```
. pwcorr B*, obs

             |    Bwarm    Bkids  Bfamily  Bnohome  Bjobsat   Bindep
-------------+------------------------------------------------------
       Bwarm |   1.0000
             |     1568
             |
       Bkids |   0.2351   1.0000
             |     1311     1483
             |
     Bfamily |   0.2289   0.5775   1.0000
             |     1314     1312     1481
             |
     Bnohome |   0.1382   0.2327   0.2850   1.0000
             |     1208     1157     1161     1352
             |
     Bjobsat |   0.0396   0.1446   0.1672   0.4659   1.0000
             |     1119     1071     1071     1033     1248
             |
      Bindep |   0.0345   0.0185   0.0546   0.1048   0.1538   1.0000
             |     1279     1217     1211     1122     1058     1438
             |
```

All that remains is to clean up and save a temporary file:

```
sort id
quietly compress
label data "Workflow example of adding analysis variables \ 2008-04-05"
notes: x-wf6-create02-binary.dta \ `tag´
datasignature set, reset
save x-wf6-create02-binary, replace
```

Creating four-category scales of positive attitudes

Next I create four-category scales that exclude neutral and that are coded in the same direction. I start by loading the data and setting up a value label (file: `wf6-create03-noneutral.do`):

```
. use x-wf6-create02-binary, clear
(Workflow example of adding analysis variables \ 2008-04-05)
. datasignature confirm
  (data unchanged since 05apr2008 19:50)
. label def Lsa_sd 1 1_SA_Pos 2 2_A_Pos 3 3_D_Neg ///
>     4 4_SD_Neg .a a_Unsure .b b_Refused .n n_Neutral
```

Starting with momwarm, I use code that is similar to wf6-create02-binary.do:

```
. * momwarm: 1=SA working mom can have warm relationship
. * C4warm:  1=SA (not reversed)
. local vin momwarm
. local vout C4warm
. recode `vin´ (1=1) (2=2) (3=.n) (4=3) (5=4), gen(`vout´)
(589 differences between momwarm and C4warm)
. label var `vout´ "Working mom can have warm relations?"
. label val `vout´ Lsa_sd
. notes `vout´: 3=neutral in source was coded .n \ `tag´
. tabulate `vin´ `vout´, m

   Working mom
 can have warm
   relations w        Working mom can have warm relations?
         kids?   1_SA_Pos    2_A_Pos     3_D_Neg   4_SD_Neg        Total
---------------------------------------------------------------------------
      1StAgree        464          0          0          0           464
        2Agree          0        712          0          0           712
      3Neither          0          0          0          0           197
     4Disagree          0          0        336          0           336
   5StDisagree          0          0          0         56            56
a_Can´t choose          0          0          0          0            26
     b_Refused          0          0          0          0             7
---------------------------------------------------------------------------
         Total        464        712        336         56         1,798

   Working mom
 can have warm      Working mom can have warm
   relations w             relations?
         kids?   a_Unsure  b_Refused  n_Neutral        Total
--------------------------------------------------------------
      1StAgree          0          0          0          464
        2Agree          0          0          0          712
      3Neither          0          0        197          197
     4Disagree          0          0          0          336
   5StDisagree          0          0          0           56
a_Can´t choose         26          0          0           26
     b_Refused          0          7          0            7
--------------------------------------------------------------
         Total         26          7        197        1,798
```

After creating the other scales, I check that they are coded in the same direction with the command pwcorr C4*. As an added check, I compare each binary variable with the corresponding four-point scale. Because of the systematic way I named variables, I can use a foreach loop:

```
. foreach s in warm kids family nohome jobsat indep {
  2.     pwcorr B`s´ C4`s´, obs
  3. }

             |    Bwarm   C4warm
-------------+------------------
       Bwarm |   1.0000
             |     1568
             |
      C4warm |  -0.8239   1.0000
             |     1568     1568
             |

             |    Bkids   C4kids
-------------+------------------
       Bkids |   1.0000
             |     1483
             |
      C4kids |  -0.8114   1.0000
             |     1483     1483
             |
  (output omitted)
```

As before, I do some bookkeeping and then save the new file.

6.6 Merging files

Another way to add variables to a dataset is by merging it with a second dataset. The most common reason for merging is when both files are for the same individuals (countries, firms, etc.), but the files contain different variables. For example, I am working on a project studying health and aging for women (Pavalko, Gong, and Long 2007) that uses data from the National Longitudinal Survey (NLS). The NLS is a huge dataset with tens of thousands of respondents, thousands of variables, and multiple panels. It took us several years to construct the variables we needed. Because several people were simultaneously working on the data, we needed to ensure that different people were not changing the same variables. Our solution was to divide the variables into seven areas and construct the datasets independently for each group of variables. For example, one person was creating a dataset with demographic and control variables while another person was creating a dataset with measures of health. We did this for half a dozen areas. For analysis, we selected variables from various files and combined them into one dataset.

Files can be combined with the `merge` command. Here I discuss the most basic features of the command, which are sufficient for many applications. For further details, see `help merge` or [D] **merge**. The `merge` command combines the dataset in memory, known as the *master dataset*, with a dataset on disk, known as the *using dataset*. With one-to-one merging, the first observation from the master dataset is combined with the first observation from the using dataset, the second with the second, and so on until one of the datasets runs out of observations. With match-merging, observations from the master dataset are linked with those in the using dataset based on an ID variable.

6.6.1 Match-merging

The syntax for match-merging is

merge *id-variable* using *filename*

where *id-variable* is a variable that is common to both datasets and that contains unique IDs. Both datasets must be sorted on *id-variable*. To illustrate match-merging, I merge control variables from wf-nls-cntrl07.dta with variables about functional limitations from wf-nls-flim05.dta. I load the master dataset and then merge using id to match observations from the two files. I start by checking the data signatures of the two files (file: wf6-merge-match.do):

```
. use wf-nls-flim05, clear
(Workflow example with NLS FLIM variables \ 2008-04-02)
. datasignature confirm
  (data unchanged since 02apr2008 13:29)
. use wf-nls-cntrl07, clear
(Workflow example with NLS control variables \ 2008-04-02)
. datasignature confirm
  (data unchanged since 02apr2008 13:29)
```

With wf-nls-cntrl07.dta still in memory, I merge the two datasets:

```
. merge id using wf-nls-flim05
```

The merge command creates the variable _merge that indicates whether an observation was found only in the master dataset (_merge==1), only in the using dataset (_merge==2), or in both datasets (_merge==3). For our example,

```
. tab1 _merge
-> tabulation of _merge
     _merge |      Freq.     Percent        Cum.
------------+-----------------------------------
          1 |         21       21.00       21.00
          3 |         79       79.00      100.00
------------+-----------------------------------
      Total |        100      100.00
```

Twenty-one observations were only in the master dataset wf-nls-cntrl07.dta, whereas the rest were in both datasets. Before proceeding, I need to determine why 21 people did not have information on functional limitations. I might do this by listing the IDs for these cases (list id if _merge==1) or by running other analyses to determine why these people had no data on their limitations. Assuming that there was no problem with these 21 cases, I would drop _merge and save the merged dataset using the following commands:

```
drop _merge
quietly compress
label data "Workflow merged NLS flim & control variables \ 2008-04-09"
local tag "wf6-merge-match.do jsl 2008-04-09."
notes: wf-nls-combined01.dta \ workflow data for chapter 6 \ `tag´
datasignature set, reset
save wf-nls-combined01, replace
```

Sorting the ID variable

Match-merging assumes that both datasets have been sorted by the ID variable. If the datasets are not sorted, you get an error:

```
. use wf-nls-cntrl07, clear
(Workflow example with NLS control variables \ 2008-04-02)
. merge id using wf-nls-flim05
master data not sorted
r(5);
```

You can use the `sort` option to automatically sort the datasets on the *id-variable*. For example,

```
. use wf-nls-cntrl07, clear
(Workflow example with NLS control variables \ 2008-04-02)
. merge id using wf-nls-flim05, sort
```

The only disadvantage to using `sort` is that merging might take a bit longer.

6.6.2 One-to-one merging

With one-to-one merging you combine the first observation from the master dataset with the first observation from the using dataset, the second observation with the second, and so on. Because no ID variable is specified, there is no requirement that observations in the source dataset correspond to those in the using dataset. Indeed, with one-to-one matching, you can combine datasets that have absolutely nothing to do with one another.

Combining unrelated datasets

In this example, I combine `wf-lfp.dta` with data on labor-force participation and `wf-acpub.dta` with data on research productivity of biochemists. I use these datasets as examples when teaching, and I think it would be convenient to combine them into the same file. What I want to do is illustrated in figure 6.13.

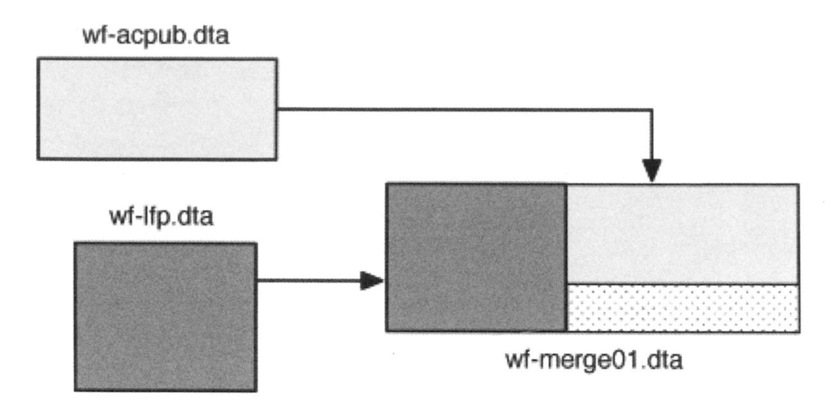

Figure 6.13. Merging unrelated datasets

The datasets that I am combining are unrelated, with `wf-acpub.dta` having fewer observations than `wf-lfp.dta`. In the merged dataset, the number of observations will equal that of the larger `wf-lfp.dta`. Once merged, the "extra observations" for variables from `wf-acpub.dta` will be assigned missing values, as illustrated by the dots in figure 6.13. I start by checking my two datasets (file: `wf6-merge-onetoone.do`):

```
. use wf-lfp, clear
(Workflow data on labor force participation \ 2008-04-02)

. datasignature confirm
  (data unchanged since 02apr2008 13:29)

. summarize

    Variable |       Obs        Mean    Std. Dev.       Min        Max
-------------+--------------------------------------------------------
         lfp |       753    .5683931    .4956295          0          1
          k5 |       753    .2377158     .523959          0          3
        k618 |       753    1.353254    1.319874          0          8
         age |       753    42.53785    8.072574         30         60
          wc |       753    .2815405    .4500494          0          1
-------------+--------------------------------------------------------
          hc |       753    .3917663    .4884694          0          1
         lwg |       753    1.097115    .5875564  -2.054124   3.218876
         inc |       753    20.12897     11.6348  -.0290001         96

. use wf-acpub, clear
(Workflow data on scientific productivity \ 2008-04-04)

. datasignature confirm
  (data unchanged since 04apr2008 17:46)

. summarize

    Variable |       Obs        Mean    Std. Dev.       Min        Max
-------------+--------------------------------------------------------
          id |       308    58654.49    2283.465      57001      62420
       enrol |       278     5.92446     2.92346          3         25
      female |       308    .3474026    .4769198          0          1
         phd |       308    3.177987    1.012738          1       4.77
         pub |       308    3.185065    3.908752          0         31
-------------+--------------------------------------------------------
 enrol_fixed |       278    5.564748    1.467253          3         14
```

Now I one-to-one merge the datasets:

```
. use wf-lfp, clear
(Workflow data on labor force participation \ 2008-04-02)
. merge using wf-acpub
. tabulate _merge

     _merge |      Freq.     Percent        Cum.
------------+-----------------------------------
          1 |        445       59.10       59.10
          3 |        308       40.90      100.00
------------+-----------------------------------
      Total |        753      100.00
. drop _merge
```

The tabulation tells me that 308 observations came from both datasets but 445 came only from `wf-lfp.dta`. Why is this? Because `wf-acpub.dta` only has 308 cases, the 445 "excess" cases are those in `wf-lfp.dta` that do not have corresponding observations in `wf-acpub.dta`. This corresponds to the section of figure 6.13 that is filled with dots. To create a dataset that combines the information from the two source datasets, I simply save the merged file:

```
. quietly compress
. label data "Workflow example of combining unrelated datasets \ 2008-04-09"
. local tag "wf6-merge-onetoone.do jsl 2008-04-09."
. notes: wf-merge01.dta \ workflow examples from chapter 6 \ `tag´
. datasignature set, reset
  753:14(117189):528693629:719271906       (data signature reset)
. save wf-merge01, replace
file wf-merge01.dta saved
```

With the combined dataset, I can replace the program that loads two datasets

```
use binlfp2, clear
logit lfp k5 k618 age wc hc lwg inc
use couart2.dta, clear
nbreg art fem mar kid5 phd ment
```

with a program that only needs one dataset:

```
use wf-merge01, clear
logit lfp k5 k618 age wc hc lwg inc
nbreg art fem mar kid5 phd ment
```

6.6.3 Forgetting to match-merge

The commands for match-merging and one-to-one merging are very similar, differing only by whether the name of the matching variable is included. For example, this command is for match-merging:

```
merge id using wf-nls-flim05
```

And this one is for one-to-one merging:

```
merge using wf-nls-flim05
```

If you meant to match-merge and forget to include the ID variable in your command, the results can be disastrous. Suppose that I have one file with bibliographic data on a sample of scientists and a second file with biographical data from the same scientists. I want to combine these into the same file, matching a scientist's biographical data with her bibliographic data. After confirming the data signatures, I run the commands (file: `wf6-merge-nomatch.do`)

```
. use wf-mergebio, clear
(Workflow biographical data to illustrate merging \ 2008-04-05)
. merge using wf-mergebib
. tab1 _merge
-> tabulation of _merge

     _merge |      Freq.     Percent        Cum.
------------+-----------------------------------
          3 |        408      100.00      100.00
------------+-----------------------------------
      Total |        408      100.00
. drop _merge
```

I have 408 cases from both files as expected. I look at the descriptive statistics and they match the results I obtained when I computed statistics for each file separately:

```
. codebook, compact
Variable    Obs Unique      Mean  Min       Max  Label
-------------------------------------------------------------------------------
job         408     80  2.233431    1       4.8  Prestige of first job
fem         408      2  .3897059    0         1  Gender: 1=female 0=male
phd         408     89  3.200564    1       4.8  PhD prestige
ment        408    123  45.47058    0  531.9999  Citations received by mentor
id          408    408     204.5    1       408  ID number
art         408     14  2.276961    0        18  # of articles published
cit         408     87  21.71569    0       203  # of citations received
-------------------------------------------------------------------------------
```

Next I look at the correlations:

```
. pwcorr job fem phd ment art cit
             |      job      fem      phd     ment      art      cit
-------------+------------------------------------------------------
         job |   1.0000
         fem |  -0.1076   1.0000
         phd |   0.3636  -0.0550   1.0000
        ment |   0.2129  -0.0100   0.3253   1.0000
         art |  -0.3534   0.0713  -0.9115  -0.2829   1.0000
         cit |  -0.2210   0.0850  -0.6700  -0.2126   0.7340   1.0000
```

The correlations among `job`, `fem`, `phd`, and `ment` look right, but the correlations between biographical and bibliographic measures are wrong. For example, the prestige of the PhD department has a strong, negative correlation with later productivity ($r = -.91$).

The problem is that I forgot to indicate the ID variable in the merge command. Because the two source files were not sorted by id, the biographical information for a given scientist is linked to a different scientist's bibliographic measures. I cannot detect this problem by looking at univariate statistics. To fix the problem, I match by id after sorting each dataset on this variable:

```
use wf-mergebio, clear
merge id using wf-mergebib, sort
```

6.7 Conclusions

This chapter has covered many of the critical steps that are involved in preparing your data for analysis. Although these steps are time consuming, they are essential. I find that the time spent cleaning my data saves time when doing the statistical analysis. Plus, if your data are not correct, your analysis will be wrong. Now that the data are clean, we can move from the workflow of data management to the workflow of statistical analysis and presentation; that is, we now consider issues involved in running analysis, fitting models, and presenting results.

7 Analyzing data and presenting results

After preparing your dataset, the only things left are statistical analysis and the presentation of findings. The time invested in naming, labeling, and cleaning your data should pay dividends by making the analysis more efficient and enjoyable. Although there might be some trips back into data management to retrieve a forgotten variable or to fix an error, these should be quick.

Building on the ideas from earlier chapters, this chapter highlights issues of special importance for statistical analysis and presentation. Sections 7.1 and 7.2 describe planning your analysis and organizing your do-files. Section 7.3 describes documentation, where the critical issue is provenance: linking every number you present to the do-file that produced that number. Section 7.4 explains how Stata programming can speed up your work, followed by section 7.5 that describes the simple but important issue of keeping track of descriptive statistics. Section 7.6 reviews problems that can limit your ability to replicate findings. Section 7.7 looks at several practical issues related to presentation ranging from constructing tables to avoiding mistakes in PowerPoint. Suitably, the chapter ends in section 7.8 with a checklist for completing a project.

7.1 Planning and organizing statistical analysis

Given all the work that goes into getting a dataset ready, it is tempting to jump into the analysis without a fully developed plan. I confess that I have done this. Soon, however, prudence overcomes enthusiasm and I make a plan to guide the analysis. A useful way to think about planning for statistical analysis is to adapt an idea from software development where work is divided into programming in the large, in the middle, and in the small (Oliveira and Stewart 2006, 59–70). Planning in the large considers the overall objectives of the work, planning in the middle divides these objectives into manageable tasks, whereas planning in the small focuses on the details necessary to complete each task.

What your plan looks like depends on many things. How specific is the research question? How well developed is the substantive literature? Are you testing specific hypotheses or exploring new possibilities? How familiar are you with the dataset and the statistical methods being used? At one extreme, your plan could be as elaborate as the one by Blau and Duncan (1967) who specified the analysis for their entire book before they had access to the data. Most often, your plan will be simpler. To give you an idea of what a plan might look like, consider a paper that I worked on with Eliza Pavalko and Fang Gong (Pavalko, Gong, and Long 2007) about the relationship between labor-force participation and health for several cohorts of women. I am not suggesting that this is the only way to write an effective plan. Indeed, my own plans vary greatly across projects. As long as your plan effectively guides your analysis and facilitates replication, it is probably a good plan.

7.1.1 Planning in the large

Planning in the large ties the objectives of your research to the potential of your data to meet those objectives. Although this part of the plan does not consider the details necessary to implement your objectives, the limitations of your data are kept in mind because it does little good to make a plan that requires data that you do not have and cannot get.

The plan for the cohort, work, and health (CWH) paper began with an abstract that described the issues we were addressing:[1]

> Social change in women's labor-force behavior in the past half century has been well documented. Coinciding with dramatic increases in women's labor-force participation are increases in percentages of women with young children in the labor force .... In this paper, we assess whether social change in women's employment has implications for women's physical health... While health benefits of employment for women are fairly clear, we know little about whether these effects have changed in concert with the changes in women's labor market experiences.

Our earlier research found several patterns that we wanted to explore further.

1. *Cohort variation by employment status versus employment category.*[2] There is little cohort variation in health by employment status, but there is interesting variation by employment category.
2. *Employment categories and health.* There are interesting relationships between employment category and health, particularly between the earliest and latest cohorts.

Motivated by these findings, we developed several research questions:

1. *Employment status and health*: Are employed women healthier than nonemployed women and does this relationship vary by birth cohort?

---

1. A more detailed version of the original plan is available at the Workflow web site (file: `wf7-plan-cwh-large.pdf`).
2. We use the term "employment status" to distinguish between women who are employed and those who are not. The term "employment category" is used to distinguish among those employed, not employed for family reasons, not employed for health, and not employed for other reasons.

2. *Employment categories and health*: Does the effect of nonemployment on health vary by the reason for nonemployment? Is the health of women who are nonemployed different than the health of employed women? Does this relationship vary by birth cohort?
3. *Explaining change in health*: Are the relationships between employment and health due to

   a. changes in the effects of employment on health?
   b. changes in the distribution of women among employment categories?
   c. the greater selectivity of the 1991 sample compared with the 1971 sample?

4. *Additional controls.* Do these relationships persist after controlling for variables such as workforce commitment, hours worked, and type of employment?

Next we planned the statistical analyses needed to explore these ideas. This is planning in the middle.

7.1.2 Planning in the middle

Middle-level plans translate the broad objectives into distinct analytic tasks and consider how the work for each task can be divided among do-files. For the CWH paper, we started with three tasks:

1. *Describe the sample and variables*: Compute descriptive statistics to describe the sample and the variables used in the analyses. Describe the distribution of health by cohort and employment status.
2. *Model health by cohort, employment status, and controls*: Estimate cohort differences in health by employment category after controlling for demographic and other variables. Compare alternative modeling approaches.
3. *Sensitivity analyses*: Examine whether the findings could be explained by sample attrition or measurement problems.

After data analysis begins, new tasks often emerge when initial analyses show that the problem is more complicated than anticipated. Other tasks merge because they are simpler than anticipated or because preliminary work shows that they are not as distinct as thought. For the CWH paper, our initial tasks grew into 16 tasks that we abbreviated as `cwh01`–`cwh16`. We began by computing descriptive statistics for all variables that we considered using:

`cwh01`: Descriptive statistics.

Within each task, the do-files were named to indicate the task they were part of and the order in which they needed to be run (e.g., `cwh01a-base-stats.do`, `cwh01b-base-graphs.do`). The second task is to fit count models predicting the number of health limitations a person had.

`cwh02`: Compare count models for number of health limitations (PRM, NBRM, and ZIP).

Because of the large percentage of zeros in the outcome, we created a binary measure that divided respondents into those with no limitations and those with at least one limitation. A logit model was fit for this outcome:

`cwh03`: Logit model for having any limitations.

The results from the logit model along with those from the ZIP model lead us to a hurdle model for counts:

`cwh04`: Hurdle model for number of limitations.

After reviewing these findings, we returned to data management to add variables that made it easier to fit models with interactions and to include data from the 1971 panel:

`cwh05`: Data management: Add interaction variables.

`cwh06`: Data management: Add data from the 1971 panel.

With the new data, we fit the models from tasks `cwh02` and `cwh04` with the 1971 panel included:

`cwh07`: Count models with 1971 data included.

Because the hurdle model provided the best fit and made the most substantive sense, we tried different approaches to parameterizing the model to make it easier to test our hypotheses and then evaluated the sensitivity of the results to variations in our specification:

`cwh08`: Hurdle model using alternative parameterizations.

`cwh09`: Sensitivity analysis of hurdle model.

Using these results, we wrote the first draft of the paper. After discussing this draft, we planned additional analyses with a few new variables. Although the new analyses could have been added to the tasks above, we created new tasks that were organized around the tables in the revised paper. When a paper is almost ready to circulate, I find it useful to add new tasks that include only those do-files needed to reproduce the analyses presented in the paper. By doing this, I can double check the analyses, and later I can easily find the analyses used in the paper. The results reported in the revised paper all came from these tasks, where the first task involved data management that simplified the later analyses:

`cwh10`: Data management: Add additional variables.

`cwh11`: Descriptive statistics for tables 1, 2, and 3.

`cwh12`: Hurdle models and predictions for figures 1–5.

`cwh13`: Supplementary analyses with the hurdle model for table 4.

We submitted the paper. After receiving a request to revise, we added variables suggested by reviewers, refined the coding of other variables, and updated our figures and tables. Because all the analyses used in the paper were included in tasks `cwh10`–`cwh13`, revisions were simple:

`cwh14`: Data management: Add work and smoking variables; revise some operational definitions.

`cwh15`: Refit models and create plots.

`cwh16`: Fit additional models for table 4.

The paper went through two more revisions before it was published. During this process, the project was put on hold for months at a time as we waited for reviews. Having the work divided into tasks made it easier to pick up the work where we had left off.

7.1.3 Planning in the small

Planning in the small implements the tasks from the middle-level plan. This involves what Oliveira and Stewart (2006, 61) refer to as the "nitty-gritty details". In the small, you decide which variables to use, how to code variables for analysis (e.g., do you want to treat two years of college as the same as having a two-year degree from a trade or professional school?), and which commands to use (e.g., should the hurdle model be fit with `logit` and `ztp` or with `hplogit`?). Planning in the small also involves organizing results within do-files, as discussed in the next section.

7.2 Organizing do-files

The number of tasks used to organize your work and the number of do-files within each task depend on how complicated your analyses are and your personal preference. I prefer more tasks and shorter do-files rather than fewer tasks and longer do-files. For one paper, I might use 100 do-files, most of which contain fewer than 100 commands. I only use longer do-files for things such as complex postestimation analysis or the construction of intricate graphs.

Here is a real-world example that illustrates why I prefer shorter do-files. I was asked for advice on adjusting for clustered observations in analyses using multinomial logit. After looking at the tables with the current findings, I asked for the do-files that generated those results. I received one file that was 753 lines long, had no comments, and produced a Stata log that was 163 pages long. The do-file had evolved over months as new analyses were added and previous analyses were revised. Somewhere in the evolution of the 753-line do-file, the results in the tables had been extracted. Unfortunately, later changes to the do-file affected those results so that I could not replicate the tables. After a great deal of frustration, we decided that the simplest thing to do was to start over. Although this example is extreme, it highlights what can happen if your do-files get too long and if you do not anticipate the need to later replicate your results.

There are several advantages to having shorter do-files.

- If you correct an error in a do-file, you need to verify that none of the later commands in the file are affected. You might accidentally change something you did not intend to or make a change earlier in the file that affects something later. With longer files, verifying that later analyses are unaffected is more difficult. For example, if you correct an error in line 243 of a 700-line do-file, you need to verify hundreds of other commands. In longer do-files, it is easier to change the wrong thing, thereby replacing one error with two.
- It is easier to keep track of corrected results from shorter do-files. Suppose that you have a long do-file that generates many pages of output. After you correct an error and rerun the program, should you print the entire log file including those parts that did not change? If you had comments written on the earlier output, these need to be transferred. If you print only the changed output, it is difficult to keep track of which output has the latest result.
- It is easier to review the results in shorter log files, especially in collaborative work. Trying to digest too much output at a time makes it more likely that you will miss errors or skip over an important finding. In collaborative work, lengthy Stata logs result in too much time spent looking for the result that someone else is discussing. I find that it works better to discuss several smaller sets of results. For example, the descriptive statistics are stapled together, analyses testing group differences are in another set, and the results from regression models are in a third.

The most common objection I hear to using multiple, shorter do-files is that it is difficult to keep track of the files and if you need to rerun them, you have to type a lot of `do` *program-name* commands. If you use the naming conventions I recommend, it is easy to keep track of your files. To easily rerun all your files, you can use a master do-file.

7.2.1 Using master do-files

A master do-file is simply a do-file that contains `do` commands to run other do-files (see page 131 for further details). For example, in a recent analysis of racial differences in sexual well-being, I used 10 do-files. To rerun these files, I ran a master do-file named `swb-all.do`:

```
capture log close master
log using swb-all, name(master) replace text
//  program:    swb-all.do
//  task:       swb \ may 2007 analyses
//  project:    workflow - chapter 7
//  author:     jsl \ 2007-03-08
//  note:       all programs required swb-00-loaddata.doi
// Task 01: descriptive statistics and data checking
do swb-01a-desc.do
do swb-01b-descmisc.do
do swb-01c-barchart.do
// Task 02: logit - sexual relationships
do swb-02aV2-srlogit.do
do swb-02b-srlogit-checkage.do
do swb-02c-srlogit-ageplot.do
// Task 03: logit - own sexuality
do swb-03a-os2logit.do
do swb-03b-os2Vos1logit.do
// Task 04: logit - self attractiveness
do swb-04a-salogit.do
// Task 05: logit - miscellaneous
do swb-05a-sr-os2-cor.do
log close master
exit
```

To rerun everything, I simply type the command:

```
do swb-all.do
```

If I want to rerun only some of the do-files, I comment out the others. For example,

```
capture log close master
log using swb-all, name(master) replace text
//  program:    swb-all.do
//  task:       swb \ may 2007 analyses
//  project:    workflow - chapter 7
//  author:     jsl \ 2008-03-07
//  note:       all programs include swb-00-loaddata.doi
/*
// Task 01: descriptive statistics and data checking
do swb-01a-desc.do
do swb-01b-descmisc.do
do swb-01c-barchart.do
// Task 02: logit - sexual relationships
do swb-02aV2-srlogit.do
do swb-02b-srlogit-checkage.do
do swb-02c-srlogit-ageplot.do
*/
// Task 03: logit - own sexuality
do swb-03a-os2logit.do
do swb-03b-os2Vos1logit.do
// Task 04: logit - self attractiveness
do swb-04a-salogit.do
```

```
// Task 05: logit - miscellaneous
do swb-05a-sr-os2-cor.do
log close master
exit
```

When actively working on a project, I like to name the master do-file `it.do`. To rerun things, I type

```
do it
```

Later I rename the file to the more cumbersome but informative `swb-all.do`.

7.2.2 What belongs in your do-file?

Page 63 described what should be included in a do-file. For statistical analyses, there are several points worth emphasizing. First, the `version` command is critical for replicating your results (discussed in section 7.6.2). Second, the `set scheme` command is necessary to create graphs that match previous graphs in style. If you do not use the same scheme, the graphs will look different. Third, you might want to include comments that interpret results and highlight key findings. To do this, you run the do-file and examine the resulting log file. Based on what you find, add comments to the do-file and rerun that file. Including interpretations within a do-file is particularly useful when collaborating. Fourth, include numbered comments that make it simpler to refer to a specific result. I began doing this when collaborating with colleagues at three universities. The analyses were complex and our emails soon involved exchanges about which part of the output someone was referring to or confusing debates when two people were discussing different parts of the same Stata log. The solution was to add numbered comments. For example, the file `swb-02b.do` might contain

```
//  #1
//  load data, select sample and variables
(commands omitted)
//  #2
//  compute means to be plotted
(commands omitted)
//  #3
//  graph of age by mean number of partners
(commands omitted)
```

If someone has a comment about the graph of the number of partners by age, he or she can write "In `swb-02b#3`, I'm concerned that the minimum age is 23. I thought it was 25." Numbering parts of a do-file is so convenient that I now do it even if the work is not part of a collaboration.

7.3 Documentation for statistical analysis

The documentation of statistical analysis is based on the principles discussed in section 2.4.1. The next section briefly reviews how these principles apply to documenting the do-files used in your analysis. Section 7.3.2 describes the important new topic of linking the statistics used in presentations and papers to the do-files that computed those statistics.

7.3.1 The research log and comments in do-files

A research log for statistical analysis might begin with an outline of the analysis plans, including a list of tasks and the do-files that I anticipate needing for each task. As the analysis proceeds, the log becomes a dated record of which do-files were run and what each file did. My comments about the do-file are usually very short, so the research log is essentially an index to the do-file. The do-files themselves contain a more detailed explanation of the work. My research log would include more detail if a problem was encountered, if the analysis involved decisions that might be confusing later, or if the file produced a result that was surprising (e.g., "Even though I did not expect this variable to be significant, it is. I verified that it is coded correctly. Explore this finding later."). The research log could also include short write-ups of the results, ideas from collaborators, and plans for future analysis. Even though my do-files indicate the dataset being analyzed (e.g., `use mydata01, clear`) and the version of Stata being used (e.g., `version 10`), I generally include this information in the research log as well. Here is an example of what a research log might look like (see figure 7.1). Parts of the log have simply been copied from comments in the do-file:

(*Continued on next page*)

jsIFLIMlog: 4/1/02 to 6/12/22 - Page 11

First complete set of analysis for FLIM measures paper

Data: *flim03.dta*

f2alt01a.do - 24May2002

Descriptive information on all rhs, lhs, and flim measures

f2alt01b.do - 25May2002

Compute bic' for each of four outcomes and all flim measures.

```
** Outcome: Can Work                    global lhs "qcanwrk95"
** Outcome: Work in three categories    global lhs "dhlthwk95"
** Outcome: bath trouble                global lhs "bathdif95"
** Outcome: adlsum95 - sum of adls      global lhs "adlsum95"
```

f2alt01c.do - 25May2002

Compute bic' for each of four outcomes and with only these restricted flim measures.

```
*    1.  ln(x+.5) and ln(x+1)
*    2.   9 counts: >=5=5  >=7=7  (50% and 75%)
*    3.   8 counts: >=4=4  >=6=6  (50% and 75%)
*    4.   18 counts: >=9=9 >=14=14  (50% and 75%)
*    5.   probability splits at .5; these don't work well in prior tests
```

f2alt01d.do - 25May2002

bic' for all four outcomes in models that include all raw flim measures (fla*p5; fll*p5); pairs of u/l measures; groups of LCA measures

f2alt01e.do - all LCA probabilities - 25May2002

Figure 7.1. Example of a research log with links to do-files

I use comments inside the do-files to explain why specific analyses were run (e.g., ZIP model used because of large percentage of 0s), which variables were used (e.g., because of limitations in source data number of children under six was not available), and why specific commands and options were used (e.g., clustering used to adjust for multiple children from same family). Although I could include this information in the research log, I find it most useful in the do-file because it is echoed to the Stata log file where I can easily refer to it when interpreting results.

7.3.2 Documenting the provenance of results

The provenance of every result should be documented.

To replicate your results, you need to know where every number you report comes from, whether it is as simple as the sample size or as complex as estimates from a multiequation model. Although it is easy to grab a number from a Stata log and use it in your paper, it is often much harder to later find the do-file that produced that number. Without knowing which do-files were used, it can be difficult or impossible to later replicate your findings or to modify your analyses if changes are needed.

There are many ways to keep track of which do-file produced which result. For years, I added tabs and notations to the printed output. This efficiently documented the source of tables and figures but did not work as well for documenting specific numbers reported in the text. If I wanted to track down a result, I had to scan through pages of output. I could also use the research log to record which do-file generated which results. Both approaches were time consuming and difficult to keep accurate. The best solution I have found is to record the origin of each number within the paper itself. For example, in this extract (figure 7.2) from a paper by Pavalko, Gong, and Long (2007), I have circled two probabilities and a test of equality that were computed by `cwhrr-fig03c-hrmemp4.do`:

> 1922-1926 cohort, employed women have fewer limitations than those who are out for family
>
> reasons, (.48 and .73, respectively (z=2.55, p<.01).) However, this gap has disappeared for the
>
> 1943-1947 cohort and, indeed, employed women have slightly more limitations (.76 for non-

Figure 7.2. Example of results reported in a paper

To document the provenance of these results, I insert the name of the do-file, the location within the do-file (i.e., `#4`), the person who ran it, and the date (see figure 7.3).

> 1922-1926 cohort, employed women have fewer limitations than those who are out for family
>
> reasons, (.48 and .73, respectively (z=2.55, p<.01 {cwhrr-fig03c-hrmemp4.do #4 jsl 17May06}).)
>
> However, this gap has disappeared for the 1943-1947 cohort and, indeed, employed women have

Figure 7.3. Addition using hidden font to show the provenance of the results

Because I do not want this information printed in drafts that I circulate, I hide it. In Word, I use a hidden font.[3] Depending on the options used, hidden text can be totally suppressed, shown on the screen but not printed, or shown on the screen and printed. In LaTeX, I include provenance information within a pair of `\iffalse` and `\fi` commands, which excludes that text from being printed. Most word processors have a similar feature.

Although you might agree in principle that it is important to document the source of results used in your research, you might also think that it takes too much time to keep track of every number. I am convinced that recording the source of results saves time. First, once you get used to doing this, it does not take long because you can paste the information from the log file. Making the text hidden only takes a few clicks. I have customized Word so that all I need to do is select the text and press Ctrl+Alt+h to make it hidden. Second, having this information readily available makes revisions much simpler. For example, the CWH paper discussed earlier received a request to revise that involved changing the coding of a variable used in all the models. Using the hidden

3. For details on using hidden fonts in Word, see the Workflow web site.

documentation, it was trivially easy to find the do-files that needed to be changed and to compute the new results.

Captions on graphs

For figures included in your paper, you can add provenance information as suggested above. I often generate dozens of graphs to explore the data or to uncover trends that are reported in the paper without showing the graph. Because it is easy to lose track of which program created a graph, I add a caption that indicates the do-file that created the graph, who wrote it, and when. For example (file: `wf7-caption.do`),

```
twoway (line art_root2 art_root3 art_root4 art_root5 articles,        ///
    lwidth(medium)), ytitle(Number of Publications to the k-th Root) ///
    yscale(range(0 8.)) legend(pos(11) rows(4) ring(0))              ///
    caption(wf7-caption.do \ jsl 2008-04-09, size(vsmall))
```

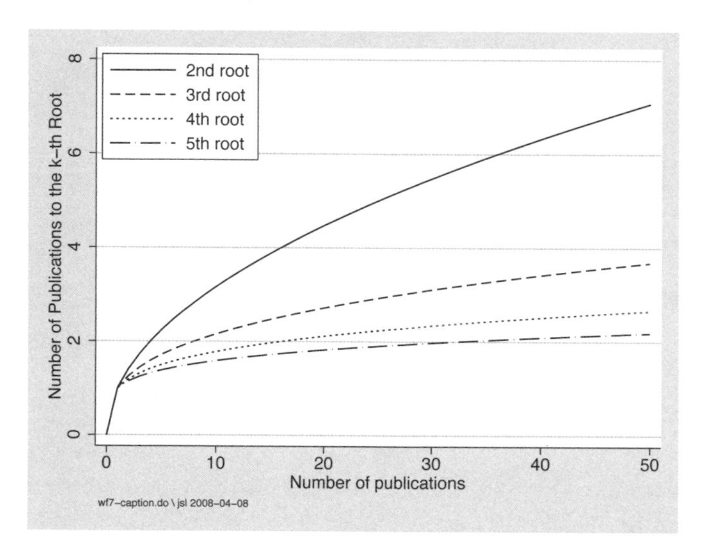

creates the above graph where the caption documents the graph's origins.

If I do not want the caption to be seen, I can crop the bottom of the graph. When producing the graph to include in an accepted paper, I can rerun the do-file with the caption removed.

7.4 Analyzing data using automation

The automation tools introduced in chapter 4 can speed up statistical analyses in many ways. Local macros can hold the names of variables making it simpler to specify commands and revise models. Loops let you easily run the same commands with multiple variables. Matrices can collect results from multiple commands that make it easier to review results and create tables. Also `include` files can be used to load a dataset and select cases the same way in multiple do-files. To illustrate these methods, I use data about gender differences in the receipt of tenure for academic biochemists (Long, Allison, and McGinnis 1993). The binary dependent variable indicates if a person received tenure with predictors such as gender, departmental prestige, time in rank, and research productivity. Each person has multiple observations corresponding to each year they were in the rank of assistant professor. Although I do not discuss the statistical commands used in the examples, you should be able to follow how the automation features are used even if you are not familiar with the specific methods.

7.4.1 Locals to define sets of variables

I generally avoid listing the same variables multiple times within a do-file. Instead, I use local macros to hold the names of groups of variables. This eliminates the possibility of changing the list inconsistently at different locations. For example, in a study of gender differences, I want descriptive statistics for the entire sample, for men, and for women (file: `wf7-locals.do`):

```
// #2
// desc statistics for men & women combined
codebook female male tenure year yearsq select articles prestige, compact
// #3
// desc statistics for women
codebook female male tenure year yearsq select articles prestige ///
    if female, compact
// #4
// desc statistics for men
codebook female male tenure year yearsq select articles prestige ///
    if male, compact
```

With these commands, if I decide to drop variable `yearsq`, I must remove it at three locations. Alternatively, I can define the local `varset` with the names of variables:

```
local varset "female male tenure year yearsq select articles prestige"
```

Then my program becomes

```
// #5
// desc statistics for men & women combined
codebook `varset´, compact
// #6
// desc statistics for women
codebook `varset´ if female, compact
// #7
// desc statistics for men
codebook `varset´ if male, compact
```

Then if I want to drop `yearsq`, I need to make the change in only one place.

Local macros also make it easier to see the links among models, which prevents errors in specification. For example, suppose that I want to fit a series of logit models that start by controlling only for gender and progressively add groups of variables. Here is how I would do this without using locals:

```
// #8a = baseline gender only model
logit tenure female, nolog or
// #8b + time
logit tenure female year yearsq, nolog or
// #8c + department
logit tenure female year yearsq select prestige, nolog or
// #8d + productivity
logit tenure female year yearsq select prestige articles, nolog or
```

While in this simple example, it is easy to verify that each model is correctly specified, with more complex models that require multiple lines to specify, it can be difficult to verify. As an alternative, I create local variables that define three groups of variables:

```
local Vtime "year yearsq"      // time in rank
local Vdept "select prestige"  // characteristics of departments
local Vprod "articles"         // research productivity
```

Next I fit four models:

```
// #9a = baseline gender only model
logit tenure female, nolog or
// #9b + time
logit tenure female `Vtime´, nolog or
// #9c + department
logit tenure female `Vtime´ `Vdept´, nolog or
// #9d + productivity
logit tenure female `Vtime´ `Vdept´ `Vprod´, nolog or
```

Using macros makes it easy to see that the models have been specified correctly. If I decide to include the third power of year in my time variables and add citations as a measure of productivity, I have to change only the two locals:

```
local Vtime "year yearsq yearcu" // time in rank
local Vprod "articles citations" // research productivity
```

Because the `logit` commands are not changed, I know that the models are correctly specified.

7.4.2 Loops for repeated analyses

In statistical analyses, I often use the same command or series of commands with multiple variables. An efficient way to do this is with loops, which I illustrate by computing a series of t tests and fitting regression models.

Computing t tests using loops

I want to test gender differences in all the variables included in a logit model predicting tenure. For the outcome `tenure` (file: `wf7-loops-ttest.do`),

```
. ttest tenure, by(female)
Two-sample t test with equal variances

   Group |     Obs        Mean    Std. Err.   Std. Dev.   [95% Conf. Interval]
---------+--------------------------------------------------------------------
  0_Male |    1741    .1315336    .0081025    .3380801    .1156419    .1474253
1_Female |    1056    .1089015    .0095908    .3116632    .0900824    .1277207
---------+--------------------------------------------------------------------
combined |    2797    .1229889    .0062111    .3284832    .1108102    .1351677
---------+--------------------------------------------------------------------
    diff |            .0226321    .0128075               -.002481    .0477451
------------------------------------------------------------------------------
    diff = mean(0_Male) - mean(1_Female)                          t =   1.7671
Ho: diff = 0                                     degrees of freedom =     2795

    Ha: diff < 0                 Ha: diff != 0                 Ha: diff > 0
 Pr(T < t) = 0.9613         Pr(|T| > |t|) = 0.0773          Pr(T > t) = 0.0387
```

For the other variables, I could run the commands

```
ttest year, by(female)
ttest select, by(female)
ttest articles, by(female)
ttest prestige, by(female)
```

Alternatively, I can use a `foreach` loop. First, I create a local with the variables to test:

```
local varlist "tenure year select articles prestige"
```

Then I loop through the variables:

```
foreach var in `varlist´ {
    ttest `var´, by(female)
}
```

This quickly generates the tests, but there is a problem. When `ttest` is run within a loop, the command is not echoed and accordingly the output does not indicate the name of the variable being tested. For example, the first pass through the loop produces this output:

```
Two-sample t test with equal variances

   Group |     Obs        Mean    Std. Err.   Std. Dev.   [95% Conf. Interval]
---------+--------------------------------------------------------------------
  0_Male |    1741    .1315336    .0081025    .3380801    .1156419    .1474253
1_Female |    1056    .1089015    .0095908    .3116632    .0900824    .1277207
---------+--------------------------------------------------------------------
combined |    2797    .1229889    .0062111    .3284832    .1108102    .1351677
---------+--------------------------------------------------------------------
    diff |            .0226321    .0128075               -.002481    .0477451
------------------------------------------------------------------------------
    diff = mean(0_Male) - mean(1_Female)                          t =   1.7671
Ho: diff = 0                                     degrees of freedom =     2795

    Ha: diff < 0                 Ha: diff != 0                 Ha: diff > 0
 Pr(T < t) = 0.9613         Pr(|T| > |t|) = 0.0773          Pr(T > t) = 0.0387
```

The solution is to use display to echo the command:

```
foreach var in `varlist' {
    display _new ". ttest `var', by(female)"
    ttest `var', by(female)
}
```

For example, the first time through the loop

```
. ttest tenure, by(female)
Two-sample t test with equal variances

------------------------------------------------------------------------------
   Group |     Obs        Mean    Std. Err.   Std. Dev.   [95% Conf. Interval]
---------+--------------------------------------------------------------------
```

(*output omitted*)

Loops for alternative model specifications

Loops can also be used to estimate a series of regressions in which each model differs by the inclusion of a single variable. For example, I can use a loop to fit a series of models with the same right-hand-side variables but different outcomes, or I can use a loop to change one of the predictors. Suppose that I want to evaluate alternative transformations of the predictor articles in a logit model for tenure. I start by creating variables that are root transformations of articles (file: wf7-loops-arttran.do):

```
1>  local artvars ""
2>  forvalues root = 1(1)9 {
3>      gen art_root`root' = articles^(1/`root')
4>      label var art_root`root' "articles^(1/`root')"
5>      local artvars "`artvars' art_root`root'"
6>  }
```

Line 1 initializes the local artvars to hold the names of the variables I am generating. If I already had these variables in my dataset, I would replace the entire loop with the local artvars containing the names of the variables. Line 2 begins and line 6 ends the loop where the local root is changed from 1 to 9 in increments of 1. To explain line 3, suppose that root equals 3. The command creates variable art_root3 equal to the 1/3 power of articles. Lines 4 adds a variable label, and line 5 adds the name of the new variable to the end of the local artvars.

Next I loop through the names collected in the local artvars to fit logit models of tenure on one of the art_root# variables along with the other predictors, where I use display to describe each model. For example,

```
foreach avar in `artvars' {
    display _new "== logit with `avar'"
    logit tenure `avar' female year yearsq select prestige, nolog
}
```

7.4.3 Matrices to collect and print results

I often run similar analyses for multiple variables and need a table to summarize the results. For example, I might want a table reporting the t tests of gender differences in the example above, or I might want a table reporting the regression coefficients and Bayesian information criterion (BIC) statistics from my experiments using different roots of `articles`. I could extract this information by hand from the Stata log, but this is time consuming, tedious, and error prone. A better solution is to collect the results in a matrix from which I can create a table. To do this, there are three steps:

1. Create a matrix with one row for each set of results and one column for each statistic.
2. Retrieve the results returned by Stata commands and place them in the matrix.
3. Print the matrix.

I provide a series of increasingly complicated examples that illustrate how to do this. If you are having problems following any of the examples, you can might want to review chapter 4 on returned results (page 90) and loops (page 102).

Collecting results of t tests

Earlier, I used a loop to compute t tests of gender differences and now I want to summarize the results in a table that will look like this:

	Women		Men		Test of equal means	
Variable	Mean	Std. Dev.	Mean	Std. Dev.	*t* test	Prob
`tenure`						
`year`						
`selectivity`						
`articles`						
`prestige`						

To collect the information, I need a matrix with five rows and six columns (file: `wf7-matrix-ttest.do`):

```
matrix stats = J(5,6,-99)
```

The `J(`*# of rows*`,`*# of columns*`,`*value*`)` function creates a matrix containing −99s:

```
. matrix list stats
stats[5,6]
     c1   c2   c3   c4   c5   c6
r1  -99  -99  -99  -99  -99  -99
r2  -99  -99  -99  -99  -99  -99
r3  -99  -99  -99  -99  -99  -99
r4  -99  -99  -99  -99  -99  -99
r5  -99  -99  -99  -99  -99  -99
```

I fill the matrix with −99 to make it easier to debug my program. When a Stata command encounters a problem, it often returns missing values for some quantities. If I place these values in the matrix, the −99s are replaced with the sysmiss. Accordingly, if I print the matrix and see missing values, I know there was a problem with the statistical analysis. If I see −99s, I know that I made a mistake when populating the matrix. Next I add row and column labels:

```
. matrix colnames stats = FemMn FemSD MalMn MalSD t_test t_prob
. matrix rownames stats = `varlist´
. matrix list stats
stats[5,6]
             FemMn    FemSD    MalMn    MalSD   t_test   t_prob
  tenure       -99      -99      -99      -99      -99      -99
    year       -99      -99      -99      -99      -99      -99
  select       -99      -99      -99      -99      -99      -99
articles       -99      -99      -99      -99      -99      -99
prestige       -99      -99      -99      -99      -99      -99
```

To populate the matrix, I use results returned by `ttest`. For example,

```
. ttest tenure, by(female)
Two-sample t test with equal variances
------------------------------------------------------------------------------
   Group |     Obs        Mean    Std. Err.   Std. Dev.   [95% Conf. Interval]
---------+--------------------------------------------------------------------
  0_Male |    1741    .1315336    .0081025    .3380801    .1156419    .1474253
1_Female |    1056    .1089015    .0095908    .3116632    .0900824    .1277207
---------+--------------------------------------------------------------------
combined |    2797    .1229889    .0062111    .3284832    .1108102    .1351677
---------+--------------------------------------------------------------------
    diff |            .0226321    .0128075                -.002481    .0477451
------------------------------------------------------------------------------
    diff = mean(0_Male) - mean(1_Female)                          t =   1.7671
Ho: diff = 0                                     degrees of freedom =     2795

    Ha: diff < 0                 Ha: diff != 0                 Ha: diff > 0
 Pr(T < t) = 0.9613         Pr(|T| > |t|) = 0.0773          Pr(T > t) = 0.0387
```

The `return list` command returns this information:

```
. return list
scalars:
                 r(sd) =  .3284832119751412
               r(sd_2) =  .3116632125613366
               r(sd_1) =  .3380801147013905
                 r(se) =  .0128074679461748
                r(p_u) =  .0386602339087719
                r(p_l) =  .9613397660912281
                  r(p) =  .0773204678175438
                  r(t) =  1.767100751070975
               r(df_t) =  2795
               r(mu_2) =  .1089015151515152
                r(N_2) =  1056
               r(mu_1) =  .1315336013785181
                r(N_1) =  1741
```

To determine what each return contains, I can check the *Saved Results* section of `help ttest`.

Now I am ready to replace the −99s in the matrix with the values computed by `ttest`. I loop through the variables and compute the required test:

```
 1> local irow = 0
 2> foreach var of varlist `tenvars´ {
 3>     local ++irow
 4>     quietly ttest `var´, by(female)
 5>     matrix stats[`irow´,1] = r(mu_2) // female mean
 6>     matrix stats[`irow´,2] = r(sd_2) // female sd
 7>     matrix stats[`irow´,3] = r(mu_1) // male mean
 8>     matrix stats[`irow´,4] = r(sd_1) // male sd
 9>     matrix stats[`irow´,5] = r(t)    // t-value
10>     matrix stats[`irow´,6] = r(p)    // p-value
11> }
```

Line 1 initiates a counter for the row of the matrix where I will put the returned results. Line 2 begins the loop through the variables in local `tenvars`. Each pass through the loop takes the next variable in `tenvars` and puts it in the macro `var`. Line 3 increments the row number. Line 4 computes the t test of `` `var´ `` by `female` where `quietly` suppresses the output. I added the `quietly` prefix after I was sure that the program was working correctly. Lines 5–10 take returns from `ttest` and places them in matrix `stats`. Let's look at line 5 carefully:

```
matrix stats[`irow´,1] = r(mu_2)
```

To the left of the equal-sign, `` matrix stats[`irow´,1] `` specifies the cell of the matrix that I want to change. On the right, `r(mu_2)` retrieves the returned `mu_2` from `ttest`, which is the mean for women. Lines 6–10 fill the other columns of the matrix for row `irow`. Before printing the matrix, I create a header with the number of men and women in the sample. First, I create locals with the number of cases

```
local n_men = r(N_1)
local n_women = r(N_2)
```

and combine this information in a local:

```
local header "t-tests: mean_women (N=`n_women´) = mean_men (N=`n_men´)"
```

Now I print the matrix

```
. matrix list stats, format(%9.3f) title(`header´)
stats[5,6]:  t-tests: mean_women (N=1056) = mean_men (N=1741)
            FemMn    FemSD    MalMn    MalSD   t_test   t_prob
  tenure    0.109    0.312    0.132    0.338    1.767    0.077
    year    3.974    2.380    3.784    2.252   -2.121    0.034
  select    5.001    1.475    4.992    1.365   -0.170    0.865
articles    7.415    7.430    6.829    5.990   -2.284    0.022
prestige    2.658    0.765    2.640    0.784   -0.612    0.540
```

where the `format(%9.3f)` option specifies three decimal digits.

Sample size? When I retrieved the sample size to create the header, I assumed that the number of cases was the same for all variables. If the sample differed across variables due to missing data, I could add two columns to `stat` for the number of valid cases for men and women for each test. I suggest you try this as an exercise.

Saving results from nested regressions

An earlier example ran a series of logits predicting tenure using different sets of control variables. Now I want to create a table with the odds ratio for `female`, its z-value, and the BIC statistics for each model. I start by creating a matrix (file: `wf7-matrix-nested.do`):

```
. local modelnm "base plustime plusdept plusprod"
. local statsnm "ORfemale zfemale BIC"
. matrix stats = J(4,3,-99)
. matrix rownames stats = `modelnm´
. matrix colnames stats = `statsnm´
. matrix list stats
stats[4,3]
            ORfemale    zfemale        BIC
    base         -99        -99        -99
plustime         -99        -99        -99
plusdept         -99        -99        -99
plusprod         -99        -99        -99
```

I run the first logit model:

```
. logit tenure female, or
  (output omitted)
```

Because `logit` does not return either odds ratios or z-values, I need to compute these quantities. I start by retrieving the estimated coefficients and placing them in matrix `b`. Because `logit` is an estimation command, the coefficients are returned with `e(b)` not `r(b)`:

```
. matrix b = e(b)
. matrix list b
b[1,2]
         female        _cons
y1  -.21454446   -1.8874666
```

Next I retrieve the covariance matrix:

```
. matrix v = e(V)
. matrix list v
symmetric v[2,2]
             female       _cons
female    .01478639
 _cons   -.00502818   .00502818
```

I use the coefficient for `female` from the first column of `b`, take the exponential to compute the odds ratios, and place that value in cell (1,1) of `stats`:

```
. matrix stats[1,1] = exp(b[1,1])
```

Next I compute $z = \widehat{\beta}_{\texttt{female}} / \sqrt{\widehat{\text{Var}}\left(\widehat{\beta}_{\texttt{female}}\right)}$ and place this value in the second column of `stats`:

```
. matrix stats[1,2] = b[1,1]/sqrt(v[1,1])
```

To compute the BIC statistic, I use `estat`:

```
. estat ic
```

Model	Obs	ll(null)	ll(model)	df	AIC	BIC
.	2797	-1042.828	-1041.245	2	2086.49	2098.363

```
Note:  N=Obs used in calculating BIC; see [R] BIC note
```

The *Saved Results* section of `help estat` indicates that the BIC statistic is in the first row and sixth column of `r(S)`. First, I move `r(S)` into a temporary matrix, and then I move BIC into `stats`:

```
. matrix temp = r(S)
. matrix stats[1,3] = temp[1,6]
```

The next models use the locals that were defined earlier:

```
local Vtime "year yearsq"
local Vdept "select prestige"
local Vprod "articles"
```

I use the same commands where the only thing that differs is the row of the matrix where I store the results:

```
logit tenure female `Vtime´, or
matrix b = e(b)
matrix v = e(V)
matrix stats[2,1] = exp(b[1,1])
matrix stats[2,2] = b[1,1]/sqrt(v[1,1])
estat ic
matrix stats[2,3] = temp[1,6]
```

After fitting all the models, I list `stats` to summarize the results from the four models:

```
. matrix list stats, format(%9.3f)
stats[4,3]
           ORfemale   zfemale        BIC
     base     0.807    -1.764   2098.363
 plustime     0.723    -2.511   1768.675
 plusdept     0.721    -2.520   1767.413
 plusprod     0.702    -2.678   1732.620
```

Using include for saving results to matrices

After each `logit` model, I used the same commands to extract the information and place it into the `stats` matrix with only the row number changing. Rather than retyping these commands, I can use an `include` file to automate the process. I create `wf7-matrix-nested-include.doi` that contains

```
local irow = `irow' + 1
matrix b = e(b)
matrix v = e(V)
matrix stats[`irow',1] = exp(b[1,1])
matrix stats[`irow',2] = b[1,1]/sqrt(v[1,1])
qui estat ic
matrix temp = r(S)
matrix stats[`irow',3] = temp[1,6]
```

Using the `include` file, the commands to fit the models and store the results are simplified to (file: `wf7-matrix-nested-include.do`)

```
//  #4
//  nested models predicting tenure
//  #4a - baseline gender only model
logit tenure female, or
include wf7-matrix-nested-include.doi
//  #4b + time
logit tenure female `Vtime', or
include wf7-matrix-nested-include.doi
//  #4c + department
logit tenure female `Vtime' `Vdept', or
include wf7-matrix-nested-include.doi
//  #4d + time
logit tenure female `Vtime' `Vdept' `Vprod', or
include wf7-matrix-nested-include.doi
```

This makes the program easier to write, and if I change the information I want to store for each model, I have to make the change only once in the `include` file.

Saving results from different transformations of articles

I can use a similar approach to collect the results from the earlier example that used different transformations of the number of articles in a model predicting `tenure` (see page 302). The local `artvars` contains the names of the variables that I want to examine:

```
local artvars art_root1 art_root2 art_root3 art_root4 art_root5
```

To summarize the results, I want a table with the standard deviation of `art_root#`, the standardized coefficient for `art_root#`, the odds ratio for a standard-deviation change, and the BIC statistic from the model with `art_root#`. Each row of the matrix `stats` corresponds to a different root transformation of `articles`. To create the matrix (file: `wf7-matrix-arttran.do`),

```
local nvars : word count `artvars'
matrix stats = J(`nvars',5,-99)
matrix rownames stats = `artvars'
matrix colnames stats = root sd b_std exp_b_std bic
```

I use the extended macro function `local : word count` to count the number of variables in local `artvars`. The advantage of using the extended macro is that if I later decide to examine the roots from 1 to 15, the size of the matrix will automatically be adjusted. The next steps are to

1. Loop through each transformation of `articles` where the local `avar` contains the name of the transformed variable.
2. Use `summarize` to compute the standard deviation of `` `avar' `` and place the result in the matrix `stats`.
3. Fit a logit model that includes `` `avar' `` along with other predictors.
4. Compute the standardized coefficient and the standardized odds ratio for `` `avar' `` and put these in the matrix.
5. Compute the BIC statistic using `estat ic` and place it in the matrix.

Here are the commands to do this:

```
 1>  local irow = 0
 2>  foreach avar in `artvars' {
 3>      local ++irow
 4>      * add root number to the matrix
 5>      matrix stats[`irow',1] = `irow'
 6>      * sd of avar
 7>      summarize `avar'
 8>      local sd = r(sd)
 9>      matrix stats[`irow',2] = `sd'
10>      * logit with avar
11>      logit tenure `avar' female year yearsq select prestige, nolog
12>      * save b*sd and exp(b*sd) for avar
13>      matrix b = e(b)
14>      matrix stats[`irow',3] = b[1,1]*`sd'
15>      matrix stats[`irow',4] = exp(b[1,1]*`sd')
16>      * save bic
17>      estat ic
18>      matrix temp = r(S)
19>      matrix stats[`irow',5] = temp[1,6]
20>  }
```

Because these commands are similar to those used earlier, I note only a few things. Line 5 adds the root number to the matrix. Lines 7–9 use `summarize` to compute the standard deviation and place it into `stats`. Line 13 creates matrix `b` with the estimated coefficients. Line 14 multiplies the estimated coefficient for `` `avar' `` by the standard deviation from line 8. Line 15 takes the exponential to compute the standardized odds ratio. Printing the results, I have all the information I need:

```
. local header "Comparing root transformations of articles in logit"
. matrix list stats, format(%9.3f) title(`header´)
stats[9,5]:  Comparing root transformations of articles in logit
                  root         sd      b_std  exp_b_std        bic
art_root1        1.000      6.576      0.361      1.434   1732.620
art_root2        2.000      1.171      0.507      1.661   1716.203
art_root3        3.000      0.646      0.581      1.788   1716.791
art_root4        4.000      0.479      0.638      1.892   1720.605
art_root5        5.000      0.403      0.683      1.979   1725.224
art_root6        6.000      0.360      0.716      2.047   1729.924
art_root7        7.000      0.333      0.738      2.092   1734.431
art_root8        8.000      0.315      0.747      2.112   1738.625
art_root9        9.000      0.303      0.746      2.108   1742.455
```

Overall, the square root looks like a good transformation to use. For details on selecting a model using the BIC statistics, see Raftery (1995).

Adding z and p

Next I want to enhance the program to add the z test and the p-value for each estimate. First, I create a larger matrix:

```
local nvars : word count `artvars´
matrix stats = J(`nvars´,7,-99)
matrix rownames stats = `artvars´
matrix colnames stats = root sd b_std exp_b_std bic z prob
```

Because `logit` does not return either z or p, I need to compute these by adding the following commands to the loop shown above:

```
19a>  matrix vc = e(V)
19b>  local ztest = b[1,1]/sqrt(vc[1,1])
19c>  local prval = 2*normal(-abs(`ztest´))
19d>  matrix stats2[`irow´,6] = `ztest´
19e>  matrix stats2[`irow´,7] = `prval´
```

Line 19a retrieves the covariance matrix, line 19b computes the z test, and line 19c computes the probability associated with the test. Lines 19d and 19e save the results to the matrix.

7.4.4 Creating a graph from a matrix

After you collect information in a matrix, you can easily graph it. For example, to plot the BIC statistics against the root transformation from the last example, I start by creating variables from the columns of the matrix `stats` (file: `wf7-matrix-plot.do`):

```
svmat stats, names(col)
```

The `svmat` command creates one variable for each column of a matrix, and the `names(col)` option indicates that the column names of the matrix should be used to name the variables. I now have the variable `root` with the root number and variable `bic` with the BIC statistic. To plot these, I use `twoway`:

```
twoway (connected bic root, msymbol(circle)), ///
    ytitle(BIC statistic) ylabel(1700(10)1750) ///
    xtitle(Root transformation of articles) xlabel(1(1)9) ///
    caption("wf7-matrix-plot.do \ jsl 2008-04-11",size(vsmall))
```

The resulting graph makes it easy to see that the BIC statistic supports the use of the second or third root transformation:

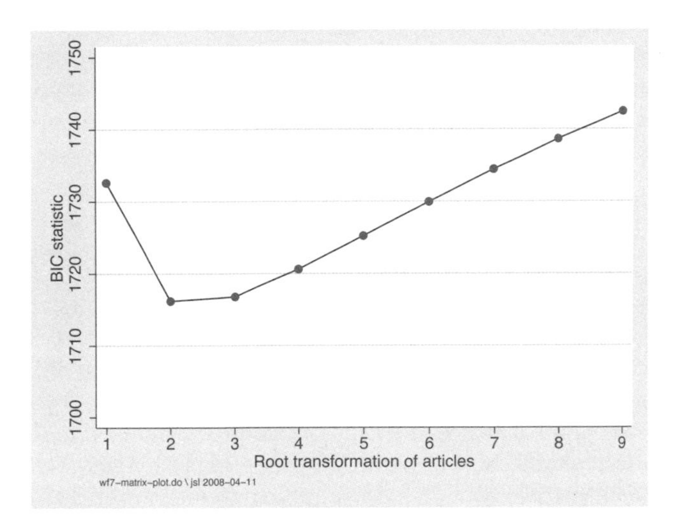

7.4.5 Include files to load data and select your sample

Suppose that I am running a series of do-files that load the same dataset and select the same observations. Each do-file could begin with these commands
(file: `wf7-include-sample.do`):

```
use wf-tenure, clear
datasignature confirm
drop if year>=11    // drop cases with long time in rank
drop if prestige<1 // drop if unrated department
```

If I later decided to drop cases if `year` is greater than or equal to 10 and `prestige` is less than or equal to 1, I need to make this change in all the do-files. When doing this, it is easy to miss one of the files or to make an incorrect change. Alternatively, I can use an `include` file:

```
//  include:    wf7-include-sample.doi
//  used by:    wf7-include-sample.do
//  task:       define sample for tenure example
//  project:    workflow chapter 7
//  author:     scott long \ 2008-04-09

//  #1
//  load data and select sample
```

```
use wf-tenure, clear
datasignature confirm
drop if year>=11 // drop cases with long time in rank
drop if prestige<1 // drop if unrated department
```

Then each do-file would contain this command

```
include wf7-include-sample.doi
```

If I decide to change the sample, I revise `wf7-include-sample.doi` and the change applies to all do-files that include it. (If I have already posted the do-files, I need to create new versions of the files.)

7.5 Baseline statistics

When fitting complex models, it is easy to forget about the most basic characteristics of your data. This can lead to wasted effort or even incorrect results. For example, last year I was working on a count model predicting the number of sexual partners a person had. Because the Poisson and negative binomial regression models did not fit, I started thinking about zero-inflated count models (Long and Freese 2006, 394–396). These did not fit either. At this point I returned to the outcome distribution and discovered what I already knew: most people had a single partner. This obvious realization lead to a very different modeling strategy. Or, consider situations in which several people are analyzing the same data and you need everyone to work with the same sample and variables. One of the easiest ways to ensure this is by comparing descriptive statistics, including the sample size. If you do not, it is possible to spend hours debating differences in results until you realize that different decisions were made about the sample used.

I like to start new analyses by double checking the descriptive statistics and including these statistics in the front of my project notebook. Consider the tenure data described above. I begin by loading the data and selecting the sample used in later analyses (file: `wf7-baseline.do`):

```
. //  #1
. //  load data and select sample
. use wf-tenure, clear
(Workflow data for gender differences in tenure \ 2008-04-02)
. datasignature confirm
  (data unchanged since 02apr2008 13:29)
. tabulate sampleis

 Sample for
     tenure
   analysis |      Freq.     Percent        Cum.
------------+-----------------------------------
      0_Not |        148        5.03        5.03
 1_InSample |      2,797       94.97      100.00
------------+-----------------------------------
      Total |      2,945      100.00
. keep if sampleis
(148 observations deleted)
```

Next I compute descriptive statistics for men and women combined:

```
. // #2
. // desc statistics for men & women combined
. codebook female male tenure year yearsq select articles prestige, compact

Variable     Obs Unique      Mean   Min   Max  Label
---------------------------------------------------------------------------
female      2797      2  .3775474     0     1  Scientist is female?
male        2797      2  .6224526     0     1  Is male?
tenure      2797      2  .1229889     0     1  Is tenured?
year        2797     10  3.855917     1    10  Years in rank
yearsq      2797     10  20.16911     1   100  Years in rank squared
select      2797      8  4.995048     1     7  Baccalaureate selectivity
articles    2797     48  7.050411     0    73  Total number of articles
prestige    2797     98  2.646591   .65   4.8  Prestige of department
---------------------------------------------------------------------------
```

Because I am looking at gender differences, I also compute descriptive statistics for men and women separately:

```
. // #3
. // desc statistics for women
. codebook female male tenure year yearsq select articles prestige ///
> if female, compact
  (output omitted)
. // #4
. // desc statistics for men
. codebook female male tenure year yearsq select articles prestige ///
> if male, compact
  (output omitted)
```

When you are actively analyzing your data, variable names and their meanings become second nature. For collaborators who have not been working with the dataset, names are often confusing. Giving everyone a log file with descriptive statistics and variable labels makes the results clearer and the discussion easier.

7.6 Replication

No matter what workflow you use, you should be able to replicate your results. There are a number of things that can make replication difficult or even impossible.

7.6.1 Lost or forgotten files

If you cannot find the dataset or do-files used for your analysis or if you can no longer read the dataset, you will not be able to replicate your results. This can occur if the data are destroyed, such as when fire destroyed the data for Wolfgang, Figlio, and Sellin's *Delinquency in a Birth Cohort* study (1972). More likely, you physically have the dataset but cannot read it, or perhaps the media are obsolete (e.g., 80-column cards), the data are in a format that is not supported by current software (e.g., OSIRIS, multipunched cards), or you might have several versions of the dataset that differ slightly. If you did not record which version of the data was used in your analysis, it can be maddeningly difficult to get answers that match exactly. Or you might know which version was used, have the do-file that is needed, but cannot find where the dataset is stored. Archiving your data and documentation are key to preventing these types of problems. Chapter 8 discusses these issues in greater detail.

7.6.2 Software and version control

Later versions of software will not necessarily give you the same results as earlier versions even when running exactly the same commands. For example, over time Stata changed the way in which `xtgee` estimated some quantities. Although these changes were improvements, they resulted in slightly different answers when using the same model to analyze the same data. The solution is to have a `version` command at the start of your do-file (see page 53 for details). For example, if you are using Stata 9.2 but your do-file includes the `version 5` command, Stata will produce results that correspond to those obtained in Stata 5.0.

7.6.3 Unknown seed for random numbers

Random numbers are used in many types of statistical analysis, such as bootstrapping standard errors, estimation by simulation, and dividing your data into training and confirmation samples. To replicate results that depend on random numbers, you must use exactly the same random numbers as used in the initial analyses. To understand what this means and how to do it, you need to know where the random numbers come from. Random numbers used by statistical packages are not truly random. Instead, they are pseudorandom numbers (PRN) that are generated by an equation. This equation transforms an initial number, called the *seed*, to create the first PRN. The first PRN is transformed by the same equation to create the second PRN and so on. If you start with the same seed, you get exactly the same sequence of PRNs. Accordingly, to reproduce analyses that use random numbers, you need to know the seed that was used. The easiest way to do this is by including a `set seed` # command in your do-file. For example, `set seed 17022` indicates that 17,022 should be used as the seed when generating numbers. Here are some examples that illustrate the importance of documenting the seed.

Bootstrap standard errors

Bootstrapping is a method to estimate the sampling distribution of your estimates by taking repeated random samples with replacement from the sample used to fit the model. These bootstrap samples are selected using pseudorandom numbers. Here is an example of how results differ when you use different seeds (see `wf7-replicate-bootstrap.do`). I start by fitting a logit model for labor-force participation:

```
. use wf-lfp, clear
(Workflow data on labor force participation \ 2008-04-02)
. datasignature confirm
  (data unchanged since 02apr2008 13:29)
. logit lfp k5 k618 age wc hc lwg inc
Iteration 0:   log likelihood =  -514.8732
Iteration 1:   log likelihood = -454.32339
Iteration 2:   log likelihood = -452.64187
Iteration 3:   log likelihood = -452.63296
Iteration 4:   log likelihood = -452.63296
Logistic regression                               Number of obs   =        753
                                                  LR chi2(7)      =     124.48
                                                  Prob > chi2     =     0.0000
Log likelihood = -452.63296                       Pseudo R2       =     0.1209
  (output omitted)
```

I use prvalue to compute the predicted probabilities of being in the labor force using 100 replications to compute the bootstrap confidence interval:

```
. set seed 11020
. prvalue, bootstrap reps(100)
logit: Predictions for lfp
Bootstrap confidence intervals using percentile method
(100 of 100 replications completed)
                                 95% Conf. Interval
  Pr(y=1_InLF|x):      0.5778   [ 0.5242,    0.6110]
  Pr(y=0_NotInL|x):    0.4222   [ 0.3890,    0.4758]
            k5        k618         age         wc         hc         lwg        inc
x=   .2377158   1.3532537   42.537849   .2815405  .39176627  1.0971148  20.128965
```

If I run prvalue again without setting the seed, I get these results:

```
. prvalue, bootstrap reps(100)
logit: Predictions for lfp
Bootstrap confidence intervals using percentile method
(100 of 100 replications completed)
                                 95% Conf. Interval
  Pr(y=1_InLF|x):      0.5778   [ 0.5361,    0.6318]
  Pr(y=0_NotInL|x):    0.4222   [ 0.3682,    0.4639]
            k5        k618         age         wc         hc         lwg        inc
x=   .2377158   1.3532537   42.537849   .2815405  .39176627  1.0971148  20.128965
```

The predicted probability is exactly the same but the confidence interval differs slightly. With a larger number of replications (1,000 is recommended), the answers will usually be very close when using different seeds.

Letting Stata set the seed

When Stata opens, it always begins with the same seed (see `help random number functions`). Accordingly, if a do-file that uses random numbers is the first program run after Stata is started, you should get the same answer every time. However, if you run your do-file during a Stata session in which other programs are using random numbers, your answers will differ depending on which commands were run first. Accordingly, I prefer to always set the seed. If you do not do this, you should record the seed that Stata is using so that you can replicate your results. Stata returns the current seed in `c(return)`. To display the seed,

```
. local seedis = c(seed)
. display "`seedis´"
X98ec336832edcfba325ca86f7001068b308a
```

To use this seed in a later program, you use the command

```
set seed X98ec336832edcfba325ca86f7001068b308a
```

Training and confirmation samples

Modeling often involves trying alternative specifications where estimates from one model are used to suggest an alternative model. When this process is automated, it is referred to as stepwise regression. The model you select based on repeated analyses of the same data can depend greatly on unique characteristics of the sample (i.e., it will not generalize to the population). One way to check this is to randomly divide your sample. Half the sample, called the training sample, is used to explore the data and to choose a model. The second half, the confirmation sample, is used to confirm the model selected with the training sample. In this example, I show how the model you select can depend on the seed used when dividing your sample (see `wf7-replicate-stepwise.do`). After setting the seed, I generate the binary random variable `train1` to divide the sample. To create this variable, I use `runiform()` to obtain uniform random numbers between 0 and 1. If the random number is less than .5, `train1` is 1, else `train1` is 0:

```
. use wf-articles, clear
(Workflow data on scientific productivity \ 2008-04-11)
. datasignature confirm
  (data unchanged since 11apr2008 10:35)
. set seed X57c74068e0f7a3200d5b8463f279bb82065a
. generate train1 = (runiform() < .5)
. label var train1 "Training sample?"
. label def trainlbl 0 "0Confirm" 1 "1Train"
. label val train1 trainlbl
```

Next I use `stepwise` to select a model using the training sample (i.e., `if train1==1`) and save the results to print later:

```
. stepwise, pr(.05): nbreg art fem mar kid5 phd ment if train1==1
  (output omitted)
. estimates store train1trim
```

Only `fem`, `ment`, and `kid5` were kept in the model. When I fit that model with the confirmation sample, the results closely match those from the training sample:

```
. quietly nbreg art fem kid5 ment if train1==0
. estimates store confirm1trim
. estimates table train1trim confirm1trim, stats(N chi2) b(%9.3f) star
```

Variable	train1trim	confirm1trim
art		
fem	-0.207*	-0.249*
ment	0.019***	0.038***
kid5	-0.138*	-0.138*
_cons	0.438***	0.340***
lnalpha		
_cons	-1.014***	-0.723***
Statistics		
N	478	437
chi2	24.417	71.336

legend: * p<0.05; ** p<0.01; *** p<0.001

Here the stepwise procedure selected a model that generalized to the other half of the sample. Alas, I had to try over 50 random splits of the data to find results that were roughly comparable in the two samples. Here is what I got the first time I tried this experiment:

```
. set seed 11051951
. generate train2 = (runiform() < .5)
. label var train2 "Training sample?"
. label val train2 trainlbl
. quietly stepwise, pr(.05): ///
> nbreg art fem mar kid5 phd ment if train2==1
. estimates store train2trim
. quietly nbreg art fem mar kid5 ment if train2==0
. estimates store confirm2trim
. estimates table train2trim confirm2trim, stats(N chi2) b(%9.3f) star
```

Variable	train2trim	confirm2trim
art		
fem	-0.304**	-0.132
mar	0.273*	0.015
kid5	-0.211**	-0.130
ment	0.033***	0.024***
_cons	0.259*	0.361**
lnalpha		
_cons	-0.722***	-1.001***
Statistics		
N	456	459
chi2	69.155	29.522

legend: * p<0.05; ** p<0.01; *** p<0.001

This example illustrates two important things. First, the seed used to generate random numbers can have a large effect on the results. Second, stepwise methods should be used with caution.

7.6.4 Using a global that is not in your do-file

In Stata, some quantities are stored as globals while others are stored as locals. Globals persist after you run a do-file, whereas locals do not. For example, suppose that `step1.do` creates the global macro `rhs` containing the names of the independent variables (see page 84 for information on global macros):

```
global rhs "k5 k618 age wc hc lwg inc"
```

In `step2.do`, I have the command

```
logit lfp $rhs
```

This fits the logit model for `lfp` on `k5 k618 age wc hc lwg inc`. If I later run `step2.do` without first running `step1.do`, the global `rhs` might not be defined. Accordingly, the command `logit lfp $rhs` will be interpreted as `logit lfp` because `$rhs` is a null string. In general, your do-files should not depend on information held in Stata's memory. This is why I suggest that you include `clear all` and `macro drop _all` at the start of each do-file.[4]

7.7 Presenting results

With the analysis completed, you are ready for the third stage of Michael Faraday's famous guide to success: *Work, finish, publish.* It would take a book as long as the current one to address the many challenges involved in presenting and publishing your results. Indeed, there are many books about scientific writing. I particularly like *The Art of Scientific Writing* by Ebel, Bliefert, and Russey (2004), which begins by noting: "Underlying all of natural science is a rather remarkable understanding, albeit one that attracts relatively little attention: Everything measured, detected, invented, or arrived at theoretically in the name of science must, as soon as possible, be made public-complete with all the details." The next 500 pages tell you how to do this, addressing topics such as approaches to writing, rules of grammar, making effective tables, copy editing, and much more. Although this book is written by chemists for chemists, I think that anyone who writes will benefit from its advice. In this section, my objectives are much more limited. I consider techniques that make it easier to collect the results you want to present, review a few ideas that will make your presentations of graphs more effective, provide some suggestions regarding PowerPoint-type presentations, and present a checklist of things to do when a draft of your paper is completed.

4. The command `clear all` does not clear macros, so `macro drop _all` is needed. In Stata 9, you must use `clear` instead of `clear all`.

7.7.1 Creating tables

Good tables are critical for effectively conveying your results. Unfortunately, creating good tables involves a lot of work that is tedious and error prone. There are many formats for tables which vary by discipline and journal. To decide how to format a table, I often start by looking at papers in the journal where I plan to submit the paper, find a table similar to what I need, and use that table as a model. You can find guidelines for creating tables in the *Publication Manual of the American Psychological Association* (American Psychological Association 2001), *The Chicago Manual of Style* (University of Chicago Press 2003), or *The Art of Scientific Writing* (Ebel, Bliefert, and Russey 2004). Regardless of the format you use, there are several principles that are important. First, make tables self-contained. The effectiveness of a table is decreased if readers need to search for information to understand the table. Second, be consistent in your use of decimal digits. Third, avoid large amounts of empty space that make it difficult to follow a row of numbers. Fourth, use labels that are clear to someone who is not actively involved in the research.

After the format of a table is chosen, I have three techniques that make it easier to create a table. First, collect results in a matrix that corresponds to the table that you want. This was illustrated in section 7.4.3. Second, paste Stata output into a spreadsheet where you can format your table; this works particularly well when you have collected results into a matrix. Third, for regression results, use Jann's (2007) `esttab` command.

Using spreadsheets

A spreadsheet can greatly simplify creating tables because it allows you to compute statistics that are not available in the Stata log (e.g., compute the odds ratios for the estimated coefficients) and to easily revise the format. During the early stages of a paper, I often keep tables in a spreadsheet and only move them into a word processor when the first circulation draft is written. Once in a word processor, the tables can be refined as needed.

The way you move results from a Stata log into a spreadsheet depends on the software you are using. Here I consider Excel, which is available for Windows and Mac OS. I start by listing the matrix that summarizes the results of t tests of gender differences:

```
. matrix list stats, format(%9.3f)
stats[5,6]
             FemMn    FemSD    MalMn    MalSD   t_test   t_prob
  tenure     0.109    0.312    0.132    0.338    1.767    0.077
    year     3.974    2.380    3.784    2.252   -2.121    0.034
  select     5.001    1.475    4.992    1.365   -0.170    0.865
articles     7.415    7.430    6.829    5.990   -2.284    0.022
prestige     2.658    0.765    2.640    0.784   -0.612    0.540
```

I select the table in my text editor, copy it, and paste the text into the spreadsheet. At this point, each cell within the box on the left of figure 7.4 contains the entire row of text and numbers (i.e., values have not been moved into individual cells).

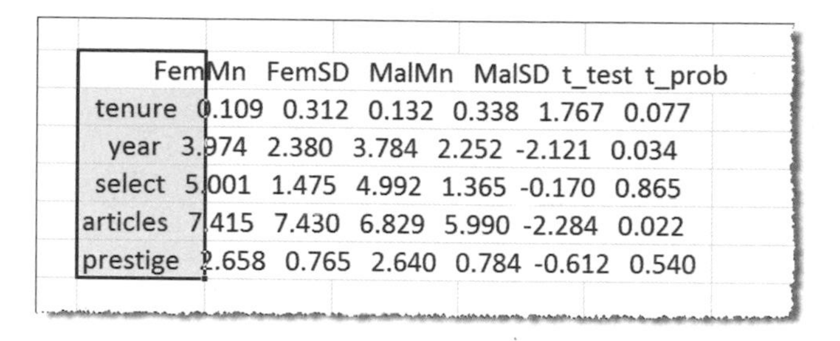

	FemMn	FemSD	MalMn	MalSD	t_test	t_prob
tenure	0.109	0.312	0.132	0.338	1.767	0.077
year	3.974	2.380	3.784	2.252	-2.121	0.034
select	5.001	1.475	4.992	1.365	-0.170	0.865
articles	7.415	7.430	6.829	5.990	-2.284	0.022
prestige	2.658	0.765	2.640	0.784	-0.612	0.540

Figure 7.4. Spreadsheet with pasted text

To convert the pasted text into numbers within cells, I use the Convert Text to Columns Wizard (how you invoke this depends on the version of Excel you are using); see figure 7.5.

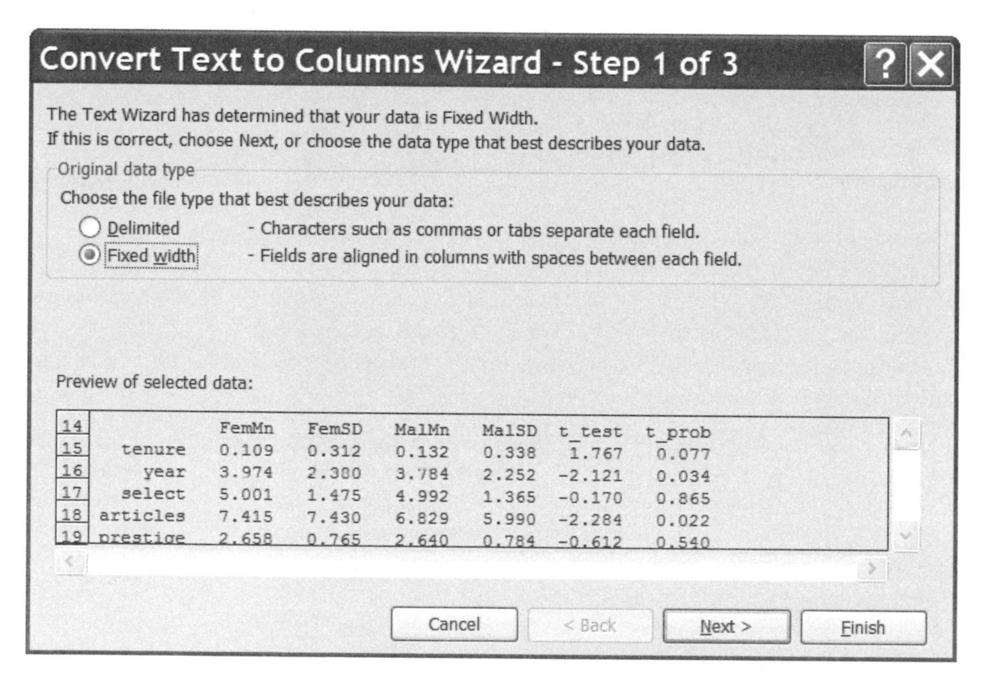

Figure 7.5. Convert text to columns wizard

Next I add header information so that I know where the results in the table came from (see figure 7.6):

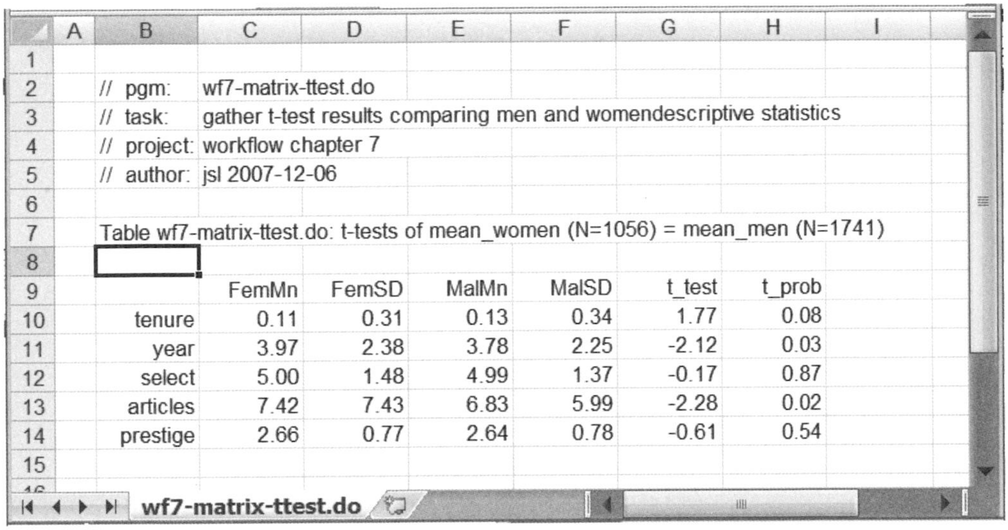

	A	B	C	D	E	F	G	H	I
1									
2		// pgm:	wf7-matrix-ttest.do						
3		// task:	gather t-test results comparing men and womendescriptive statistics						
4		// project:	workflow chapter 7						
5		// author:	jsl 2007-12-06						
6									
7		Table wf7-matrix-ttest.do: t-tests of mean_women (N=1056) = mean_men (N=1741)							
8									
9			FemMn	FemSD	MalMn	MalSD	t_test	t_prob	
10		tenure	0.11	0.31	0.13	0.34	1.77	0.08	
11		year	3.97	2.38	3.78	2.25	-2.12	0.03	
12		select	5.00	1.48	4.99	1.37	-0.17	0.87	
13		articles	7.42	7.43	6.83	5.99	-2.28	0.02	
14		prestige	2.66	0.77	2.64	0.78	-0.61	0.54	
15									

wf7-matrix-ttest.do

Figure 7.6. Spreadsheet with header information added

I can now use the formatting tools in Excel to revise the table.

Regression tables with esttab

Jann (2007) has written several extremely useful commands for creating tables with regression results. His `estout` command is powerful and flexible, but as Jann (2007) notes "`estout`'s syntax is not as intuitive and user-friendly as it could be." Unless you are creating a lot of tables, the investment in mastering this command is probably not worth it. Fortunately, Jann (2007) wrote a new pair of commands, `esttab` and `eststo`, that are easy to use yet retain the flexibility of `estout`. For example, I can use these commands to create a table for the nested regressions in an earlier example. After each logit, the command `eststo` stores the estimates so they can be used to create a table (file: `wf7-tables-esttab.do`):

```
//  #3a - baseline gender only model
logit tenure female, nolog or
eststo
//  #3b + time
logit tenure female `Vtime', nolog or
eststo
//  #3c + department
logit tenure female `Vtime' `Vdept', nolog or
eststo
```

Using the default options for `esttab`, I easily create a basic table:

```
. esttab
```

	(1) tenure	(2) tenure	(3) tenure
female	-0.215	-0.324*	-0.327*
	(-1.76)	(-2.51)	(-2.52)
year		1.805***	1.818***
		(11.21)	(11.23)
yearsq		-0.129***	-0.130***
		(-9.35)	(-9.35)
select			0.141**
			(3.12)
prestige			-0.262**
			(-3.15)
_cons	-1.887***	-6.927***	-7.002***
	(-26.62)	(-15.59)	(-13.30)
N	2797	2797	2797

```
t statistics in parentheses
* p<0.05, ** p<0.01, *** p<0.001
```

With a few simple options, I can fine-tune the format:

```
. esttab, eform nostar bic label varwidth(33) ///
>     title("Table 7.1: Workflow Example of Jann´s esttab Command.") ///
>     mtitles("Model A" "Model B" "Model C") ///
>     addnote("Source: wf7-tables-esttab.do")
```

Table 7.1: Workflow Example of Jann´s esttab Command.

	(1) Model A	(2) Model B	(3) Model C
Scientist is female?	0.807	0.723	0.721
	(-1.76)	(-2.51)	(-2.52)
Years in rank		6.079	6.161
		(11.21)	(11.23)
Years in rank squared		0.879	0.878
		(-9.35)	(-9.35)
Baccalaureate selectivity			1.151
			(3.12)
Prestige of department			0.770
			(-3.15)
Observations	2797	2797	2797
BIC	2098.4	1768.7	1767.4

```
Exponentiated coefficients; t statistics in parentheses
Source: wf7-tables-esttab.do
```

Even better, `esttab` lets me save the table in formats that can be imported into a spreadsheet or a word processor. For example, I can create a LaTeX table with the commands

```
esttab using wf7-estout.tex, eform nostar bic label varwidth(33) ///
    mtitles("Model A" "Model B" "Model C") ///
    addnote("Source: wf7-tables-esttab.do")
```

Table 7.1 is the resulting table.

Table 7.1. Example of a LaTeX table created using `esttab`

	(1) Model A	(2) Model B	(3) Model C
Scientist is female?	0.807	0.723	0.721
	(-1.76)	(-2.51)	(-2.52)
Years in rank.		6.079	6.161
		(11.21)	(11.23)
Years in rank squared.		0.879	0.878
		(-9.35)	(-9.35)
Baccalaureate selectivity.			1.151
			(3.12)
Prestige of department.			0.770
			(-3.15)
Observations	2797	2797	2797
BIC	2098.4	1768.7	1767.4

Exponentiated coefficients; t statistics in parentheses
Source: wf7-tables-esttab.do

To import a table to Word, I would save the table with the extension `.rtf` rather than `.tex`. To import the table into Excel, I would save it with the extension `.csv`. For full details, I recommend Jann's web site at htpp://repec.org/bocode/e/estout/.

7.7.2 Creating graphs

There is vast literature on what makes graphs effective and ineffective. This topic is beyond the scope of this book, but I highly recommend Tufte (2001), Cleveland (1993, 1994), and Wallgren (1996). If you know what you want your graph to look like and need help in finding the commands in Stata to create the graph, I suggest Mitchell (2008), which includes hundreds of examples. In this section, I assume that you know what you want your graph to look like and you know the commands needed to create the graph. I focus on evaluating how your graph looks when presented in different media (e.g., projectors, printed, web). The most important rule for presenting a graph is simple:

Try it before you present it.

If you plan to project your graph, test how the image projects. If you plan to print it in black and white, try printing it. If you are putting graphs on the web, look at them on the web, and so on. Do not assume that how a graph appears on your monitor is a good indication of how the graph will look when printed or projected. If you attend many talks, you have almost certainly heard comments such as: "Unfortunately, the colors look the same on the screen, but the bar on the left is red and the one on the right is blue." Or "This graph is included in your handout, although unfortunately it is hard to tell the lines apart because they are in black and white instead of color."[5]

There are two problems that commonly occur and can be easily addressed. First, graphs that look fine on your monitor lose critical detail when you print them in black and white or project them to a screen. Second, labels that are clear when printed full size become illegible when reduced in size or shown on a projector.

Colors, black, and white

Color can be wonderful for conveying information. However, even if your graph looks perfect on your monitor, you need to check how it will appear when printed. Here is what can happen. You create a colored graph that looks great on your monitor (file: `wf7-graphs-colors.do`):

```
graph bar (mean) Mbg (mean) Wbg, over(Vbg) ///
  legend(label(1 Men) label(2 Women)) ytitle("Percent Tenured") ///
  ylabel(0(3)15) legend(label(1 Men) label(2 Women)) ///
  bar(1,fcolor(red)) bar(2,fcolor(green))
```

You send the file to the publisher or have black and white copies made to hand out. To your chagrin, the graph looks like figure 7.7.

5. Tables are not immune to this problem. If you project a table that is clear in print, the numbers are often too small to be read on the screen. Here, too, the rule is to try it before you present it.

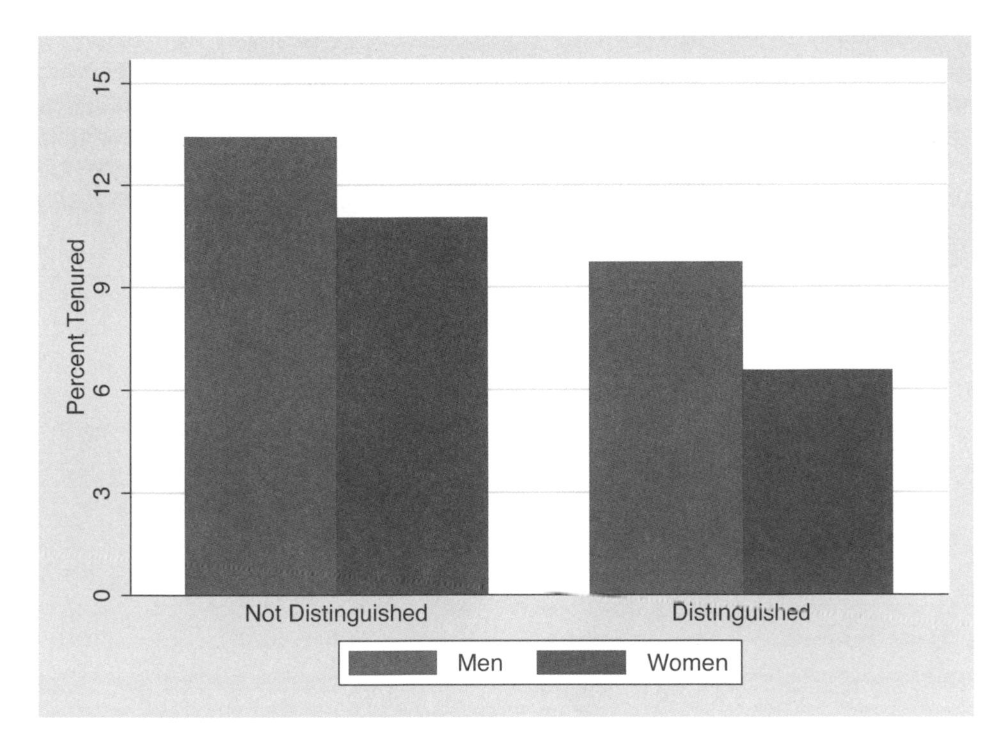

Figure 7.7. Colored graph printed in black and white

Colors that were easily distinguished on a monitor look nearly identical when printed in black and white. If you want to see the color version of this graph, check the Workflow web site.

You also need to consider how colors from your monitor translate into colors that are projected. Depending on the brand of the projector, how the color space was set up for the projector, and the way color is managed in the computer driving the projector, great colors on your screen can look terrible or indistinguishable when projected. For example, you might have a colored background with a contrasting color for the lettering that is stunning on your monitor, but the colors merge when projected. Or your carefully chosen colors to distinguish groups all look that same when projected. For presentations where you are not sure of the projector, create graphs that do not depend exclusively on color. For example, you might use colored lines to distinguish two groups, but the lines for one group could be solid while the other is dashed. You also need to be careful with shades of gray. Shades that appear distinct on your monitor might be similar when printed or projected. The bottom line is, when using color and shading in graphs, check how the graph appears on your final output source.[6]

6. For details on the reasons why colors look different with different media, see Fraser, Buntin, and Murphy (2003).

Font size

You should test your graphs in the size they will appear in presentation. For example, do not rely on a full-page printout of your graph to evaluate how it will look when reduced to 2"×3" in the published paper. Suppose that you create a graph that is five inches wide when printed in draft form. The journal decides that the printed graph should be 2" wide. The lettering that was legible at 5" is too small to easily read when reduced to 2". This is illustrated with the graphs in figure 7.8 (file: `wf7-graphs-fontsize.do`):

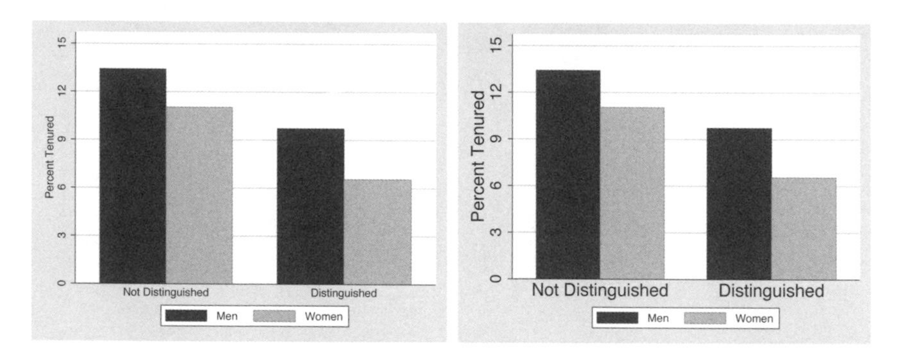

Figure 7.8. Graph with small, hard to read text and a graph with readable text

The graph on the left uses the default font size while the graph on the right uses Stata's `vlarge` font size. Here is the new program:

```
graph bar (mean) Mbg (mean) Wbg, over(Vbg, label(labsize(vlarge))) ///
    legend(label(1 Men) label(2 Women)) ytitle("Percent Tenured", size(vlarge)) ///
    ylabel(0(3)15, labsize(large)) legend(label(1 Men) label(2 Women)) ///
    bar(1,fcolor(gs4)) bar(2,fcolor(gs13)) legend(size(vlarge))
```

To determine the font size required, you need to experiment. To see a list of font sizes use `help textsizestyle`. Not only is it important to use larger fonts when graphs are printed at a smaller size, but you might need larger fonts when graphs are projected.

7.7.3 Tips for papers and presentations

Here are a few suggestions that I have found useful when preparing, sharing, and presenting results.

Papers

There are several simple things that help avoid problems when circulating drafts of your paper. First, include the date, the filename, and the person who has control of the file on the cover page of each draft. This helps prevent multiple, inconsistent versions of the same paper and avoids a collaborator mistakenly revising the wrong version of the paper.

Second, if you are distributing an electronic copy of a paper, send a PDF file instead of the word processor file (unless you are asking someone to revise the paper). With many word processors, a document prints differently with different printers or versions of the software, while PDFs usually look and print the same way for everyone. Third, do not send large files as attachments, which can fill people's email quotas. Instead, put the file on the web and send the URL or place the file on the LAN and send the location. Fourth, when asking people to send comments, consider adding line numbers to the paper. This lets people refer to a specific line (e.g., "on line 426 you missed a reference to Mitroff (2003)") rather than a location on a page (e.g., "on page 16, about half way down..."). In Word, search help for "line numbers". In LaTeX, the `lineno` package can be used.

Presentations

In recent years, presentations using overheads and handouts have largely been replaced by projected presentations using software such as PowerPoint from Microsoft, Keynote from Apple, or Beamer (Tantau 2001) using LaTeX. I am not the first to notice that this has not been all for the better. Although beamed presentations can be extremely effective, they can also be deadly. Tufte (2006) has been a vocal critic:

> Imagine a widely used and expensive prescription drug that claimed to make us beautiful but didn't. Instead the drug had frequent, serious side effects: making us stupid, degrading the quality and credibility of our communication, turning us into bores, wasting our colleagues' time. The side effects, and the resulting unsatisfactory cost/benefit ratio, would rightly lead to a worldwide product recall.

I have found several sources that can help you make your beamed presentations more effective. Tufte's (2006) *The Cognitive Style of PowerPoint* can help you avoid some egregious problems, even though Tufte does not appear to see any hope for this media. Peter Norvig, Director of Research at Google, makes many of the same points more humorously with his *The Gettysburg PowerPoint Presentation* (http://www.norvig.com/Gettysburg/). For specific suggestions on how to make an effective presentation, I recommend chapter 5 of Tantau's (2001) *User's Guide to the Beamer Class*, even if you do not use Beamer. Here are five of his points that are particularly important:

1. Create a presentation that fits the time you have.
2. Never use a smaller font to fit more on a page.
3. Never include things that you will not discuss.
4. Use colors carefully, maximize contrast, and avoid shaded backgrounds.
5. Test your presentation.

These suggestions imply several things. First, you cannot put detailed information in a table that will be projected. If you need people to see the fine print in a table, distribute paper copies of the tables. Second, because projectors do not show colors in the same way that your computer monitor does and presentations are often made in rooms with bright light, avoid subtle shading and use high contrast.

Although overheads worked with any overhead projector, this is not true with projected presentations. If you use PowerPoint 2007, the presentation will not work on a machine that has PowerPoint 2003. I recommend saving your presentation in older formats unless you are sure that the computer you will use for your presentation has the latest software. I also suggest saving your presentation as a PDF file which should work on almost any computer you use. As a backup, I like to save my presentations on the web as a PDF file. This provides a backup copy of my talk and I can make the presentation using Acrobat Reader or most browsers.

7.8 A project checklist

The completion of a project or a major stage in the project (e.g., a draft of a paper) is a good time to address some bookkeeping chores: verify the work, organize the files, and make backups. If I have followed my own advice while working on the project, this should not take long. Here is a list of things to do.

1. Verify that the research log is complete and clear. If something is confusing now, it is likely to be very confusing when you really need the information.
2. Verify that you can replicate your results. Move the do-files to a new directory and run the master do-files. You do not want to overwrite your current Stata logs because they will be very useful if you have problems when you try to rerun the do-files. Often, I find small problems that can be readily fixed.
3. Verify that all results in the paper, book, or presentation are linked to a do-file through hidden text within the paper.
4. Post supplementary analyses on the web. It is easy to postpone doing this and it is always harder to do it later.
5. Clean-up stray files. I invariably accumulate files ranging from PDFs of articles to multiple variations of tables and graphs. Delete files that are not needed and put the rest of the files in the directory where they belong.
6. Archive the do-files, log files, datasets, documentation, and drafts of your paper. This is discussed in the next chapter.

7.9 Conclusions

This chapter discussed workflow issues that occur when your data are cleaned and ready for statistical analysis. First, planning, organizing, and documenting are just as important in this stage of your work as they are when you are preparing your data for analysis. Second, documenting the provenance of each number you present is essential. Without this documentation, replication is likely to be very difficult if not impossible. Third, automation tools in Stata can make data analysis more efficient and simplify the process of transferring your results into tables and figures. Fourth, your goal in data analysis is to present your results to others. The key principle of a presentation is to test your presentation in the same format and conditions under which the presentation will occur. Finally, the completion of a presentation or draft of a manuscript is an ideal time to verify that everything is documented, organized, and archived. Archiving and backing up are now considered in greater detail.

8 Protecting your files

A recent story in the New York Times (Schwartz 2008) began "LISTEN. Do you hear it? The bits are dying." The same day at the National Science Foundation, a review panel evaluated proposals for the $100 million DataNet program, an initiative to develop the next generation of tools for the preservation and access of scientific data. As digital data accumulate, people are discovering to their chagrin that digital data can be more fragile than paper and easier to misplace. The 10-year-old snapshot is still in the album, but the hard drive with last summer's JPGs crashed and the pictures were lost. I know exactly which binder holds analyses from a paper published in 1978, but I need to search several hard drives for the unprinted log files from a paper published two years ago. In 30 years, will it be as easy to access those digital logs as it is to read the printed output from 1978? If I left the printout on a shelf for 30 more years, someone could probably still read them to confirm my findings. Will the same be true for my digital log files? Will it be possible to use a USB drive? Will files be destroyed by a hardware failure or corrupted by a virus? Will newer software read the old file formats? Such concerns affect everything that is stored digitally.

Preservation of files is a critical part of the workflow of data analysis. You want to prevent the loss of files that you are actively working on, to maintain files from completed work that you might need later, and to preserve critical data and analyses for future generations of scientists. My goal is to help you develop a realistic plan for protecting your files. A successful plan needs to consider both the risk of losing a file and the probability of following the plan:

$$\begin{aligned}\Pr(\text{File loss}) = &\{\Pr(\text{File loss using plan}) \times \Pr(\text{Follow the plan})\} \\ &+ \{\Pr(\text{File loss without using plan}) \times \Pr(\text{Ignore the plan})\}\end{aligned}$$

A simpler plan that is religiously followed is more likely to protect your data than a more elaborate plan that is inconsistently applied. Accordingly, the workflow I suggest balances ease of use with the degree of protection. The procedures are simple enough to fit into my daily work and robust enough that my chances of data loss are small. The workflow illustrates general principles and provides a framework that you can adapt to your needs and your willingness to take an active part in backing up files. Fortunately, with current technologies, you can have a very effective workflow for protecting files that requires only minimal effort. The key is to have a comprehensive plan, keep files organized, and automate the process.

The chapter begins by distinguishing among levels of protection that range from simply making a duplicate copy to preserving an important dataset for the next 100 years. Deciding what form of protection you need involves considering the trade-off between the cost of losing files and the cost of preserving files. The most basic tool for preserving files is to have multiple, duplicate copies, but the best thing is to not lose files in the first place. Accordingly, I discuss how files are lost and ways to minimize risks. To provide a context for understanding why any workflow for data protection must include redundancies and anticipate unlikely events, I discuss Murphy's law and what it implies for protecting files. Next I outline a basic workflow designed for a "typical" data analyst who is more interested in substantive analysis than data preservation. Although this workflow might not deal with all your requirements, it provides a basic structure that you can adapt. The chapter ends by discussing long-term, archival preservation.

Before proceeding, I must add three warnings. First, I hesitate to recommend how you should protect your files because these files are valuable and I cannot be responsible for their loss, but file preservation is too important in the workflow of data analysis to ignore. Although the suggestions in this chapter have worked for me, I cannot guarantee that they will prevent you from losing files. Second, before you rely on any method to protect your files, you should test it with files that are not critical. Do not abandon your current method until you are sure that the new method works. Third, the technology for data preservation is changing rapidly. By the time you read this, there could be better ways to protect your files. The Workflow web site will add new information as it becomes available.

8.1 Levels of protection and types of files

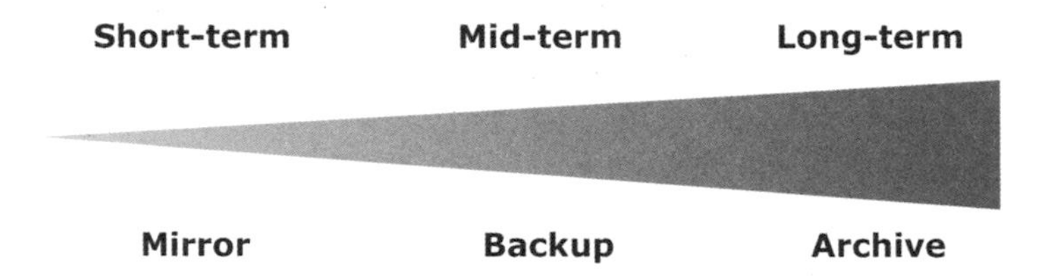

Figure 8.1. Levels of protection for files

The workflow that I present considers three levels of protections (see figure 8.1). Short-term protection focuses on making sure that the files you are using today will be there tomorrow. Protection involves continuously making duplicate copies of your files, referred to a mirroring. Short-term protection protects against the failure of your hard drive, a computer virus, or accidental deletion. Mid-term protection involves protecting files that you have finished working on but that you might want to use in the years to come. I refer to this as backup. As technology changes, you will need to move these files to new media and data formats, but you are not concerned with preserving the files beyond your own interest in them. Long-term protection, referred to as archiving, seeks to maintain the information in perpetuity. Archiving is extremely difficult, requiring constant concern with migrating files to new media and formats and the availability of documentation that is accessible to anyone interested in the files.

Deciding on a workflow for file preservation requires evaluating the costs of losing files against the costs of preserving files. If a file is mistakenly deleted, how much time and money are needed to replace the file? If you cannot replace it, what are the consequences? Although you never want to lose files, how much time and money can you invest protecting them? Answers to these questions vary for different types of files and depend on the type of work you do. Table 8.1 summarizes costs for four major classes of files.

Table 8.1. Issues related to backup from the perspective of a data analyst

	Type of File			
	Working	Posted	Archival	System
How do you recover a lost file?	Redo recent work	Redo older work	Download files	Reinstall the software
What is the cost of recovering the file?	Minor delays	Potentially lots of work	Minor inconvenience	Minor inconvenience
For how long are you preserving the file?	1–3 years	3–10+ years	100+ years	1–3 years
How difficult is it to preserve the file?	Some work	More work	(Substantial)	Trivial
Should you consider media & format?	Very little	Some	(Critical)	Very little

Note: Items in parenthesis apply if you are concerned with archival preservation.

Working files include documents, do-files, log files, datasets, reprints, and other files that are part of ongoing analyses. Because you are actively working on these files, if they are lost you could probably redo the work. Rather than risking the lost time to redo your work, it makes sense to protect these files. Because the files change every day you work, you need to make duplicate copies of these files frequently. However, because you are concerned only with short-term protection, you do not need to worry about storage media becoming obsolete or file formats becoming unreadable.

Posted files are working files that you have completed and will no longer change. For example, after you create a dataset and verify that it is accurate, the dataset and do-files should be posted. If you later find problems, new files are created but the old ones are left as is. Because posted files are essential for replicating your work, you need to preserve these files for longer than working files. Accordingly, you may eventually need to migrate the posted file to newer storage media and file formats. For critical files, you might want to store the information in more than one format. For example, datasets could be saved in Stata format, ASCII, and SAS XPORT, which is the U.S. Food and Drug Administrations standard. SAS XPORT files can be created with Stata's `fdasave` and read with `fdause`. I would also treat user-written ado-files installed in the `PLUS` and `PERSONAL` folders (page 349) as posted files. These files are not part of official Stata, and you will need them to replicate your work.

Archival files include datasets, codebooks, articles, and other files obtained from organizations committed to preserving and distributing digital information. For example, ICPSR distributes and preserves datasets in the social sciences, whereas JSTOR preserves and distributes published research. If you lose an archival file, it should be easy to obtain another copy. Although you might want to back up these files along with your posted files, this is a matter of convenience rather than necessity.

System files include the operating system (e.g., Mac OS, Windows) and software (e.g., Stata, Stat/Transfer). These files can be replaced by reinstalling the source DVDs or by downloading the files. If one file is corrupted or deleted, it is a minor inconvenience to reinstall the program. If the boot drive of your computer fails, the cost is greater. If you want to recover quickly from losing your boot drive, you can back up the drive including hidden files using utilities such as *Norton Ghost* for Windows or *Carbon Copy Cloner* for Mac OS. Unless you work under strict deadlines, the time required to back up system files might not be worth it. The only system files I back up are for software that is essential for reading critical files or for reproducing analyses.

8.2 Causes of data loss and issues in recovering a file

Although the workflow presented below protects files by making duplicate copies, you want to avoid losing files in the first place, a topic that is now considered.

Deleted and lost files

Files can be lost if they are mistakenly deleted. Most operating systems let you recover deleted files, a feature called Recycle Bin in Windows and Trash in Mac OS X, as long as you make the recovery before the original space used by the file is used by newer files. The `\- Hold then delete` folder suggested in chapter 2 also serves this function but is more dependable because you have complete control over when files disappear. More elaborately, Mac OS X 10.5 includes Time Machine, which backs up files on a daily basis, allowing you to "go back in time" to recover a file that was deleted. This works well as long as your backups fit on one drive and that drive does not malfunction.

A file can also be literally lost—it is not deleted, but you cannot find it. This is the proverbial needle in a haystack, a growing problem as people accumulate more files on larger and cheaper hard drives. The way to avoid misplaced files is to keep things organized, carefully choose filenames, and use a utility that searches by name and content.

You can also lose files if you think they are backed up, but they are not. Before relying on the backups made by network administrators, verify how long the backups are kept and what the procedures are for recovering files. These backups might only be available to restore files after a catastrophic hardware failure soon after the backup was made, rather than to retrieve a single file. Further, some organizations are destroying backups because they might contain sensitive information that could be used for identity theft. If you rely on backups made by others, verify how long the backups are kept and what the procedures are for file recovery.

Corrupted files

A file is corrupted when the information within the file is stored incorrectly due to a write error when recording the data, to deterioration of the media (e.g., a disk goes bad), or a virus. Even one incorrect bit in a 100-megabyte file can make the file unreadable. There are several ways to prevent this problem. First, when a file is copied, do a bit comparison to verify that the source file and copy are exactly the same (see page 337 for further information). When you disconnect a USB or Firewire drive (e.g., an external hard drive, a memory stick), eject it as recommended by your operating system (e.g., right-click on the drive icon and select eject) instead of simply pulling out the plug. Local IT staff tell me that this is the most common reason they encounter for people losing files. Keep your virus software up to date. Files can also become corrupted as media age. Files on a CD left in the sun may last only a few weeks. Hard drives should be replaced after five years. Because files can be corrupted by write errors caused by power fluctuations, use an uninterruptible power supply (UPS) if you live in an area where there are frequent brown-outs or power failures.

Hardware failures

Data are also lost or sometimes corrupted when hardware fails. Failure rates follow a "bathtub curve" with high rates of early-life failures followed by lower rates until the hard drive begins to wear out and failure rates increase. New drives can fail because of a flaw in their manufacturer, as I recently learned when the boot drive on my 3-month-old computer died. Old drives fail because they wear out. Although hard drives are typically rated with a mean time to failure (MTTF) of over one million hours (114 years), Schroeder and Gibson (2007) found observed rates in the range of 2–4% per year.

There are several simple things that help prevent hardware failure. First, turn off your computer by exiting all programs and shutting down the computer as suggested by the operating system rather than simply turning off the power. Second, use a surge protector. If you have unstable electrical power, use an uninterruptible power supply (UPS). Third, make sure your computer has plenty of ventilation and remove dust from the fan; however, a recent study at Google (Pinheiro, Weber, and Barroso 2007) found that heat is not as much of a problem as previously reported. Fourth, do not move your computer when the hard drive is reading or writing. Although hard drives are remarkably robust, if you bump a drive at the wrong time, you can lose a file or the entire disk. Finally, if a drive develops a hum or squeal, replace it immediately.

Obsolete media and formats

Obsolete media and formats are an easy way to lose access to your files. Although this is not a concern with working files, files that have not been used for even a few years can be at risk. My recent experience illustrates the problem. When analyzing data from a 10-year-old study of patients, we discovered that our dataset was missing a critical variable. After a lengthy search, we determined that the file with the missing variable was on a backup tape. Because we no longer had a drive that could read the tape (i.e., obsolete media), we hired a firm specializing in this problem. They recovered the file for $1,000. Then we discovered that the file was in a format no longer supported by current software (i.e., an obsolete format). It took a $1,000 in staff time and months of delay to reinstall an old operating system and the software needed to read the file.

If you do not have access to the equipment needed to read your storage media, the files on those media are effectively lost. Once common media, such as ZIP disks, can disappear quickly. To prevent such loss, you need to migrate your files from older storage media to new ones as technology changes. When you get a new computer, make sure that you can still access the media used with your old computer. If not, find a way to transfer the files before getting rid of the old computer. To gain a better appreciation on how rapidly media appear and become obsolete, look at the online *Chamber of Horrors: Obsolete and Endangered Media* (Kenny and McGovern 2003–2007).

Even if you can copy a file from the backup media to your computer, you need software that can decode the file. It does no good to "preserve the bits" if you have lost their content. For example, if a dataset is in the once common OSIRIS format, but your software cannot read OSIRIS, the data are lost until you find a program that can decode the files. To illustrate the magnitude of the problem, a recent story from BBC News (2007) reported that the British Museum is at risk of losing 508,000 encyclopedias worth of digital information stored in formats that are no longer commercially supported. There is no easy solution to this problem because archival formats have not been fully established. To prevent this type of loss, store critical files in several formats. For datasets, save data in Stata format, but also in ASCII, SAS Transport format using `fdasave` (because it is now the standard for the FDA), and perhaps a few other formats. For text files, save your file in the format you are using (e.g., Word's `.doc`) but also in other formats such as Rich Text Format or PDF.

Recovering lost files

Even if you are careful, you are likely to lose a file sometime. When this happens, hopefully you have a backup copy. But, sometimes a file is deleted or corrupted without a backup. When this happens, take your time so that in your rush to recover the file you do not delete more files or make it harder to recover the lost file. If you have deleted a file that has not been backed up, do not write new files to the drive until you have tried to recover the file. Adding new files can make it impossible to recover a deleted file. If you lose a file because of a problem with your computer or a possible virus, do not connect your backup drive, disk, or tape until you are sure what the problem is.

A colleague once lost a file to a software malfunction and then lost the backup when the same software corrupted the backup. Unless you are sure what the problem is, ask your IT support staff for help. If all else fails, you can use a commercial data-recovery service (search the web for "data recovery"). These businesses specialize in recovering data from damaged disks but are very expensive and most require payment even if they fail to recover your files.

8.3 Murphy's law and rules for copying files

Any workflow to protect files must take Murphy's law very seriously. Murphy's law states: "If anything can go wrong, it will." Much can be learned from the context in which this famous law emerged.[1] In 1949, at Edwards Air Force Base, Murphy's Law was named after Captain Edward A. Murphy who was an engineer working on a project to see how much sudden deceleration a person could stand in a crash. Dr. John Paul Stapp, who rode a sled that created 40Gs of deceleration and lived, noted that the good safety record on the project was due to a fundamental belief in Murphy's Law and extraordinary efforts to circumvent it. He coined Stapp's Ironical Paradox: "The universal aptitude for ineptitude makes any human accomplishment an incredible miracle." When it comes to saving your work, expect things to go wrong, expect that you will delete the wrong file at the worst possible time, and expect a hose to be left on in the room above your computer. If you expect the worst, you might be able to prevent it. Based on this notion and the sometimes painful experience of others, the following rules should be followed no matter what approach you take to preserving your files.

Rule 1: Make at least two copies

Have at least two copies plus the original of all files.

Rule 2: Store the copies at different locations

Many disasters affect everything at a location, so store copies in different buildings. External drives can be stolen, so keeping all backup drives in the same room is bad idea. Similarly, fire and water damage often affect everything in a room.[2]

Rule 3: Verify that copies are exact duplicates

When copying files, use software that does a bit comparison of the source files and the copies. Most copy programs do not verify that the source file and the copy are

1. My history of the law is based on http://www.murphys-laws.com/murphy/murphy-true.html, which reproduces the article "Murphy's Law" from the March 3, 1978 issue of *Desert Wings*.
2. Disasters can occur in places that seem very well protected. ICPSR recently suffered damage when maintenance left a hose running on the floor above their server room (M. Gutmann 2005, pers. comm.)

exactly the same. Without bit verification, you might think you have an exact copy of your file when in fact it is a corrupted version of the file.

8.4 A workflow for file protection

I suggest a two-part workflow for protecting your files, as illustrated in figure 8.2.

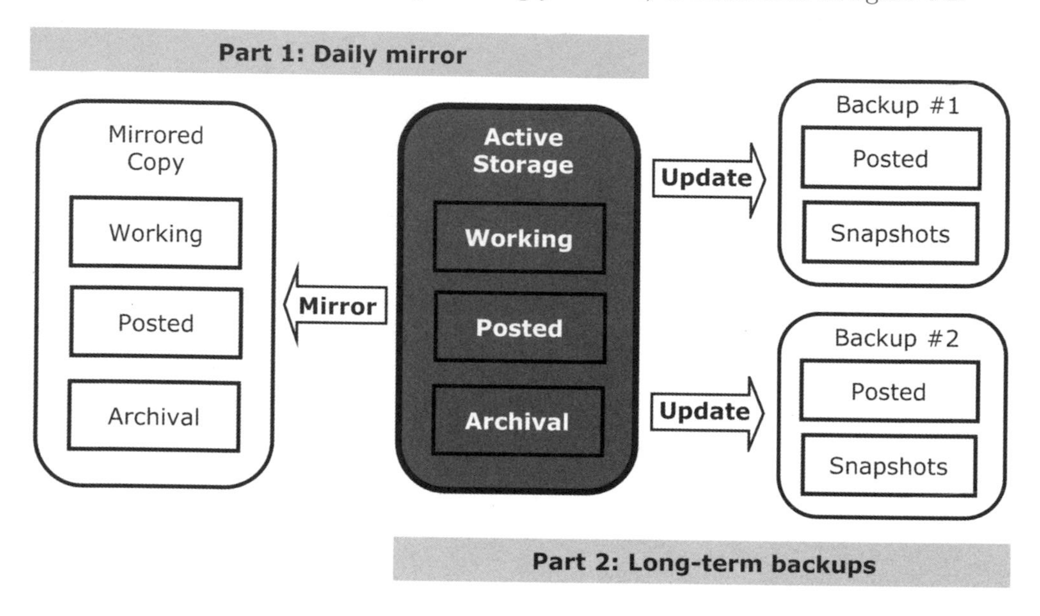

Figure 8.2. A two-part workflow for protecting files used in data analysis

Each rounded rectangle is a storage device. For simplicity, I refer to these as hard drives, but they could be other storage devices. The rectangles within each drive represent the types of files I want to protect. The shaded rectangle represents my active storage. These are the files that I have created and that I am using today. Part 1 of the workflow creates duplicate copies of active storage, referred to as a mirror, to provide short-term protection. Part 2 provides mid-term protection of posted and project files by making two copies of the files, referred to as backups.

Part 1: Mirroring active storage

For active files, the greatest risk is the catastrophic loss of all files due to hardware failure, a virus, or accidental deletion. To protect against such loss, I save duplicate copies of the files at least once every day. The duplicate files are referred to as a mirrored copy because all changes to files in active storage are reflected in the mirror. There are two basic approaches to creating a mirrored copy. With continuous mirroring, changes to files in active storage are transferred to the mirror within a few seconds.

The disadvantage of continuous mirroring is that if I accidentally delete a file, it is immediately deleted on the mirror so that I cannot use the mirror to recover the file. More sophisticated programs for mirroring files move copies of changed files to a separate folder as a fail-safe. Periodic mirroring occurs only at times you select. You might update the mirror as the last thing you do each day before logging off. Because you only change the mirror periodically, you can recover files that changed since the last update to the mirror. The disadvantage is that if the drive with active storage crashes, your mirror will not include copies of files that changed since the last update of the mirror. If you want the best of both continuous and periodic mirroring, you can create both a continuous mirror and a periodic mirror, although this increases the complexity of the workflow.

Configurations for mirroring active storage

The specific way that you mirror active storage depends on how many computers you use, what storage devices are convenient, and the software you use. Here are some alternative configurations, all of which provide excellent protection.

Single computer mirror: If you always work on the same computer, it is easiest to keep your active files on an internal hard drive and to mirror files to an external drive, a LAN, or other storage connected to the computer.

Multiple computer mirror with LAN: If you use two or more computers that are connected to the same LAN, you can keep active files on the LAN and use the internal hard drives on each computer to hold mirrored copies of the active files from the LAN. Because you are using multiple computers, you have extra security since files are mirrored multiple places. If the computers you use do not allow you to save files to the internal drive (e.g., in a public computing lab), you can create a mirror on an external or portable drive that you carry with you. Alternatively, you can use the next configuration.

Multiple computer mirror with portable drive: The best solution for the way I work is to use a portable drive for active storage and carry the portable with me as I move among computers. For years, I transferred new or changed files between computers using a ZIP disk or memory stick but regularly forgot to transfer the file I needed or was confused about which computer had the latest version of a file. By using a portable drive for active files and the internal drives on each computer to hold my mirror, I always have the latest version of my files (unless I forget the portable). On each computer's internal drive, I have a directory `\- Portable mirror` that is used for a continuous mirror when the portable is attached to that computer. When I leave work, I eject the portable drive, put it in my briefcase, and take normal precautions not to drop the briefcase. At home, I plug in the portable drive and follow the same procedures. When traveling, I take the portable drive with me and make a mirror to the internal laptop drive. If I travel without a laptop, I take the portable drive with me and use a memory stick to make duplicates of new files I create.

Multiple computer mirror using the Internet: If you use multiple computers that are on the Internet, you can use Microsoft's Foldershare or a similar program to mirror files over the Internet. Suppose you use one computer at work and another at home. When using the computer at work, Foldershare continuously synchronizes your active files to your home computer. When you are home, you use the files on that computer as your active storage and Foldershare mirrors changes to your work computer. I have experimented with Foldershare, and it is easy to use and efficient with a few exceptions. First, the beta software limits you to 10,000 files per library and 10 libraries. Second, if the Internet connection goes down, the other computer is turned off, or the software has a problem, you need to work without a mirror until the problem is fixed. Third, if you use a laptop that is not always connected to the Internet, you must remember to update files to the laptop before you use it. Fourth, if you work on computers that do not have Foldershare installed, you will not have access to your files.

Other configurations can work equally well depending on how you work and what technology you have available.

Part 2: Offline backups

The second part of the workflow makes backup copies of posted files and snapshots (defined below) of files from active projects.

Posted files

Posted files include the do-files, log files, datasets, text files, and others that you have used in your writing or shared with others so that they should no longer be changed. The rule for posted files is simple: Once a file is posted, it should never be changed. See chapter 2 (page 22) and chapter 5 (page 125) for further details on posting. While I continue to work on the project, I keep posted files from that project in my active storage, but because these files are part of results that I have distributed or shared with coauthors, I want to give them an extra level of protection against hardware failures, viruses, or accidental deletion. About once a month or when I finish a major task, I copy newly posted files from active storage to a backup drive. If my archival files are not very large, I often back them up along with the posted files.

The key to backing up posted files is to automate the process so that the backup software can determine which files need to be copied. If you must manually determine which files need to be copied, your workflow will not work unless you are very patient and exceptionally thorough. The easiest method I have found is to put posted files in subdirectories named `\Posted`. Suppose that I have the following directories on my computer (where for simplicity I use a simple directory structure):

```
\Projects
    \COGA
        \Posted
        \Work
    \EPSL
        \Posted
        \Work
\Workflow
    \Posted
    \Work
```

My backup program only copies files from `\Posted` directories and automatically reproduces the directory structure from active storage:

```
\Posted files
    \Projects
        \COGA
            \Posted
        \EPSL
            \Posted
    \Workflow
        \Posted
```

If I add another `\Posted` folder to my active storage, the next time I run the backup utility it will create the same folder within `\Posted` and will copy the newly posted files.[3]

This plan for backing up posted files only works if your backup program can automatically select all files located in directories named `\Posted`. On my computer, it takes my backup software about a minute to search 4,000 folders to find which files need to be backed up. If your software cannot select files to back up in this way (and not all backup programs can), you need to organize your files some other way that makes it easy to determine which files to back up. For example, you could create one folder, `\Posted`, and create subdirectories for each project. I do not find this approach convenient, but you might.

Before proceeding, I want to emphasize that this plan only works if you stick to the posting principle: Once a file is posted it cannot be changed. For years, I made the mistake of changing, renaming, or relocating "posted files". Then I needed to back up the files again. What about the files I had backed up before with different names? If I deleted them, old do-files would not work and my research logs would point to a file that no longer exists. If I did not delete them, I had multiple files with the same information but different names. This process wasted time and was not reliable. The solution was to follow the simple principle for posting a file: After a file is posted, it can no longer be changed. If I find a mistake in a posted file, I can create a new file with a new name, but I cannot change the original file. If I decide that I no longer need a posted file, I can delete it but cannot change it.

---

3. The software I use can copy only files in directories that contain a particular phrase, such as "posted". Accordingly, it will copy files from `\Posted`, `\Text-posted`, etc, but not from `\Work`. I also use the rule that if a directory name ends with `+` (e.g., `\Data+`), that directory is treated as posted. See the Workflow web site for details on software.

Snapshots

I also periodically take "snapshots" of all files, posted or not, associated with a project. For example, when I completed the first draft of the workflow book on January 14, 2007, I backed up all the files in `\Workflow` to `\Snapshots\Workflow\2007-01-14`. If I later need a file that was used for this draft but that was not a posted file and was later deleted (e.g., a figure I did not use), I check here. I also take snapshots before I start cleaning files at the end of a project. That way if I make a mistake, I can easily recover the file. My snapshot folder might look like this:

```
\Snapshots
    \Workflow
        \2006-12-12
            (copy of all workflow files on this day)
        \2007-01-14
            (copy of all workflow files on this day)
    \EPSL
        \2004-06-02
            (copy of all epsl files on this day)
        . . . . . .
```

I might keep the snapshots for a long time or delete the files after a few weeks.

Although I keep at least two copies of files located on different devices stored in different locations, I do not worry about keeping the different drives continuously synchronized. Although keeping them synchronized would be better, I find that this is impractical. Instead, I synchronize the drives every few months. When a backup drive is full, I buy a new drive. With the rapidly increasing capacity of hard drives, the new drive is usually at least twice as large so it can hold all my current files and have capacity for at least that many more files. Using bit verification, I copy all files from the current device to the new one. The old drive is stored as "deep backup".

Configurations for backup storage

You should have at least two copies of each file stored on media located in different physical locations to protect against local disasters such as water damage or theft. The drives are only plugged in when making copies or recovering files, so they are not vulnerable to power surges or viruses. Here are some configurations you can consider for your backup storage.

Backups using external drives. My preferred approach is to use multiple external drives to hold the two backup copies of files. These drives are inexpensive, have fast read/write speeds, and are easy to move. When I want to synchronize the drives, I carry the drive I store at home to work and bring it back home after the two drives are synchronized.

Backups using a LAN. If you have sufficient space on your LAN, you can store one copy on the LAN and the other on an external drive or an internal drive (assuming that drive is not used for active storage).

Backups using enterprise mass storage. If your organization provides enterprise storage, such as tape backups, one copy could be stored there and the other copy on another device.

Backups on the Internet. One backup copy could be saved over the Internet with a company that sells storage. As of mid-2008, "unlimited" storage costs less than $5 a month. The catch is that transfer speeds limit you to copying a maximum of about 7 GB of files a day (i.e., two weeks to copy 100 GB, one year to copy 2.5 TB). After initial copies are made, only changed files are transferred. If you have a hard disk failure, it could take weeks to restore files over the Internet. You can also pay to have DVDs made and mailed. The software for these sites only allows simple criteria for selecting files, so selecting only files in `\Posted` folders would not be possible. To find companies that sell Internet storage, do a web search of "Internet storage", "online backup", or "online storage". Keep in mind that if the company selling storages goes out of business, you could lose your backups.

Backups on DVDs. DVDs have the advantage that once you write to them, the files cannot be deleted (assuming you avoid the multiple-write DVDs, which provide more expensive and less stable storage). DVDs, however, are slow, they hold only about 5 GB so you end up with a lot of disks to organize, and quality DVDs are no longer cheaper per GB than hard drives.

8.5 Archival preservation

Important files get lost, sometimes surprisingly so.[4] On November 22, 1963, President John F. Kennedy was assassinated. A few days later, the National Opinion Research Center (NORC) conducted a national opinion survey known as the Kennedy Assassination Study (KAS). After the tragedy of September 11, 2001, NORC decided that the KAS survey should be replicated to allow comparisons of these two tragedies. The KAS codebook was found in the NORC library, a new survey was constructed, and interviews began on September 13. The original KAS data could not, however, be found in the NORC archives, nor were they found in other data archives, such as the Roper Center and the Inter-university Consortium for Political and Social Research. The search then began for the original 80-column punch cards, with hopes that they would be found in NORC's 24,000 cubic feet of storage. Three thousand, six hundred and sixty-nine boxes of cards were listed on the storage inventory, but KAS was not on the list. Someone noticed that while the inventory listed the contents of 3,669 boxes of cards, there were 8,348 boxes of cards in storage. A retired staff member remembered a memo from 1987 that included bar codes for the KAS cards. Amazingly, the 11 boxes of cards were found. Next NORC had to locate a card reader. A firm in New York was hired and the cards were hand delivered. The card reader was literally a little rusty, causing further delays. When the cards were finally read, the multiple-punch data on the cards were

4. This example is based on Smith and Forstrom (2001).

not correctly interpreted.[5] Two months later, the data were decoded. Unfortunately, the variable names simply indicated the card number and column position (e.g., `c3c14`) and no value or variable labels were available. Adding this information caused further delays. Smith and Forstrom (2001, 14) summed up the experience:

> The lessons from the KAS experience are simple but important. Survey data must be sent to survey archives like the Roper Center and ICPSR where the documentation and data will be preserved, backed-up, periodically updated as technologies change, indexed, and made routinely and easily accessible to researchers. Failure to archive studies is poor science and a disservice to other contemporary researchers and those in the future.

A second example of lost data also involves a defining event in modern history.[6] On July 20, 1969, half a billion people watched Neil Armstrong walk on the moon. Macey (2006) summarized the event: "The heart-stopping moments when Neil Armstrong took his first tentative steps onto another world are defining images of the 20th century: grainy, fuzzy, unforgettable." Indeed, the images were of poor quality, obtained by pointing a television camera at the monitor that was receiving the transmissions from the moon. The original images were of too high a resolution to show on TV. Over the next 30 years, the moon tapes moved around, although where they ended up is uncertain. In 2002, a technician from Australia's Honeysuckle Creek ground station found a tape in his garage that seemed to be from the moon landing. Although his tape was not of the walk on the moon, it initiated a search for the moon tapes that had once been stored at the National Records Center. Documents were found showing that 26,000 boxes of tapes had been requested by Goddard during the 1970s and 1980s, but no trace of the tapes was found at Goddard (Kaufman 2007). In 2006, NASA admitted that the tapes were lost. Dolly Perkins who led the failed search explained (Kaufman 2007): "Maybe somebody didn't have the wisdom to realize that the original tapes might be valuable sometime in the future. Certainly, we can look back now and wonder why we didn't have better foresight about this." Following NASA's admission, news stories appeared around the world discussing how NASA lost this historic and scientifically valuable data. Macey (2006) at the *Sydney Morning Herald* wrote a story titled "One giant blunder for mankind: how NASA lost moon pictures." Remarkably, the Australian film producer Peter Clifton heard about the tapes on TV and remembered that in 1979 he had purchased two moon tapes to use in a rock film about Pink Floyd's album *The Dark Side of the Moon* (Egan 2006). His two tapes were recovered from a Sydney vault. In October of 2006, nearly 100 moon tapes were found in the basement of a lecture hall at Curtin University of Technology in Perth, Western Australia (Amalfi 2006). It remains to be seen if these are the right tapes and whether the remaining tape drive at Goddard still works.

5. Multiple punches are a way in which one column of a card can record information on multiple variables. This allows one card to hold much more information but requires special processing to decode the information.
6. This example is based on articles by Macey (2006), Egan (2006), Amalfi (2006), and Kaufman (2007).

I find that most data analysts have not thought much about archival data preservation. I had not until I spent four years on the Council of the Inter-university Consortium of Political and Social Research (ICPSR), a nonprofit organization dedicated to preserving data. When I joined the council, I thought of backing up and archiving as the same thing. I made two copies of files, one on tape and one on disk, and thought I had archived my files. I learned that archival storage is much more complicated, as the two examples above illustrate. The hard part in archiving files is anticipating changes in file formats and physical media, and making sure that the files are documented in a way that is clear to someone who is not associated with the original work. When it comes to archiving, there is no difference between losing a file because it was deleted or losing the knowledge about what the file contains. A good source of information on these issues is *The Guide to Social Science Data Preparation and Archiving* (ICPSR 2005, available at http://www.icpsr.umich.edu/access/dataprep.pdf) and the *Digital Preservation Management Tutorial* (http://icpsr.umich.edu/dpm/).

The best way to archive datasets is to let someone else do it! I highly recommend depositing your original data with an organization that specializes in data preservation, such as ICPSR for the social sciences. For analysis files from published papers, you should consider depositing the files at the journal's archive if they have one (Freese 2007). Not only does this ensure that the data are preserved, but it gives other researchers access to the data for replication and additional research.

You should start thinking about archiving data when you start a project, not at the end of the project. I have projects that I began before I understood what was required to archive data. I kept careful documentation that I could use, but this information was not in a form that others could use. As a consequence, I have data I would like to archive at ICPSR, but it would take a great deal of time to get the data and documentation in a form that others could use. If I had been thinking about preservation from the start, archiving the data would have been much simpler.

8.6 Conclusions

Procedures for preserving the files used in cleaning, analyzing, and presenting your data are a critical part of the workflow of data analysis. Like filing taxes, having your teeth cleaned, changing oil in your car, and removing leaves from your gutters, preserving your files is something you need to do. Most people realize this, but very few people appear to do it systematically. The best way to avoid data loss is to have a comprehensive plan for protecting your files and following it conscientiously. Hopefully, this chapter will help you find a manageable workflow for protecting files and will encourage you to make this a priority. If you adapt the suggestions in this chapter to your own needs, you should be able to do a very good job protecting your files with a minimum of effort. If you begin planning how to archive data as it is being collected, you can save a lot of time and increase the chances that your data will be preserved.

Finally, this chapter focused on procedures for copying files to multiple locations. It assumes that you know what your files are! This seems easy, but versions of files can quickly get out of hand, especially in collaborative projects. In a recent article on data preservation (Schwartz 2008), Dr. Margaret Hedstrom echoed the issues of organization and provenance that I discussed in earlier chapters: "Which architectural drawings of the many versions generated for a project were actually used to erect the building, and what was the chain of decisions that led to the brick-and-mortar result?" Preserving your files is much easier and more effective if you keep files organized and carefully named.

9 Conclusions

This book makes explicit the workflow that I discovered through my own mistakes; learned from colleagues, teachers, and collaborators; and read in sources ranging from Donald E. Knuth's classic *Art of Computer Programming* (1997; 1998a; 1998b) to Peter Krogh's *The DAM Book: Digital Asset Management for Photographers* (2005). Although many books discuss statistical analysis, I do not know of any devoted to the workflow of data analysis from the import of data through the presentation of results. Too often, people learn their workflow by first making mistakes and then looking for ways to avoid future problems. I hope that this book helps you avoid those mistakes, most of which I have already made at least once.

Many of the topics in this book, such as naming variables and adding labels, may seem far removed from your substantive research goals. Research begins with an exciting idea that you hope will contribute to our knowledge of the world. Between the idea and the published results there is a lot of undifferentiated heavy lifting. This is the phrase used by Jeff Bezos, the founder of Amazon.com, when discussing the challenges of establishing a startup company (Steinberg 2006): "Undifferentiated heavy lifting is all the work that must be done in order to bring your product to the marketplace, but which has nothing to do with your product itself." *The Workflow of Data Analysis* is about the heavy lifting that goes on in data analysis and ways to minimize the time spent doing this lifting. Still, even with the best workflow over half the work, probably much more, in a research project involves getting the data ready. An efficient workflow can reduce the time you spend doing data management and can produce a dataset that is easier to analyze. With carefully chosen names, effective labels, and clean data, the time you spend in statistical and graphical analyses will be more productive and more enjoyable.

It is easy to take the goal of developing an effective workflow too far and in the process defeat the objectives of improving your research productivity while increasing the replicability of your work. If you spend all your research time perfecting the way in which your data are prepared or your files are organized, nothing gets done. At some point, the time it takes for one more iteration at improving names and refining labels is not worth the time that could have been used for analyzing the data and writing the results. On the other hand, if you spend too little time developing an effective workflow, you will be inefficient and risk obtaining incorrect results or results that cannot be replicated. An effective workflow is about being efficient and about obtaining correct results that can be replicated. An effective workflow involves a balance among competing demands.

If you are like me, you will be tempted to violate the principles of a good workflow. Do I really need to document every step? Why not wait until next week to back up files? Because I used the dataset in only one do-file, can I just fix the problem and keep the same name? For a simple research note, do I need to document the source of every number? Sometimes making exceptions works fine. More often, shortcuts take longer and there is no way to predict which shortcuts really are faster. The simplest thing is to stick with a sound workflow and make no exceptions. An effective workflow does not add much time to your work and in the long run will save you time. The key is to decide on a workflow and stick with it.

Do you need to follow the workflow that I suggest? Absolutely not. There are many workflows to choose from and many excellent ways to conduct data management and statistical analysis. As long as the fundamental goals of accuracy, replicability, and efficiency are met, you should feel comfortable with the procedures and standards that you develop. Still, there are two general advantages to using the workflow in this book. First, having a fully documented workflow is extremely useful. While writing the book, I was surprised by how often I referred to the working draft to remind myself of the standard way of doing things or of a sequence of commands that I rarely use but that make a hard job easier. Written standards made my work more consistent and easier. In collaborations, especially with more than two people, there are tremendous advantages to standardization and written procedures. Second, in writing this book, I was forced to think systematically about all the steps of the workflow, to create consistent standards, and to make my workflow explicit. This took a lot of time, indeed more than I anticipated, but it also made my workflow better. Still, if you have a better way or a different standard than I have presented, by all means use it. I hope you will let me know so that I can improve my own work.

Finally, this book was motivated by my work as a social statistician, not by an interest in how to name and label things. I find data analysis to be far more engaging and challenging than data preparation. However, I find data analysis to be more productive and fun when my data are well named, fully labeled, documented, and cleaned. Data analysis is hard enough without being handicapped by data that are hard to use or that include problems unrelated to the goals of your research. The challenges facing a data analyst are increasingly complex as a result of advances in computational power and storage capacity, the availability of increasingly large and complex datasets, and the growing choices among more and more challenging methods of analysis. To manage these complex datasets and methods, I believe it is essential to develop a workflow that increases efficiency, ensures accuracy, and facilitates replication. I hope that this book is a step in that direction.

A How Stata works

The more you use Stata the more you can benefit from understanding some of the finer points about how the program works. This understanding can help you install user-written programs, deal with problems that occur from running Stata on a network, make the interface easier to use, increase the memory available for data, and many other things. To this end, section A.1 describes the files used by Stata and where they are located. Section A.3 reviews ways to customize Stata, ranging from the fonts used to the amount of memory available. Section A.4 describes other resources for learning about Stata.

A.1 How Stata works

The Stata program consists of two major types of files. First, there is the Stata executable. This file contains the compiled program that is the core of Stata. This file might be named `wstata.exe`, `wsestata.exe`, or `wmpstata.exe`. When you click on the Stata icon or start the program in some other way, the operating system runs this file. Second, much of Stata is contained in ado-files, which are programs that use features in the executable to add new commands to Stata. For example, the `nbreg` command fits the negative binomial regression model. This command is not part of the executable but is contained in `nbreg.ado`, which is written using commands in the executable as well as commands contained in other ado-files. Stata 10 includes nearly 2,000 ado-files. Together, the commands included in the executable and in the ado-files written by StataCorp are referred to as official Stata to distinguish them from commands written by users.

A clever feature in Stata is that when you run a command, say, `summarize` to compute summary statistics, you cannot tell whether the command is part of the executable or an ado-file. This means that users can write their own commands and use them just like official Stata commands. Indeed, Stata users have written many excellent programs that are discussed in the *Stata Journal* and on Statalist
(http://www.stata.com/statalist/). For example, the commands `esttab` and `eststo` discussed in chapter 7 are not part of official Stata but were written by Ben Jann (see Jann [2007]).

Stata directories

Most of the time, you do not need to know where the Stata executable and ado-files are located. However, sometimes this information is critical, especially when something goes wrong or you are writing your own ado-files. The `sysdir` command tells you where Stata stores files. On my computer,

```
. sysdir
   STATA:  D:\Stata10\
 UPDATES:  D:\Stata10\ado\updates\
    BASE:  D:\Stata10\ado\base\
    SITE:  D:\Stata10\ado\site\
    PLUS:  D:\Stata10\plus\
PERSONAL:  D:\Stata10\personal\
OLDPLACE:  D:\Stata10\personal\
```

The `STATA` directory contains the executable. The remaining directories are for ado-files; as a group, these directories are referred to as the *ado-path*. The `BASE` directory contains ado-files written by StataCorp that were placed there when Stata was initially installed. Periodically, StataCorp updates these files to fix problems or to add features. These updates are placed in the `UPDATES` directory. Suppose that you type `nbreg`. Because this command is not part of the executable, Stata checks if `nbreg.ado` is in the `UPDATES` directory. If not, it checks in `BASE`. If it is not there, it checks other system directories on the ado-path. On a network, `SITE` might contain files that were installed by your site administrator. If you install unofficial ado-files (i.e., those that are not distributed by StataCorp), they are saved in `PLUS`. For example, the ado-files from the Workflow package are saved here. `PERSONAL` is for ado-files that you have written, as discussed in chapter 4. `OLDPLACE` is where personal files were located in earlier versions of Stata. For additional information, including how to install updates, see [U] **17 Ado-files**.

The working directory

When you start Stata, you are assigned a working directory. To understand the concept of a working directory think of the word processor that you use. When you want to open a document, a dialog box comes up that shows you files from a default directory. In Windows, this might be the directory `My Documents`. That directory is the working directory—the default location for reading and writing files. The same holds for Stata: the working directory is where Stata looks for your do-files and datasets and where Stata will save the files you create, such as new datasets or log files. If you type a command to load a dataset and do not specify a directory, Stata looks in the working directory. For example, with the command

```
use wf-lfp, clear
```

Stata looks for `wf-lfp.dta` in the working directory. If you include a directory as part of a command, Stata looks in that directory instead of the working directory. For example, with the command

```
use d:\data\wf-lfp, clear
```

Stata will look for the dataset in the d:\data directory. Similarly, if you type the do mypgm.do command, Stata looks for the do-file in the working directory. If you type the do d:\workflow\work\mypgm.do command, Stata looks for the do-file in d:\workflow\work\. In chapter 3, I explain why I suggest that you almost always read and write files to the working directory, rather than hardcoding a directory name as part of the command.

To determine what your working directory is, in Windows, you use the cd command:

```
. cd
e:\data
```

In Mac OS X or Unix, you use the pwd command instead:

```
. pwd
~:data
```

You can change the working directory with the cd command. For example, writing do-files for this book, I used the e:\workflow\work directory. To make that directory my working directory, I used the command:

```
cd e:\workflow\work
```

To change to the working directory used for the CWH project, I type the command

```
cd e:\cwh\work
```

The ls command is a synonym for dir.

A.2 Working on a network

If you are running Stata on a network, which is often the case if you are working in a computing lab, I suggest that you begin your session by running cd or pwd to determine your working directory. On some networks, the working directory will be the private directory associated with your username. Or the working directory could be a directory on the local computer such as d:\data. If the working directory is a public directory, you risk losing your work because other users can delete your files. Also the files in the default working directory might be automatically deleted when you log off. If you want to save your work, make sure that the working directory is a directory that you control, such as your private directory on the network or a USB drive attached to the computer. If it is not, change to it with the cd command.

The second issue that you may need to consider is how to install ado-files, such as those that are part of the Workflow package. On many networks, you will not have permission to write files to Stata's system directories because these are protected to avoid a user from mistakenly or maliciously changing something that is essential for Stata to work properly. Although this is a reasonable thing to do, it prevents you from installing programs written by others or updating the official ado-files. For example, if

I run Stata from a network and try to install the Workflow package, I get the following error message:

```
cannot write in directory Y:\Stata10\plus\n
```

The error occurs because Stata is trying to copy the file `nmlab.ado` to a directory that is write protected. The solution is to tell Stata to install the files to a folder that is not restricted. One way to do this is to create a new PLUS directory using the following steps:

1. Check what the current PLUS directory is with `sysdir`. You might need to know this if you encounter a problem:

   ```
   . sysdir
      STATA:   Y:\STATA10\
    UPDATES:   Y:\STATA10\ado\updates\
       BASE:   Y:\STATA10\ado\base\
       SITE:   Y:\STATA10\ado\site\
       PLUS:   Y:\Stata10\ado\plus\
   PERSONAL:   L:\
   OLDPLACE:   C:\ado\
   ```

   The PLUS directory is `Y:\Stata10\ado\plus\`, which is a restricted directory located on the `Y:` drive.

2. Create a directory that will become the new PLUS directory. You can do this with your file manager or the `mkdir` command in Stata. For example, to create `L:\adoplus`, I use the command

   ```
   mkdir L:\adoplus
   ```

   where `L:` is my personal drive on the network (i.e., I can read and write files there).

3. After you have verified that the folder was properly created, you use `sysdir` with the `set` option to redefine the location of the PLUS directory:

   ```
   sysdir set PLUS L:\adoplus
   ```

4. Now you run `sysdir` again to verify that the PLUS directory has been changed.

   ```
   . sysdir
      STATA:   Y:\STATA10\
    UPDATES:   Y:\STATA10\ado\updates\
       BASE:   Y:\STATA10\ado\base\
       SITE:   Y:\STATA10\ado\site\
       PLUS:   L:\adoplus\
   PERSONAL:   L:\
   OLDPLACE:   C:\ado\
   ```

After making these changes, you should be able to run the `findit workflow` command and follow the prompts to install the package. You will have to follow these steps each time you open Stata on the network. A convenient way to do this is to create a do-file with the commands.

Because networks can be set up in many different ways, it is possible that these steps will not work for you. If that is the case, you should check with a local consultant.

A.3 Customizing Stata

If you use Stata regularly, you might want to customize Stata to make your work more efficient. Here I consider the customizations that I find most useful. If you are working on a network, you may not be able to change some of these options.

A.3.1 Fonts and window locations

I find the default fonts used by Stata to be too small for most monitors. You can change the font by right-clicking on the window whose font you want to change. Select the Fonts option and choose the font, style, and size that you like. In Windows, I prefer the Fixedsys or Lucida console fonts. You can also change the size and position of the windows by clicking and dragging. After you have things the way you like, you can make your choices the defaults by selecting **Edit** > **Preferences** > **Manage Preferences** > **Save Preferences...**.

A.3.2 Commands to change preferences

Some preferences can be changed only by running a `set` command. Some of these commands change your preferences permanently, meaning that you need to change the setting only once and it will persist in later Stata sessions. Other preferences need to be set each time you launch Stata. Fortunately, as discussed in the next section, you can include these commands in your `profile.do` file so that they will be run each time you start Stata.

There are dozens of options that you can change. To see what these options are and what the settings are for your computer, use the `query` command. The settings are listed in blue, which means you can click on the blue word and a Viewer window will open with details on that option. Here are the options that I find most important for an efficient workflow. For further details, see `help set` or [R] **query** and [R] **set**.

Options that can be set permanently

For some `set` commands, you can specify the `permanently` option so that the change will be remembered the next time you run Stata.

(Continued on next page)

Format used for log files

Log files can be text files or SMCL files. To set the option, type

set logtype { text | smcl } [, permanently]

Because I prefer text files, I run the command

```
set logtype text, permanently
```

Scrollback buffer in the Results window

The number of lines of output that can be retrieved by scrolling the Results window is controlled by the command

set scrollbufsize #

where 10000 ≤ # ≤ 500000.

Unless you have very little computer memory, I suggest

```
set scrollbufsize 500000
```

This command is permanent even though it does not have a `permanently` option.

Maximum size of matrices

The largest matrix that Stata will use is controlled by the command

set matsize # [, permanently]

where 10 ≤ # ≤ 11000 for Stata/MP and Stata/SE, and
where 10 ≤ # ≤ 800 for Stata/IC.

The default values are 400 for Stata/MP and Stata/SE, and 200 for Stata/IC.

If you are fitting complex models, the default `matsize` might be too small. If you have enough memory, I recommend setting `matsize` to its maximum. For Stata/MP and Stata/SE, I suggest

```
set matsize 11000, permanently
```

and for Stata/IC, I suggest

```
set matsize 800, permanently
```

Memory

Stata keeps all data in memory, rather than reading data from disk every time it needs the information. If your dataset is large, you might run out of memory. The amount of memory that Stata uses is controlled by the command

set memory #[b|k|m|g] [, permanently]

where # is specified in terms of bytes, kilobytes, megabytes, or gigabytes. If you routinely get an out of memory error, run **query** to find out how much memory has been set. Then increase it until the out of memory error is resolved. Once you find out how much memory you need, you can make this permanent. For example,

```
set memory 10m, perm
```

Options that need to be set each session

These set commands need to be run each time Stata is started. They can be put in your profile.do or myprofile.do (discussed in section A.3.3).

Line size

The length of a line before wrapping is controlled by the command

set linesize #

I prefer

```
set linesize 80
```

A.3.3 profile.do

When Stata is launched, it looks for a do-file named profile.do. If profile.do is found in one of the system directories, Stata runs it. This allows you to customize Stata by including commands in the profile.do that you want to run each time you start Stata. Although you should consult [GS] *Getting Started with Stata* for full details or type help profile, the following example shows you some things that I find useful:

```
// these settings must be run each time Stata is loaded
   set linesize 80
// function key definitions (discussed below)
   global F3 "codebook, compact"
// change to the working directory I want to use
   cd d:\stata_start
   exit
```

myprofile.do for a network

If you are working with a networked version of Stata, you probably are not allowed to change the `profile.do`. As an alternative, you can create a personal profile file, perhaps called `myprofile.do`, that you store in your default working directory. When you load Stata and are in your working directory, type `do myprofile`.

Function keys

Stata allows you to associate strings of text with the function keys F2–F9. When you press one of these keys, that string is inserted into the Command window. To set this up, you create global macros named F2–F9. When you press a function key, Stata checks if the global has been defined. If it has, the content of the global is pasted into the Command window. For example, to assign `codebook, compact` to the F3 key, I create the global:

```
global F3 "codebook, compact"
```

After this global is defined, if I press F3, `codebook, compact` is written in the Command window. If I want F3 to work this way every time I load Stata, I can add this command to my `profile.do` (or `myprofile.do` if I am working on a networked version of Stata).

If you use a different working directory for each of your projects, functions keys are a convenient way to change directories. For example,

```
global F4 "cd e:\workflow\work"
global F5 "cd e:\analysis\coga\work"
global F6 "cd e:\teaching\s650\work"
```

A.4 Additional resources

I have only scratched the surface of what you can do in Stata. If you are interested in learning more, I suggest the following resources:

1. The Stata site http://www.stata.com/support/ has information and links to many resources for learning about Stata, including books, courses, the Statalist listserver, meetings, and other web sites with information on Stata. This is the best place to begin if you are looking for more information.
2. The Statalist is a very active listserver hosted at the Harvard School of Public Health. For details, go to http://www.stata.com/statalist/. This is a great way to get help if you have a problem or do not know how to do something.
3. StataCorp offers web-based courses. I highly recommend NetCourse 151, *Introduction to Stata Programming*. I have recommended this course to many users and all of them, regardless of their expertise, found it to be extremely useful. You do not have to be a programmer to benefit from this course.

4. The following sections of the Stata manual are particularly useful for learning more about how Stata works: [U] *User's Guide* includes both introductory materials and advanced material. Look through the table of context and read those sections that look useful. [P] *Programming Reference Manual* has more advanced materials on writing ado-files.

References

Amalfi, C. 2006. Lost Moon landing tapes discovered. *Cosmos online*, November 1. http://www.cosmosmagazine.com/features/online/818/lost-moon-landing-tapes-discovered.

American Psychological Association. 2001. *Publication Manual of the American Psychological Association*. Washington, DC: American Psychological Association.

BBC News. 2007. Warning of data ticking time bomb. *BBC News*, July 3. http://news.bbc.co.uk/1/hi/technology/6265976.stm.

Blau, P. M., and O. D. Duncan. 1967. *The American Occupational Structure*. New York: Wiley.

Chavez, C. 2007. *Religious switchers and their sexual behavior: Examining the role of religious affiliation*. Master's thesis, Indiana University.

Cleveland, W. S. 1993. *Visualizing Data*. Summit, NJ: Hobart Press.

———. 1994. *The Elements of Graphing Data*. Summit, NJ: Hobart Press.

Cragg, R. H. 1967. Work, finish, and publish' the chemistry of Michael Faraday 1791–1867. *Chemistry in Britain* 3: 482–486.

Davis, J. A., T. W. Smith, and P. V. Marsden. 2007. General Social Surveys, 1972–2006. File ICPSR04697-v2. Chicago, IL: National Opinion Research Center [producer]; Ann Arbor, MI: Inter-university Consortium for Political and Social Research [distributors].

Ebel, H. F., C. Bliefert, and W. E. Russey. 2004. *The Art of Scientific Writing: From Student Reports to Professional Publications in Chemistry and Related Fields*. New York: Wiley.

Egan, C. 2006. One Small Step in Hunt for Moon Film World Didn't See. *Sydney Morning Herald*, August 20, 2006. http://www.smh.com.au/news/national/one-small-step/2006/08/19/1155408073519.html.

Fraser, B. 2005. *Real World Camera Raw with Adobe Photoshop CS2*. Berkeley, CA: Peachpit Press.

Fraser, B., F. Buntin, and C. Murphy. 2003. *Real World Color Management*. Berkeley, CA: Peachpit Press.

Freese, J. 2007. Replication standards for quantitative social science: Why not sociology? *Sociological Methods & Research* 36: 153–172.

ICPSR. 2005. *Guide to Social Science Data Preparation and Archiving: Best Practice Throughout the Data Life Cycle*. 3rd ed. Ann Arbor, MI: Inter-university Consortium for Political and Social Research. http://www.icpsr.umich.edu/access/dataprep.pdf.

International Social Survey Program. 2004. International Social Survey Program: Family and Changing Gender Roles III, 2002. Computer file. ICPSR version. Cologne, Germany: Zentralarchiv fur Empirische Sozialforschung [producer], 2004. Cologne, Germany: Zentralarchiv fur Empiriscche Sozialforschung/Ann Arbor, MI: Inter-university Consortium for Political and Social Research [distributors], 2004.

Jann, B. 2007. Making regression tables simplified. *Stata Journal* 7: 227–244.

Kanare, H. M. 1985. *Writing the Laboratory Notebook*. Washington, DC: American Chemical Society.

Kaufman, M. 2007. The Saga of the Lost Space Tapes: NASA Is Stumped in Search for Videos of 1969 Moonwalk. *The Washington Post*, January 31, 2007. http://www.washingtonpost.com/wp-dyn/content/article/2007/01/30/AR2007013002065.html.

Kenny, A. R., and N. Y. McGovern. 2003–2007. Digital Preservation Management: Implementing Short-term Strategies for Long-term Problems. Online tutorial developed for the DPM workshop series by Anne R. Kenney and Nancy Y. McGovern, et al., at Cornell University Library (2003–2006) with funding from the National Endowment for the Humanities (NEH) and hosted by the Inter-university Consortium for Political and Social Research (ICPSR). Available at http://www.icpsr.umich.edu/dpm/dpm-eng/eng_index.html.

Knuth, D. E. 1997. *The Art of Computer Programming. Volume 1: Fundamental Algorithms*. 3rd ed. Reading, MA: Addison–Wesley.

———. 1998a. *The Art of Computer Programming. Volume 2: Seminumerical Algorithms*. 3rd ed. Reading, MA: Addison–Wesley.

———. 1998b. *The Art of Computer Programming. Volume 3: Sorting and Searching*. 2nd ed. Reading, MA: Addison–Wesley.

Krogh, P. 2005. *The DAM Book: Digital Asset Management for Photographers*. Sebastopol, CA: O'Reilly Media.

Long, J. S. 2002. *From Scarcity to Visibility: Gender Differences in the Careers of Doctoral Scientists and Engineers*. Washington, DC: National Academy Press.

Long, J. S., P. D. Allison, and R. McGinnis. 1993. Rank advancement in academic careers: Sex differences and the effects of productivity. *American Sociological Review* 58: 703–722.

Long, J. S., and J. Freese. 2006. *Regression Models for Categorical Dependent Variables Using Stata*. 2nd ed. College Station, TX: Stata Press.

Long, J. S., and E. K. Pavalko. 2004. Comparing alternative measures of functional limitations. *Medical Care* 42: 19–27.

Macey, R. 2006. One giant blunder for mankind: how NASA lost moon pictures. *Sydney Morning Herald*, August 5, 2006. http://www.smh.com.au/news/national/one-giant-blunder-for-mankind-how-nasa-lost-moon-pictures/2006/08/04/1154198328978.html.

Mackenzie, D. 2008. Cryptologists Cook Up Some Hash for New 'Bake-Off'. *Science*, vol. 319, March.

Mitchell, M. 2008. *A Visual Guide to Stata Graphics*. 2nd ed. College Station, TX: Stata Press.

Oliveira, S., and D. Stewart. 2006. *Writing Scientific Software—A Guide to Good Style*. New York: Cambridge University Press.

Pavalko, E. K., F. Gong, and J. S. Long. 2007. Women's work, cohort change, and health. *Journal of Health and Social Behavior* 48: 352–368.

Pescosolido, B., J. K. Martin, J. S. Long, and T. W. Smith. 2003. Stigma and Mental Illness in Cross-National Perspective. NIH Grant Number R01TW006374 from the Fogarty International Center, the National Institute of Mental Health and the Office of Hehavioral and ocial Science Research to Indiana University–Bloomington. July 10, 2003–June 30, 2008.

Pinheiro, E., W.-D. Weber, and L. A. Barroso. 2007. Failure trends in a large disk drive population. In *Proceedings of the 5th USENIX Conference on File and Storage Technologies (FAST '07)*. San Joe, CA: USENIX.

Raftery, A. E. 1995. Bayesian model selection in social research. *Sociological Methodology* 25: 111–163.

Royston, P. 2004. Multiple imputation of missing values. *Stata Journal* 4: 227–241.

Schroeder, B., and G. A. Gibson. 2007. Disk failures in the real world: What does an MTTP of 1,000,000 hours mean to you? Presentation to FAST '07: 5th USENiX Conference on File and Storage Technologies, San Joe, CA, February 14–16, 2007. Pittsburgh, PA: Carnegie Mellon University. Downloadable from http://www.sagecertification.org/events/fast07/tech/schroeder/schroeder_html/index.html.

Schwartz, J. 2008. In storing 1's and 0's, the question is $. *The New York Times*, April 9, 2008. http://www.nytimes.com/2008/04/09/technology/techspecial/09store.html.

Smith, T., and M. Forstrom. 2001. In praise of data archives: Finding and recovering the 1963 Kennedy assassination study. *IASSIST Quarterly* Winter: 12–14.

Steinberg, D. H. 2006. Web 2.0 podcast: A conversation with Jeff Bezos. December 20, 2006. http://www.oreillynet.com/pub/a/network/2006/12/20/web-20-bezos.html.

Tantau, T. 2001. *User Guide to the Beamer Class, Version 3.07.* http://latex-beamer.sourceforge.net (accessed March 11, 2007).

Tufte, E. R. 2001. *The Visual Display of Quantitative Information.* 2nd ed. Cheshire, CT: Graphics Press.

———. 2006. *The Cognitive Style of PowerPoint: Pitching Out Corrupts Within.* 2nd ed. Cheshire, CT: Graphics Press.

University of Chicago Press. 2003. *The Chicago Manual of Style.* 15th ed. Chicago: University of Chicago Press.

Wallgren, A. 1996. *Graphing statistics and data: Creating better charts.* Thousand Oaks, CA: Sage.

Wolfgang, M. E., R. M. Figlio, and T. Sellin. 1972. *Delinquency in a Birth Cohort.* Chicago: University of Chicago Press.

Author index

Subject index

Symbols

If a term includes a symbol (e.g., _N), the term is listed under the letter it begins with. For example, _N is listed under N in the index; `tag´ is listed under T.

P

Q

R

T

X